Energy from the Desert

This fourth volume in the established Energy From The Desert series examines and evaluates the potential and feasibility of Very Large Scale Photovoltaic Power Generation (VLS-PV) systems, which have capacities ranging from several megawatts to gigawatts, and to develop practical project proposals toward implementing the VLS-PV systems in the future.

Comprehensively analysing all major issues involved in such large scale applications, based on the latest scientific and technological developments and by means of close international cooperation with experts from different countries.

From the perspective of the global energy situation, global warming, and other environmental issues, it is apparent that VLS-PV systems can:

- contribute substantially to global energy needs;
- become economically and technologically feasible soon;
- contribute significantly to the global environment protection;
- contribute significantly to socio-economic development.

Energy policies around the world are gradually changing direction to focus less on nuclear energy with the expectation to turn to denuclearization entirely with the negative impacts of nuclear energy, while in parallel the importance of and expectations for renewable energy technologies are increasing drastically as possible energy infrastructure, as well as environmental friendly technology.

This book recognises that very large scale solar electricity generation provides economic, social and environmental benefits, security of electricity supply and fair access to affordable and sustainable energy solutions and that VLS-PV systems must be one of the promising options for large scale deployment of PV systems and renewable energy technologies.

Keiichi Komoto is a senior manager in the field of renewable energy for Mizuho Information and Research Institute.

Christian Breyer is managing director of the Reiner Lemoine Institute, Berlin, focused on reaching 100 % renewable energy supply.

Edwin Cunow is an engineer with two diplomas, internationally acting as consultant and has over 27 years of experience in photovoltaics.

Karim Megherbi is currently the founder and President of Hélios Energie, a French company active exclusively in Sub-Saharan Africa and the Caribbean.

David Faiman is a professor of physics at Ben-Gurion University of the Negev and director of Israel's National Solar Energy Center.

Peter van der Vleuten has an extensive background in photovoltaics and has been president of Free Energy Europe and Free Energy International.

Energy from the Desert

Very Large Scale Photovoltaic Power – State of the Art and Into The Future

EDITED BY

Keiichi Komoto

Christian Breyer

Edwin Cunow

Karim Megherbi

David Faiman

Peter van der Vleuten

Routledge
Taylor & Francis Group
LONDON AND NEW YORK

earthscan
from Routledge

First published 2013
by Routledge
2 Park Square, Milton Park, Abingdon, Oxon, OX14 4RN

Simultaneously published in the USA and Canada
by Routledge
711 Third Avenue, New York, NY 10017

Routledge is an imprint of the Taylor & Francis Group, an informa business

British Library Cataloguing in Publication Data
A catalogue record for this book is available from the British Library

Library of Congress Cataloging-in-Publication data
Very large scale PV power : state of the art and into the future /
edited by Keiichi Komoto, Christian Breyer, Edwin Cunow,
Karim Megherbi, David Faiman, Peter van der Vleuten.
p. cm. – (Energy from the desert ; 4)
1. Photovoltaic power systems. I. Komoto, Keiichi.
TK1087.V47 2013
621.31´244–dc23 2012018942

ISBN13: 978-0-415-63982-8 (hbk)
ISBN13: 978-0-203-08140-2 (ebk)

Typeset in Bembo
by HWA Text and Data Management, London

Contents

II

Foreword

Looking at recent developments, the role of solar photovoltaic electricity in the future energy supply can be observed to increase constantly in present energy scenarios, policies and, above all, the real markets. Looking at the topic from an International Energy Agency (IEA) perspective, three recent publications mark the increasing role of renewable energy in general and photovoltaics in particular: the *IEA Technology Roadmap Solar Photovoltaic Energy* (2010), the *IEA Energy Technology Perspectives* (2010) and the *IEA Solar Energy Perspectives* (2011). The common message of these publications regarding solar energy and photovoltaics particularly is that this technology *can* provide significant contributions to the future electricity supply. Looking at the greater energy picture, climate change issues, the development of fossils fuels and the Fukushima nuclear accident have increased the perception that renewable energy not only can but will *have to* contribute strongly to the future energy supply. Furthermore, the recent rapid growth of photovoltaic markets all around the world, together with dramatically reduced costs, provides the confidence that photovoltaics *will* be able to play the expected role. These times are thus crucial for the future development of this young energy technology which is facing the challenges of rapid implementation, fast industry growth, increasing market dynamics and fierce global competition.

This new IEA PVPS Task8 book on Very Large Scale Photovoltaic Power Generation Systems comes at a time when photovoltaic solar electricity is about to achieve generation costs close to or at the costs of the retail electricity market. As a result of this development, utility scale photovoltaic applications are observed to be increasing in various countries with high solar irradiation. This trend marks the beginning of a transition from a policy driven to a more business oriented market.

The work of IEA PVPS Task8 is an excellent example of the forward looking nature of international cooperation within the IEA PVPS Implementing Agreement. When the project started in 1999, very large scale photovoltaic power systems were seen as a futuristic, long-term concept which many doubted would ever become reality. Today, only 13 years later, we can see many examples of large and very large scale photovoltaic power systems being planned and realized. The vision of Professor Kosuke Kurokawa who inspired the work of IEA PVPS Task8 has become reality much faster than expected.

This new edition of *Energy from the Desert* provides a very useful update on all critical elements, issues and solutions related to very large scale photovoltaic power systems, with concrete examples of what has been realized so far around the world. Another highlight of this new edition is the comparison with other large scale solar energy concepts such as Concentrated Solar Power (CSP) and Concentrated Photovoltaics (CPV). Finally, the book also serves as a comprehensive compendium of the many publications relevant for large scale photovoltaic power systems.

Therefore, the publication of this book is very timely for the future expected growth of large scale photovoltaic power systems and will help to satisfy the increasing need for information in this growing field. I would like to thank Keiichi Komoto for leading the work of IEA PVPS Task8 and all the experts for their dedication and contributions necessary in making this publication possible. I am sure that this book will have many interested readers who will find the insights provided most inspiring for their future work.

S. Nowak

Stefan Nowak
Chairman IEA PVPS Programme
May 2012

Preface

"It might be a dream, however ...". This was our motivation when Task8 was established in 1999. Since then, we have published extensive reports as the series 'Energy from the Desert', focusing on very large scale photovoltaic power generation (VLS-PV) systems. The books show that VLS-PV is not a simple dream but is becoming realized and well known all over the world.

During our work, large scale PV systems increasingly count as a realistic energy option and have started to appear around the world; 100 MW scale PV systems are becoming a reality.

Since the nuclear accident in Japan caused by the Great East Japan Earthquake on 11 March 2011, energy policies around the world have been more likely to change direction to less nuclear energy or denuclearization eventually. In parallel, the importance and expectation of renewable energy technologies are increasing drastically as possible energy infrastructure, as well as environmental friendly technology.

Not only as a Japanese, but also as Operating Agent of PVPS Task8, I strongly felt that the direction of this publication and the next step of Task8 activity should be reconsidered. Discussions for preparing this publication within the Task8 group might be the most fruitful discussions since Task8 was established.

Some people are of the opinion that our concept, VLS-PV, is being overtaken by market developments. In a way, this might be correct. However, states/governments all over the world consider solar power plants as a viable option for their energy policy.

We have recognized that very large scale solar electricity generation provides economic, social and environmental benefits, security of electricity supply and fair access to affordable and sustainable energy solutions and that VLS-PV systems must be a promising option for large scale deployment of PV systems and renewable energy technologies.

We believe solar energy can realize our bright future.

河本桂一

Keiichi Komoto
Operating Agent, Task8

Task 8 Participants

Keiichi Komoto, OA
Mizuho Information & Research Institute, Inc. (MHIR), Japan

Kosuke Kurokawa, OA-alternate
Tokyo Institute of Technology (Tokyo Tech), Japan

Masanori Ishimura, Secretary
New Energy and Industrial Technology Development Organization (NEDO), Japan

Michele Appendino
Solar Ventures, Italy

Aurélien Bertin
Hélios Energie, France

Christian Breyer
Reiner Lemoine Institut gGmbH, Germany

Edwin Cunow
LSPV Consulting, Germany

Tomoki Ehara
E-konzal, Japan

David Faiman
Ben-Gurion University of the Negev, Israel

Fabrizio Donini Ferretti
Chora Finance, France

Gianluca Gigliucci
ENEL, Italy

Thomas N. Hansen
Tucson Electric Power Company, USA

Herb Hayden
Southwest Solar Technologies, Inc., USA

Kay Herget
Bosch Solar Energy AG, Germany

Xu Honghua
Electrical Engineering Institute, Chinese Academy of Sciences, China

Masakazu Ito
Tokyo Institute of Technology (Tokyo Tech), Japan

Francesco De Lia
ENEA, Italy

Christof Koerner
Siemens AG, Germany

John S MacDonald
Day4Energy, Inc., Canada

Karim Megherbi
Hélios Energie, France

Fabrizio Paletta
RSE, Italy

Jinsoo Song
Korea Institute of Energy Research (KIER), Korea

Roberto Vigotti
Solar Ventures, Italy

Peter van der Vleuten
Free Energy Consulting, The Netherlands

Namjil Enebish, Observer
National Renewable Energy Center, Mongolia

List of Contributors

Kenji Araki, Daido Steel Co., Ltd., Japan (Section 4.1)
Aurélien Bertin, Hélios Energie, France (Section 9)
Christian Breyer, Reiner Lemoine Institut GmbH, Germany (Sections 2.2, 3, 10)
Edwin Cunow, LSPV Consulting, Germany (Section 6)
Tomoki Ehara, E-konzal, Japan (Sections 2.1, 13)
Namjil Enebish, National Renewable Energy Center, Mongolia (Section 12)
David Faiman, Ben-Gurion University of the Negev, Israel (Sections 4.2, 8, 11.2)
Kay Herget, Bosch Solar Energy AG, Germany (Section 15)
J.C.C.M. Huijben, Eindhoven University of Technology, The Netherlands (Section 14)
Masakazu Ito, Tokyo Institute of Technology, Japan (Sections 4.1, 5)
Keiichi Komoto, Mizuho Information & Research Institute, Inc., Japan (Sections 1, 2.2, 4.1, 15)
Antonluca Loteta, ENEL, Italy (Section 6)
Karim Megherbi, Hélios Energie, France (Sections 7, 11.1)
Rob Raven, Eindhoven University of Technology, The Netherlands (Section 14)
A.A. Solomon, University of California at Berkeley, USA (Section 8)
Geert Verbong, Eindhoven University of Technology, The Netherlands (Section 14)
Erik van der Vleuten, Eindhoven University of Technology, The Netherlands (Section 14)
Peter van der Vleuten, Free Energy Consulting, The Netherlands (Sections 1, 14)

About the Editors

Keiichi Komoto has contributed to VLS-PV activity since its preliminary stages. He currently works as a senior manager in the field of renewable energy for Mizuho Information and Research Institute, and is focusing on photovoltaic technology from environmental and socio-economic viewpoints.

Christian Breyer is managing director of the Reiner Lemoine Institute, Berlin, focusing on reaching 100 % renewable energy supply. He has worked in research and the market development responsible for PV-based technology screening, economic analyses like PV grid-parity and PV fuel-parity and specializes in hybrid PV systems. He is member of international working groups EU PV Technology Platform WG Developing Countries, Alliance for Rural Electrification, IEA-PVPS Task 8 and a founding member of DESERTEC Foundation.

Edwin Cunow is an engineer with two diplomas and has over 27 years experience in photovoltaics. He joined Siemens Solar in 1986 and built up the system engineering and procurement group. In 1996, his group engineered an innovation which set the basis for all MW PV plants after: the worldwide first Central MW PV generator with a special "MW PV module" and a Central MW Inverter Unit. In 1997, he was appointed as project manager for all Megawatt projects and continued the job after Shell took over Siemens Solar. Since 2006 he is a freelancer offering worldwide his expertise of currently about 160 MW as Owners Engineer, Lenders Engineer and Technical Advisor. Additionally he held a lot of special training courses in selected countries on the professional design of large scale grid-connected PV power plants.

Karim Megherbi is currently the founder and President of Helios Energie, a French company active exclusively in Sub-Saharan Africa and the Caribbean. Helios Energie develops large photovoltaic power plants connected to the grid, for national utilities and major power consumers.The key mission of the company is to provide electricity at a competitive price compared with current and future energy prices, without any feed-in-tariff nor subsidies from local states. Karim worked previously in the financial and development sectors. He occupied various positions, such as Founding Partner/Financial Advisor at Aurion, an advisory firm recently acquired by Astris Finance; Project Manager at Dexia in the structured finance department, focusing on Energy and Renewables; and consultant in several countries in Africa for the United Nation Development Program (UNDP) and NGOs. Karim has graduated from HEC French School of Management and holds a Masters Degree in Economics of natural resources and environment.

David Faiman is a professor of physics at Ben-Gurion University of the Negev and director of Israel's National Solar Energy Center. He was a founder member, 36 years ago, of the university's Jacob Blaustein Institutes for Desert Research at Sede Boqer, whose Department of Solar Energy and Environmental Physics he heads.

Peter van der Vleuten has an extensive background in photovoltaics: as president of Free Energy Europe he gained experience with the development, manufacture and worldwide marketing of thin film silicon modules, while his presidency of Free Energy International involved him in strategy and finance in the field of solar electricity, as well as in joint research and development projects. Since 2007, he acts as an international consultant and he started several new ventures in the field of solar technology. He has been involved in work on VLS-PV for desert applications from 1996 and has enjoyed contributing significantly to the IEA Task 8 project, being eager to see the first gigawatt projects appear

Acknowledgements

This report was accomplished with the kind support of various organizations and people around the world.

Our activity, Task8, started in 1999. Since then we have published three technical reports, as the series 'Energy from the Desert'. We have organized various international symposiums, workshops and seminars in countries around the world, e.g. Japan, Korea, Mongolia, Australia, Israel, Germany, Greece, Italy and the United Kingdom. Through these activities, we obtained various kinds of expert advice and new members. Based on these experiences, we started our fourth phase of activity in 2009.

We thank Masamichi Yamamoto (NEDO, Japan), who provided a great deal of support as the Operating Agent country member of the IEA PVPS Executive Committee.

The contribution from Japan, as Operating Agent country, was strongly supported by members of the Japanese domestic committee for VLS-PV: Ichiro Araki (GMG Corporation), Kenji Araki (Daido Steel), Hiroo Konishi (NTT Facilities), Taku Nishimura (University of Tokyo), Kenji Otani (AIST) and Noboru Yumoto (YN International).

We thank Michael Epstein for valuable discussions concerning CSP, and Samuel J. Bader for help in computing Table 4.11 in Chapter 4.

We thank Vasilis Fthenakis, Operating Agent of IEA PVPS Task12: PV Environmental, Health and Safety (E, H & S) Activities, for his valuable comments on Chapter 5.

We thank the Paris Diderot University, the Institut de Physique du Globe de Paris and the Agence Nationale de la Recherche Technologique, in particular Jean-Pierre Frangi, for their support on Chapter 9. We also thank Marion Perrin and Xavier Le Pivert from the French Alternative Energies and Atomic Energy Commission, and Vladimir Momberger, from Hélios Energie, for the fruitful discussions.

We thank Stephan Rieke (Solar Fuel) and Michael Sterner (Fraunhofer IWES) for valuable discussions concerning the renewable power methane issue in Chapter 10.

We thank the invited writers of Chapter 14: J.C.C.M. Huijben, Erik van der Vleuten, Rob Raven and Geert Verbong of the System Innovations and Sustainability Transitions Group, School of Innovation Sciences, Eindhoven University of Technology for their valuable contribution to our work.

Finally, the Task8 members thank the IEA PVPS Executive Committee and the participating countries of Task8, for giving them valuable opportunities to perform this task.

II

Executive summary

A.1 INTRODUCTION AND OVERVIEW: POTENTIAL OF VLS-PV

Objectives

The purpose of this study is to examine and evaluate the potential and feasibility of Very Large Scale Photovoltaic Power Generation (VLS-PV) systems, which have capacities ranging from several megawatts to gigawatts, and to develop strategies toward implementing the VLS-PV systems in the future (see Figure A.1).

Our study has comprehensively analysed all major issues involved in such large scale applications, based on the latest scientific and technological developments and by means of close international cooperation with experts from different countries [1, 2, 3].

From the perspective of the global energy situation, global warming, and other environmental issues, it is apparent that VLS-PV systems can:

- contribute substantially to global energy needs;
- become economically and technologically feasible soon;
- contribute significantly to global environment protection;
- contribute significantly to socio-economic development.

Also, it is concluded that desert regions contain an abundant and inexhaustible source of clean energy and that very large scale solar electricity generation provides economic, social and environmental benefits, security of electricity supply and fair access to affordable and sustainable energy solutions.

During the past 10 years, MW-scale PV systems have been increasing substantially in the world and 100 MW-scale PV systems are becoming reality. Further realistic discussions and plans of over 100 MW-scale and GW-scale PV systems are going on in some regions. However, to accelerate and implement real VLS-PV projects, decision-makers should be informed about the feasibility of such projects in and around desert regions in an appropriate manner.

VLS-PV basically means generation of electricity by means of solar irradiation at utility scale. In our specific case, we have studied utility scale generation of solar electricity in desert regions. By nature, desert regions are rich in sunshine, water is scarce and they are sparsely populated. Living conditions in deserts are harsh and they are often considered as unproductive land.

Solar electricity generation at utility scale can change this situation substantially. This will not only provide an infinite, sustainable and inexhaustible source of energy, but also offer the possibility for establishing a source for socio-economic growth in the region. Deserts have the potential to become the future powerhouses of the world.

Based on the viewpoints mentioned above, this report will show the overview and potential of VLS-PV, guidelines for VLS-PV systems, technical options and strategic options

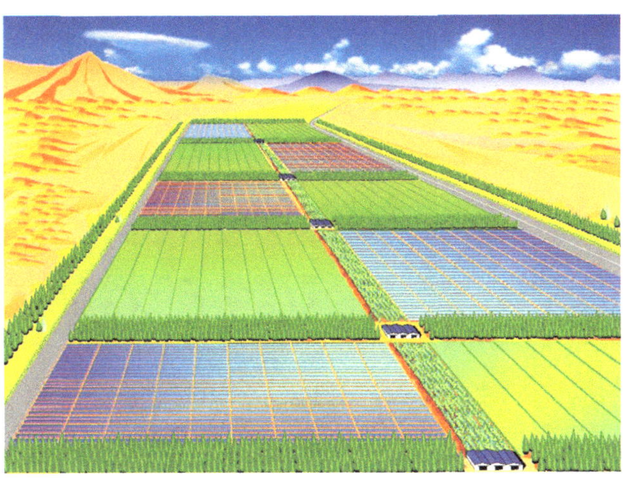

Figure A.1 An image of a VLS-PV system in a desert area

for implementing VLS-PV, and will propose our next steps into the future.

Proposed scenarios for solar and renewable energy

It has been repeatedly pointed out by scientists that human beings will face several sustainability issues sooner or later if appropriate measures are not taken promptly, especially from the environmental point of view. It is clear that the anthropogenic environmental burden will be greater than the capacity of the Earth as the global population and its economic activities expand at the current pace.

There are many projections for global energy supply but the roles of PV and other renewable energy vary greatly among the scenarios. Those are heavily dependent on underlying assumptions and projection approaches. In general, PV is expected to increase drastically in most scenarios but it might take a few decades to be a main energy source in the global energy mix without innovations.

PV has already achieved more than some projections presented in the past, for example, *IEA: World Energy Outlook*, and should continuously go over projection in order to become a main energy supply. Achievement of the projection by *IEA: Energy Technology Perspectives* [4] and *PV Roadmap* [5] toward 2050 (see Figs. A.2 and A.3) should be the minimum level, and then the VLS-PV roadmap toward 2100 should be targeted as we proposed in our previous edition [3]. Also, many countries/regions have set ambitious targets for solar and renewable energy.

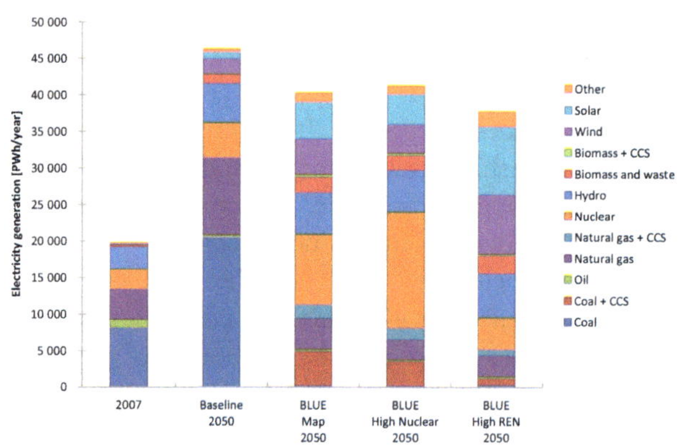

Figure A.2 Projection of electricity generation in different scenarios in ETP [4]

An excellent example of how such a fully renewable powered energy system could work has been analyzed for the Desertec project focused on a cooperation of Europe (EU) and North Africa and Middle East (MENA). The EU-MENA Desertec project gained pace by the Desertec Industrial Initiative (Dii) led by industry giants [6] and might become a blueprint for similar interregional cooperation in other parts of the world reaching a global power grid. As an industrial initiative, Dii assumes the role of a facilitator. It will ensure that development projects come to fruition, but not make any investments itself either by constructing or operating power plants. The focus of the planning phase is to establish the appropriate framework and create a lasting structure for renewable energy which will make long-term commitments to solar energy plants (solar PV and solar thermal) and wind farms, as well as integrated networks, attractive to public and private investors. The Desertec vision has become a real project for a sustainable energy supply within the EU-MENA region and gained a lot of momentum for harvesting the precious and nearly unlimited energy resources of the deserts of the world. Very large scale PV power plants are expected to become a cornerstone of such an energy system based on a real sustainable fundament.

In the Mediterranean region, it is recognized that current trends on energy supply and demand are not sustainable and will strain energy security. The following actions are necessary immediately as alternative countermeasures:

- Increase renewables in power generation and in final use
- Demand-side management policies

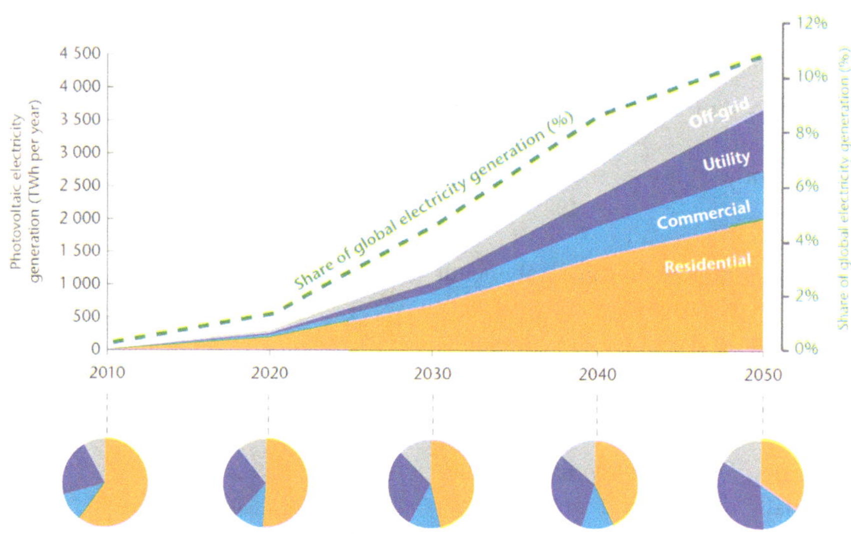

Figure A.3 PV roadmap presented by IEA [5]

- Encourage energy efficiency and energy conservation
- Remove progressively subsidies
- Invest in research, technology and innovation.

In addition to the Dii, the Mediterranean Solar Plan (MSP) and relative initiatives for deploying renewable energy in the Mediterranean region are ongoing. The MSP is one of the key projects proposed within the Union for the Mediterranean, and intends to increase the use of solar energy and other renewable energy sources for power generation, improve energy efficiency and energy savings, develop electricity grid interconnections and foster and encourage the transfer of know-how and technology towards developing countries. The final target is the development, by 2020, of 20 GW of new renewable energy installed capacity in the Southern and Eastern Mediterranean countries.

On the other hand, in the Asian region, the Asian Development Bank (ADB) announced the launch of the Asia Solar Energy Initiative (ASEI) [7], targeted to assist in identifying, developing, and implementing 3 000 MW of solar power in Asia and the Pacific over the next 3 years. The ASEI aims to create a virtuous cycle of solar energy investments in the region, toward achieving grid parity, so that ADB developing member countries optimally benefit from this clean, inexhaustible energy resource. To meet its target, the ASEI will use an integrated, multipronged approach featuring three interlinked components:

- Knowledge management: The Asia Solar Energy Forum
- Project development: 3 000 MW of solar capacity
- Innovative finance: The Asia Accelerated Solar Energy Development Fund.

The ASEI has the potential to not only drive down costs and to usher in innovation in the solar space, but also to demonstrate replicable models to sustain solar capacity additions in Asia and the Pacific. It is also expected to serve as a model for other regions in the world, such as northern Africa, Latin America, and the Caribbean, which are similarly placed. There has been keen support for the ASEI and ADB leadership in this area. Beyond the next 3 years, measures of ASEI's success will be the widespread reach of solar energy generation programs across developing member countries (DMCs) in Asia and the Pacific, and the extent of mainstreaming of solar energy as an option for generating capacity addition in all DMCs with significant potential for solar energy in the region. ADB is committed to seeing this success in the future of Asia and the Pacific.

Global potential of solar energy resource

Energy resource availability is decisive for all power technologies. The basis for all solar energy technologies is the amount and the composition of the global horizontal irradiation (GHI). Different solar system concepts have been developed for optimizing the energy yield. Solar resources adapted to systems' requirements are derived and compared with each other.

Global horizontal irradiation is more or less homogenously distributed on Earth (Figure A.4). Variations of GHI for most populated regions in the world are typically within a range of a factor of two, e.g. Central Europe (~ 1 100 kWh/m^2/y) and Arabian Peninsula (~ 2 200 kWh/m^2/y). NASA Surface meteorology and Solar Energy (SSE) 6.0 dataset is taken for all calculations [8].

Absolute and relative diffuse irradiation is strongly coupled to regional climatic conditions, in particular to daily and seasonal occurrence of clouds. Most deserts on Earth are located between 15° and 40° N/S. Diffuse irradiation in deserts is quite low and ranges typically between 15 percent and 30 percent. Diffuse irradiation is lost for solar power technologies only capable of converting direct solar irradiation, thus sites of low diffuse irradiation are most appropriate for these technologies. Solar system technologies increasing insolation on active surfaces by tracking the Sun's position are most beneficial for lower diffuse irradiation.

A major advantage of solar PV technology is the need for no moving parts; this is still the case for fixed optimally tilted systems. To give up such a great advantage very good reasons are needed; these might be given for specific tracking systems. 1-axis north-south horizontal continuous tracking systems follow the Sun over the day. 1-axis north-south optimally tilted tracking systems are better adapted to each site latitude. The PV system of highest annual irradiation is the 2-axes non-concentrating tracking system. A large plant area is needed due to two-directional shading effects and a higher mechanical stress and cost is effective due to more moving parts and engines. 2-axes concentrating tracking variation is only needed for high concentrating PV (HCPV) systems and solar thermal electricity generation (STEG).

Summarizing the results, 1-axis horizontal north-south continuous tracking PV systems might be competitive for fixed optimally tilted PV systems. An increase in irradiation of 20–30 percent for sunny and very sunny regions might justify higher capital and operational expenditures, especially for plant operators achieving further advantages by more full load hours of the PV plant [9]. 2-axes non-concentrating and concentrating tracking PV systems further increase the irradiation on the module surface; however the higher complexity of such systems significantly reduces the financial benefits of these approaches. In the case of HCPV systems, significantly higher power conversion efficiencies in reference to non-concentrating PV systems result in about 35 percent higher electricity generation potential per area unit, leading to a beneficial cost potential of HCPV systems.

In total, solar resource is available in all populated regions in the world. Fixed optimally tilted PV systems

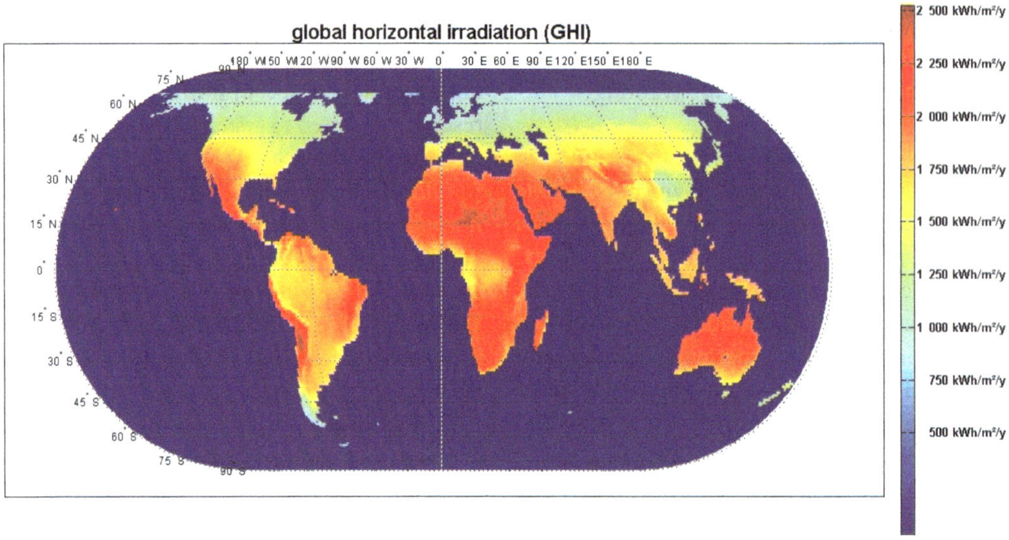

Figure A.4 Annual global horizontal irradiation (GHI)

are the standard PV power plant system types, however depending on local conditions, adapted tracking systems increase the annual yield per installed capacity and lower the power generation cost.

PV market potential for 2020

Photovoltaic energy technology has the potential to contribute to the global energy supply on a large scale. This potential can only be realized if sustainable and highly competitive PV economics are achieved. An integrated economic PV market potential assessment is presented consisting of grid-parity and fuel-parity analyses for the on-grid markets and an amortization analysis for rural off-grid PV markets. All analyses are mainly driven by cost projections based on the experience curve approach and growth rates for PV systems and electricity and fossil fuel prices for the currently used power supply. Finally the outcome of the derived economic PV market potential has been transformed to pessimistic, realistic and optimistic estimates of a cumulated installed PV capacity by the end of 2020. These installed capacity estimates are compared with PV scenarios published by leading organizations in the field of energy and PV scenarios.

The economic PV market potential derived by the grid-parity approach is estimated to about 980 GW up to 3 930 GW and in the most likely case to about 2 070 GW by the year 2020, if an advanced economic storage system for PV electricity is available [10]. The market estimate according to the fuel-parity approach leads to about 900 GW up to 1 500 GW fully economic potential for PV power plants in the year 2020 [10]. The total addressable economic market potential for rural off-grid PV applications is estimated as about 70 GW [10]. An integrated economic PV market potential has been performed to avoid double counting. The integrated assessment leads to an overall sustainable economic PV market potential of about 2 800 GW to 4 300 GW in the year 2020 [10].

The outcome is depicted in Figure A.5 for an integrated sustainable economic market potential estimate for PV systems on the basis of a learning rate of 20 percent for modules and inverters and 15 percent for the other balance of system (BOS) components.

Only a fraction of the market potential will be realized in time, thus PV market expectations are derived resulting in cumulated installed PV capacity for the year 2020 in a pessimistic (600 GW installed capacity in 2020 and an annual growth rate of 20 percent), realistic (1 000 GW in 2020 and growth rate of 30 percent) and optimistic market view (1 600 GW in 2020 and growth rate of 40 percent) [10]. Only EPIA and Greenpeace can imagine such a fast progress in PV development. Other leading organizations like the IEA, WWF and EWG expect about one third or even less compared with the derived pessimistic market expectation.

Due to the complementarity of solar PV and wind power, market expectations for PV need not be lowered. Moreover, being complementary to wind power, PV together with wind power might become the backbone of the global energy supply in the coming decades [11, 12]. However, hybrid PV-Wind systems are very likely to start to strongly influence the global power system in the next decade by sophisticated hybrid concepts, like renewable power methane technology. The fast progress of sustainable energy technologies seems not to be included in current energy scenarios of leading international organizations. Plenty of indications are given that several organizations would make no mistake in revising their scenarios in the field of PV. In conclusion, PV is on its way to become a highly competitive energy technology.

Trends in large-scale PV systems

Installation of large-scale (MW-scale) PV systems has been increasing substantially year by year throughout the world. The starting point and milestones for this development

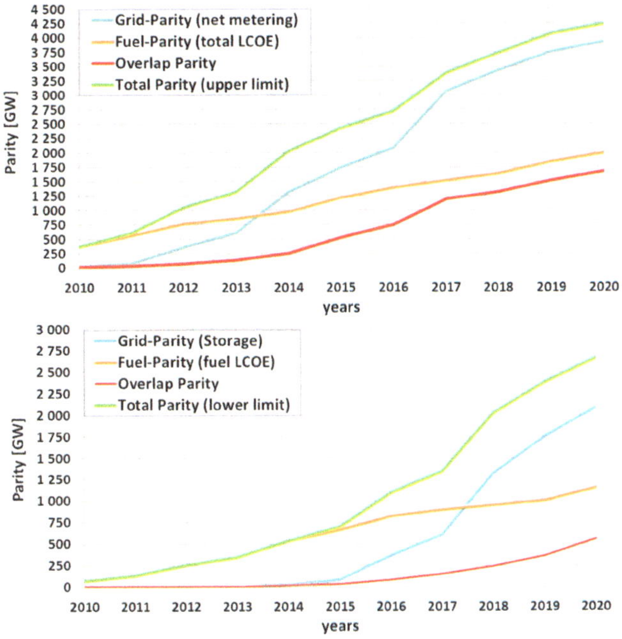

Figure A.5 Total economic market potential for PV systems in the 2010s in the upper (top) and lower limit (bottom)

have been the feed-in-tariff schemes in European countries. These schemes created an attractive market for developers, investors, etc. and a new business model of MW-scale PV systems was established.

In 2011, the number of MW-scale PV systems is over 1 500, and that of PV systems over 20 MW is more than 80. Among these, the largest PV system, Perovo I-V PV power plant in Ukraine is 100 MW, constructed in December 2011. The second largest PV system, Sarnia PV power plant in Ontario, Canada is 97 MW_{AC} (80 MW_{DC}), and the third largest PV system, Montalto di Castro PV power plant in Lazio, Italy is 84 MW.

Summing up MW-scale PV systems installed by 2011, the total capacity of MW-scale PV systems has reached more than 8 GW, and this occupies a significant share of the cumulative installed capacity of all PV applications in the world.

It is obvious that MW-scale PV systems are already a major PV application. As the capacity of the MW-scale PV systems expands year by year, the maximum capacity is achieving 100 MW and some PV plants which have a capacity of hundreds of MW are being prepared. Indeed, a GW-scale PV plant consisting of several 100 MW-scale PV systems is expected to be realized within a few decades. In the near future, VLS-PV systems will become an option for many large remote desert regions in the world where conventional power plants have never been built.

Comprehensive comparison between solar-powered VLS technologies

For utilizing the huge potential of solar energy globally, we can employ a plurality of solar-powered technologies, e.g. not only PV, but also concentrator photovoltaics (CPV), concentrated solar power (CSP) and even wind. It is therefore worthwhile to review the advantages of each from the standpoint of how they can enable solar energy, in all of its manifest varieties, to play a major part in mankind's future.

The developed world has a highly-developed electric power infrastructure that is currently powered mainly by *heat* from either fossil fuel or nuclear sources. CSP, because it too generates heat, fits more smoothly into that infrastructure than does any other renewable energy technology. To this end, another virtue of VLS-CSP is that the heat it generates is easier to store for short periods (i.e. in order to overcome the effect of passing clouds, or for peak load shifting) than electricity. In fact, owing to the relatively large thermal mass of the heat transfer fluid, CSP automatically contains a certain amount of built-in storage. Yet a third virtue of VLS-CSP is that it can power conventional turbines. There are accordingly no problems with the removal of harmonics of the sort that can arise in PV and wind systems when an inverter may function in a less than perfect manner. Last, but by no means least, the fact that the original Luz parabolic trough CSP systems are still operative provides so-called *bankability* to this kind of technology, probably more readily than to any other. Disadvantages of CSP at the present time include the need for cooling water (for efficient operation); an as-yet incomplete knowledge about the efficacy of thermal storage – without which, some form of fossil backup is necessary; and, of course, the need for a "sunny" climate.

For the developing part of the world, VLS-PV is most likely to be the technology of choice because of (a) the increasing local familiarity with PV technology owing to the ever more widespread employment of small stand-alone systems; and related to this (b) its relative simplicity when compared with CSP systems. Inter-connected VLS-PV systems, when strung out across different time zones in the developing world, will go a long way to overcoming the storage problem that is inherent to PV technology. In many regions, such systems will have an advantage over similarly located VLS-CSP systems by virtue of another feature of their simplicity: only the addition of power lines (without the need for additional supply concerns such as cooling water and perhaps backup fuel) being needed to create an electricity grid. The downside of PV, and also of wind, is the intrinsic lack of storage that, in contrast to inter-connected VLS-PV plants, will have to be developed if such plants are to perform as major power sources in isolation.

CPV has the advantage over PV that it requires typically 3 orders of magnitude less PV material. This has economic implications on both system cost and on the size needed for a fabrication plant (also 3 orders of magnitude smaller per MWp); on the continued availability of possibly rare materials; and on the energy payback time of the PV cells. Another advantage is that the actively-cooled variety of CPV plants can provide thermal energy for a variety of associated processes (desalination, air-conditioning, etc.). The prospect of employing the thermal output for additional electricity production is poor at the output temperatures presently allowed by CPV receivers. However, with improved materials, it is possible that VLS-CPV plants might in the future provide both PV and CSP power. The downside of CPV is that, as in the case of PV and wind, appropriate storage systems will have to be developed and, as in the case of CSP, "sunny" climates are necessary.

Wind has the advantage that it is currently the most mature of the VLS technologies. Its disadvantages are (like PV and CPV) the need for storage. An additional disadvantage is the large amount of land area required for very large scale plants unless dual land usage (e.g. for farming) is possible.

We shall therefore eventually have four kinds of very large scale solar technologies, e.g. VLS-PV, CPV, CSP and Wind, with which to be able to replace fossil fuel. These technologies may be regarded as *complementary* to one another rather than in competition. Ultimately, one may envisage a worldwide grid of perhaps ultrahigh voltage DC power lines (or possibly ambient-temperature superconducting cables) fed by the various VLS technologies, each located in climatic regions appropriate to its most favorable economic performance. In summary, at the present stage of their respective developments, all VLS renewable energy technologies look as if they will be complementary rather than competitive with one another.

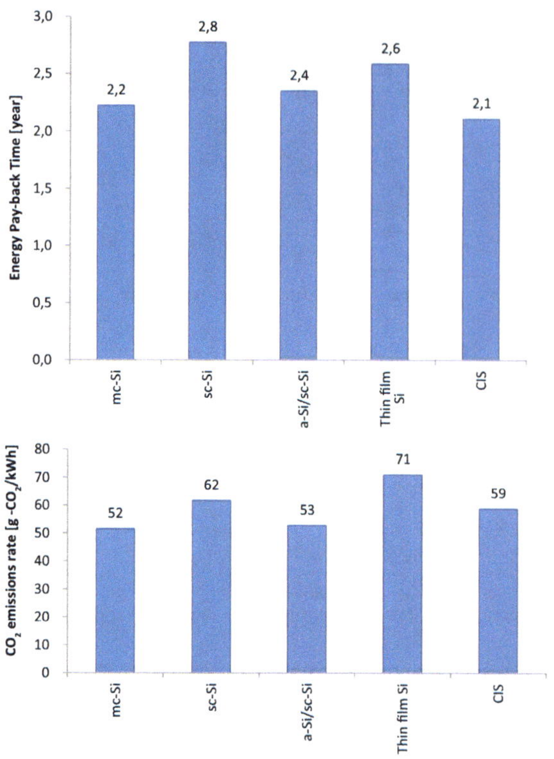

Figure A.6 Energy pay-back time (top) and CO_2 emissions rate (bottom) of the VLS-PV systems

Environmental aspects of VLS-PV

The Life-Cycle Analysis (LCA) methodology is an appropriate measure to evaluate the potential of VLS-PV systems in detail, because the purpose of this methodology is to evaluate its input and output from cradle to grave.

The LCA of VLS-PV systems were carried out by support of IEA PVPS Task12: PV environmental health and safety. Five types of PV modules were evaluated: single crystalline silicon (sc-Si), multi-crystalline silicon (mc-Si), amorphous silicon/single crystalline silicon hetero junction (a-Si/sc-Si), amorphous silicon/micro crystalline thin-film silicon (thin-film Si), and copper indium di-selenide (CIS).

Figure A.6 shows the energy pay-back time and the CO_2 emissions rate. The results of energy pay-back time are 2,1 to 2,8 years. This means that they can recover energy consumption in less than 3 years. CO_2 emissions rates are 52 to 71 g-CO_2/kWh. These results are higher than the estimates published in papers by Fthenakis *et al.* [13] or Mason *et al.* [14]. The reasons for this are the efficiency of PV modules, the assumptions on manufacturing that PV modules and inverters would be produced in Japan, and the inclusion of long transmission lines and heavy mounting hardware, which are based on Japanese standards, produced in China.

(Both figures assume 2 017 kWh·m^{-2}·yr^{-1} irradiation in Hohhot in China. BOS and 100 km transmission lines on Japanese construction standard are produced in China, and PV modules and inverters are produced in Japan. PV modules data are from NEDO. Module efficiencies are 13,9% (mc-Si), 14,3% (sc-Si), 16,6% (a-Si/sc-Si), 8,6% (thin-film Si), 10,1% (CIS).)

In addition, as an indicator of sustainability, Ecological Footprint (EF) analysis was carried out. The EF is a measure of how much biological productive land area humanity uses to produce resources and to absorb waste. On the other hand, the Earth's available capacity to produce useful biological materials and to absorb the waste is defined as Biocapacity (BC) (see Figure A.7). If the EF exceeds the BC (i.e. the ecological balance (BC – EF) is negative), this ecosystem is not sustainable. If we are able to reduce the EF or increase the BC, we can reduce the negative number of the ecological balance. The method of calculating the EF of PV systems and fossil fuel energy and the BC of the greening to reduce the ecological balance has been developed.

A.2 ENGINEERING AND FINANCIAL GUIDELINES FOR VLS-PV SYSTEMS

Technical and engineering guidelines

From the point of view of engineers, VLS-PV power plants have outstanding advantages, mainly: relatively easy identification of suitable sites, flexibility of planning and engineering, easy energetic integration into local grids and increasing commercial viability. These benefits can

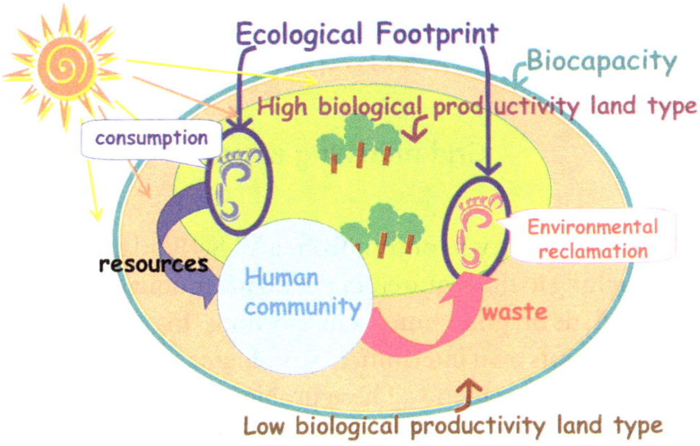

Figure A.7 The Ecological Footprint and Biocapacity

especially be used by emerging economies having very good solar conditions.

Assessment and evaluation of VLS-PV takes place via economic and financial issues, which are always in coherence with a basic understanding of technical issues. The technical engineering guidelines facilitate the access to a basic technical understanding of VLS-PV. They describe the current state-of-the-art for this special power plant technique. They are typically produced by governmental owned utilities and Ministries of Energy as well as Ministries of Environment.

The guidelines introduce and exemplify criteria for selection of appropriate power plant sites. After the identification of a site, discussions with the owner of the local medium or high voltage grid must be established. Issues of discussion are the capacity of the grid in regard to the power from the VLS-PV plant and technical requirements. The desired result is a guarantee of a grid-friendly feed-in operation.

The principle approach to planning and design is shown as Figure A.8.

The design engineer's professional qualification is an essential precondition for securing a high technical quality of engineering. The professional selection of key components and their combination is a fundamental condition. This requires experience and knowledge of the worldwide market prices for several key PV components. During the engineering phase the requirements of the later long-lasting operational phase can be influenced. This is especially valid for the power plant monitoring system. The monitoring system allows recording and visualization of all important operating states, signals and stores them. Individual requirements can be carried out by adequate software adaptation. Dependent on the size of the plant or the intention of the operator, VLS-PV can be remotely controlled and operated or operated by local personnel in a service building. The requirements of the software should also be checked and balanced with the local utility; it sometimes demands a control access to the plant.

Necessary safety conditions of personnel and plant are fulfilled by consideration and realization of set quality criteria and valid local regulations, national and international safety standards.

In an ideal case the country already has its own qualified personnel for planning and engineering VLS-PV. Then the national economy would be independent from services abroad. A domestic content then would contribute to a national value added. Since this is sometimes not the case, a know-how transfer can eliminate this disadvantage. Independent experienced and qualified consultants can teach and train engineers to the necessary technical state-of-the-art level.

Financial guidelines

A project is considered as "successfully developed" once the financial closing stage is reached. Indeed, at that point onward, the investors and the lenders are committed to disburse their funds for financing the plant's construction, which will start shortly after the financial closing.

To reach this stage, the project has to be approved by credit and investment committees, i.e. by both banks and investors. In case of approval, technical and legal due diligences will be conducted by independent auditors mandated by the financers. The result of such audits will

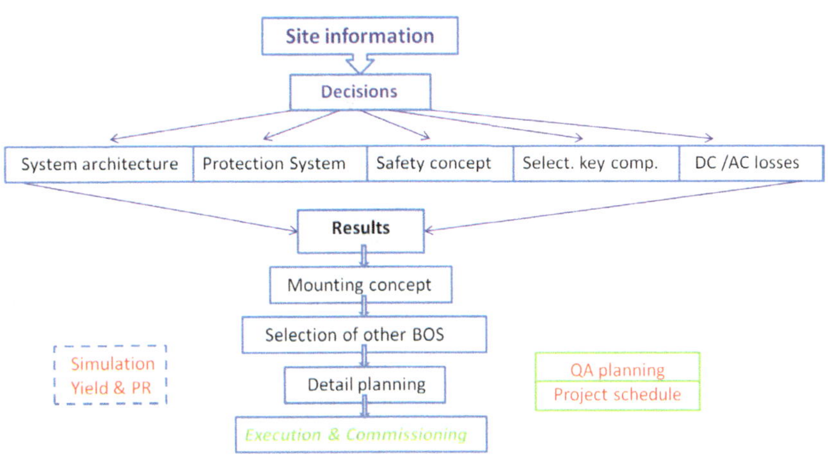

Figure A.8 Structure of planning and designing approach

condition the availability of funds for later stages of the project.

The feasibility study should anticipate the requirements of the credit and investment committees and the contents of due diligences. A high quality level, pluri-disciplinary feasibility study is therefore necessary to secure the financing. It should address, together with the technical aspects, all the key financial and legal issues.

Figure A.9 shows the key milestones in the development of a VLS-PV transaction.

In the VLS-PV transaction, the development costs and risks are quite high and the studies are complex. In this respect, the support of governments and development agencies should be as great as possible, directed toward removing the most difficult barriers. Complete studies should point out such barriers and suggest a pluri-disciplinary approach and appropriate management to guide these players.

Unfortunately, such studies are not yet available because, for most national support mechanisms, the value of PV is not at stake. The promoters that develop projects with the greatest value in current markets will not be awarded any extra benefit. On the contrary, as developing such projects is more expensive and time-consuming, promoters will tend to completely avoid developing high value projects, searching only for the cost reduction.

We believe that, given the great cost reduction of the last few years, future markets, based on value, will rapidly emerge and that PV will definitely meet with the conventional energy sector before 2020 in order to take its natural and specific place in the energy mix of all countries.

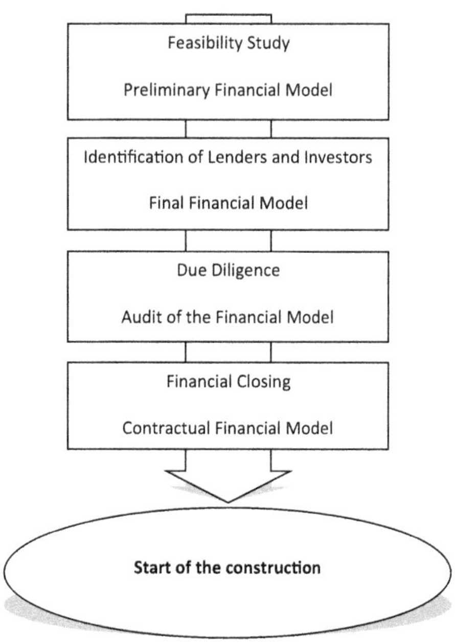

Figure A.9 Key project development milestones – evolution of the business model [15]

A.3 VLS-PV TECHNICAL OPTIONS AND APPLICATIONS

Grid matching issues

The intermittency in output from a VLS-PV plant renders its matching to the requirement of a conventional electricity grid an issue of quite complex proportions. In the first place, the grid has its own intermittency problems, associated with customer requirements. These problems are handled via a combination of economic considerations and the *controlled input* from three basic kinds of power plant: (1) Base-load plants, which have limited ramping capabilities. They operate on a continuous basis for extended periods of time with relatively infrequent shut-downs for maintenance. (2) Intermediate-load plants, which have improved ramping capabilities. They can be started up in the early morning and shut down in the evening. (3) Peak-load plants which are capable of ramping from zero to full power in typically 10 minutes and can undergo smaller ramping steps with extreme rapidity. They are used to follow the moment-to-moment requirements of the grid. Each of these plant types has its own economic considerations based on: fuel cost, frequency of maintenance, age of plant, cost of replacement, and moment-to-moment selling price of the electricity.

When it comes to attempting to input the power from a VLS-PV plant, a number of additional problems arise that result in new grid matching challenges. First, on clear days, a VLS-PV plant performs something like an intermediate-load plant in that it starts up in the morning and shuts down at night. However, unlike fossil-fuelled plants, a VLS-PV plant cannot be ramped up and down at will. Moreover, its peak output, which occurs at noon, will generally not coincide with the time of day at which grid requirements reach their peak. Thus, for example, if a VLS-PV plant is sized to provide the grid's peak load on a given summer day, it will produce more power at noon than can be used. Secondly, even on clear days, the peak output of a VLS-PV plant depends upon the time of year and its specific design. For example, a fixed-tilt plant will typically exhibit its peak output in the spring season, when noontime solar angles are close to normal incidence and ambient temperatures have not yet reached their summer highs. Now a consequence of the relatively benign springtime ambient temperatures is that air-conditioning loads have not yet reached their summer maxima. Therefore, if a VLS-PV plant is sized to meet the peak demand in summer, it will produce more power in spring than can be used. Yet a third challenge is the intermittency of clear days, a solution to which requires a combination of some form of weather forecasting, a facility for storage, and an appropriate grid operation strategy. Otherwise, precisely the same amount of backup conventional generating capacity will have to be maintained for the provision of power when solar input is not available. Finally, the effect of clouds on less than perfectly clear days can lead to grid instabilities unless the output of the VLS-PV plant is suitably smoothed.

A series of papers [16–21] have analyzed the hourly requirements of the Israel electricity grid during the year 2006, and simulated the outputs of various kinds of VLS-PV plant (static, 1-axis, 2-axes and CPV); alone, in combination, in combination with wind energy generation systems (WEGS); localized, and distributed over the Negev Desert.

It was found that very little improvement in grid penetration may be expected by distributing VLS-PV plants around the Negev. It is not clear whether this is a general result, or one that is specific to small geographical regions like the Negev. However, a positive result is that geographical distribution of VLS-PV plants significantly reduces the additional ramping requirements that a single plant would impose upon the grid. This has important implications regarding the lifetime of the non-solar plants in the grid and on their frequency of maintenance. Also, the addition of a suitably sized WEGS plant will be able to increase the rate of annual grid penetration from a pure VLS-PV to approximately 30 percent of the annual requirements. This may appear to be the best that one can do without storage.

The study then analyzed the required properties of storage and appropriate grid operating strategies for increasing grid penetration. It was concluded that up to approximately 90 percent of annual needs could be provided by VLS-PV, storage, and a revised grid operating strategy.

Although these results may not be quantitatively applicable to other regions, many of the results should be approximately correct in other desert areas. Furthermore the overall methodology will certainly be of wide applicability and the various details discussed here all need to be taken into account for optimizing VLS-PV plants wherever they may be.

VLS-PV intermittence and stationary storage

Use of stationary storage for VLS-PV systems will be influenced by the power output intermittence, impacting the grid requirement matching strategy. Photovoltaic power output intermittence prediction encompasses planning, forecasting and nowcasting. Nowcasting anticipates intermittence from the present time up to few hours ahead, forecasting covers up to 10 days, and planning refers to longer time periods. It helps VLS-PV systems to be operated in the same way as conventional power plants.

Regarding planning, yearly to seasonal intermittence is obtained through use of satellite-derived methods or ground-based measurements, and is part of important warranted items in PV project EPC contracts (Engineering, Procurement and Construction). Yet infra-yearly intermittence is not commonly assessed, because it is not commonly mandatory in Power Purchase Agreements (PPA).

Numerical Weather Prediction (NWP) models are used in planning to predict the values of a large parameters range, including solar radiation parameters and temperature, the two main factors impacting PV production. Such models being probabilistic, their use for short-term application is limited.

NWP models cover from the synoptic scale (eastern and central Europe, US and Canada), where they typically provide irradiation forecasts every 3 hours up to 72 hours (3 days) and every 6 hours up to 168 hours (7 days), at the regional and local scale, where they tend to higher spatial resolution, from 4 to 25 km, and at higher temporal resolution, especially for Rapid Update Cycle (RUC) models, allowing hourly predictions.

Model Output Statistics (MOS) techniques, based on statistical methods, use NWP model outputs to obtain solar radiation parameters and temperatures, or to directly predict the PV power output.

Regarding forecasting and nowcasting, daily to very short-term intermittences are linked to the variation of meteorological parameters: temperature, pressure, wind speed and direction aerosols content and water vapor content mainly. They are translated into radiation extinction when dealing with short-term intermittence.

These parameters are taken into consideration to perform classification of solar radiation distribution in typical days, which is important for predicting cloudy days where maintenance operations would be preferred. Classification is usually performed using four classes: clear-sky day, intermittent clear-sky day, intermittent cloudy day and cloudy day.

Nevertheless, such classification is usually performed with a single station, neglecting the smoothing effect when dealing with large PV installations. The smoothing effect is more effective during intermittent regimes, because perturbation does not impact all the plant at the same time.

As an example, a simple calculation technique presented by Hoff and Perez [22] shows that getting a VLS-PV scale will decrease the 1 min expected relative power output variation from 60 percent with a 4,59 MW plant to 10 percent with a 1 GW plant.

Cloud motion models, contrary to NWP ones, are deterministic models, which predict cloud cover and therefore radiation extinction based on cloud advection.

Satellite-based models provide better forecasts than NWP ones under a 5 hours perspective. As they provide estimations for infra-hourly variations, imagery techniques, using generally a Sky Imager, provide better information for very short term and sub-kilometer predictions.

VLS-PV systems power output intermittence and prediction can be performed at various time scales, with the use of databases, and mathematical and statistical tools to predict the power output variations. Persistence models, i.e. forecast equals instantaneous measured parameters with adjustment of solar position, are used in benchmarking the prediction models.

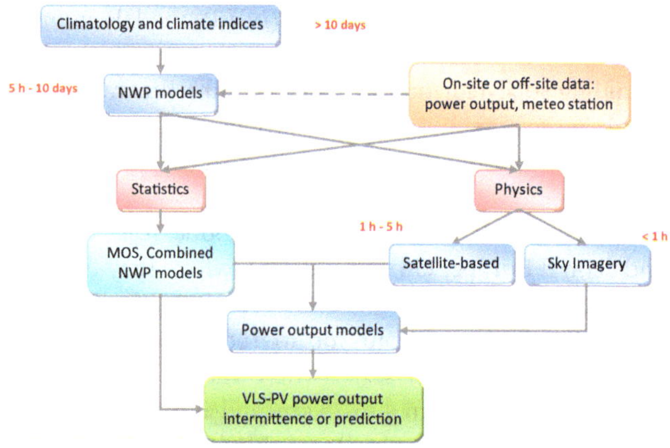

Figure A.10 PV systems power output intermittence and prediction at various time scales

Figure A.10 summarizes the temporal chain of VLS-PV systems power output intermittence and prediction; models are in blue boxes.

Stationary storage is divided into three sectors: chemical, mechanic (including here kinetic, magnetic and electrostatic) and thermal storage.[1] Their grid applications can be of two types: grid services (power quality control through fast delivery of mainly large amount of power and bridging power applications) and energy management (store electricity when the kWh price is low or when unused and inject back when the kWh price is high or to provide guaranteed electricity during identified periods).

Batteries, such as sodium-sulfur (NaS), redox-flow (RFB), nickel-cadmium (NiCd), nickel-metal hybrid (NiMH), zebra ($NaNiCl_2$), valve-regulated lead-acid (VRLA) and lithium-ion (Li-ion) usually considered for grid services applications, tend to be designed for energy management purposes. There are four implementation problems: self-discharge and depth of discharge (DoD), energy density, environmental impacts and maintenance, dismantling and recycling.

Kinetic (flywheel), magnetic (Superconducting Magnetic Energy Storage – SMES) and electrostatic (Supercapacitor Energy Storage – SCES) storage are designed for fast delivery of power, and therefore are used in specific grid services applications, as power quality for crucial applications.

Potential energy storage, Pumped-Hydro Energy Storage (PHES) and Compressed Air Energy Storage (CAES), are the most important installations that can be implemented on one site, and are today reserved for large energy management. Small scales are targeted; for PHES in coastal areas and for CAES through adiabatic operation. It will limit the development times and help get rid of specific sites.

Cost comparison between technologies is hard to achieve because of the variety of pricing regarding the project context. Therefore it is proposed to perform technology cost comparison on the basis of a targeted usable capacity (linked to the yield) and through collection of parameters for Capex and Opex calculation:

For Capex:

- Investment costs:
  - Procurement, assembly and tests
  - Installation and commissioning
- Replacement/refurbishment costs:
  - Replacement numbers over lifetime (depending on the annual operability and the number of cycles)
- Cost proportion compared with initial investment.

For Opex per year:

- Spare parts
- Man-hours
- Consumables.

Predicting intermittence will decrease the need for reserves, progressively bringing PV production from unpredictable to guaranteed. Integrating photovoltaics on grids will require adequacy between available production and demand and participation in grid balance strategies.

Production variation can be performed with coupling to other processes, with energy dumping or with energy storage. Such storage can be implemented by the utility itself or by the operator.

Storage on grid can improve grid stability, by virtually aggregating produced and/or consumed electricity on specific grid spots (homes, small plants, industries, etc.), and then improve flexibility faster than power line creation. Private entities have invested in this sector, bringing a constraint to the storage availability and the associated flexibility. Batteries are the best candidates for this kind of storage, as it remains in MW-scale. Battery limitations must be carefully considered.

Onsite storage refers to the ability of a VLS-PV system to participate in voltage and frequency regulation and to perform load leveling to fit grid requirements and maximize the grid integration. Batteries have specifications that fit with voltage and frequency regulation purposes, but careful attention must be paid to the size of such systems. Large storage capacities will be required for load leveling, and therefore PHES and CAES will be the best candidates.

Onsite storage will therefore not rely on a specific solution, but on a combination of solutions, allowing the VLS-PV system to be considered as a conventional power plant.

Renewable power methane

A 100 percent renewable power supply on a low cost basis is prerequisite for sustainable global development. Solar and wind resources are abundantly available on Earth, enabling the use of PV and wind energy technologies on a large scale in most regions in the world. A global energy supply scenario based on PV and wind power supported by an appropriate energy storage infrastructure is investigated. Results for the degree of complementarity of PV and wind power supply are presented [23] and the need for appropriate energy storage solutions is emphasized.

The renewable power methane (RPM) storage option is presented and discussed for the integration option of hybrid PV-Wind-RPM power plants. Renewable power methane storage enables bidirectional linking of power and gas networks and represents a competitive seasonal storage option. Due to a comparably low efficiency of the full RPM process, the cost of producing RPM is rather high. Therefore, the levelized cost of electricity (LCOE) of the input power needs to be as low as possible. PV and wind power will reach quite competitive LCOE by the end of the 2010s, are abundantly available and show a high degree of complementarity in time. However, both technologies are still fluctuating. Thus combining low cost PV and wind power with the balancing RPM storage to hybrid PV-Wind-RPM power plants represents a new power option for a 100 percent renewable energy supply [24].

Based on the LCOE approach and on cost assumptions for the year 2020, hybrid PV-Wind-RPM power plant economics are derived on a global scale. By the end of the 2010s, the economics of hybrid PV-Wind-RPM power plants are very promising in all regions of good solar and wind resource quality. Hybrid PV-Wind-RPM power plants might represent the fundamental centrepiece of sustainable and low cost power supply in the years to come (Figure A.11).

Estimates for the global energy supply potential of hybrid PV-Wind-RPM power plants show both rapidly increasing competitiveness and low distances between the centres of demand and least cost energy supply, which is complemented by abundant solar and wind resource availability. By the year 2020, about 90 percent of mankind might be supplied by 100 percent renewable power fully competitive to fossil fuel prices of about 150 $USD/barrel_{eq}$

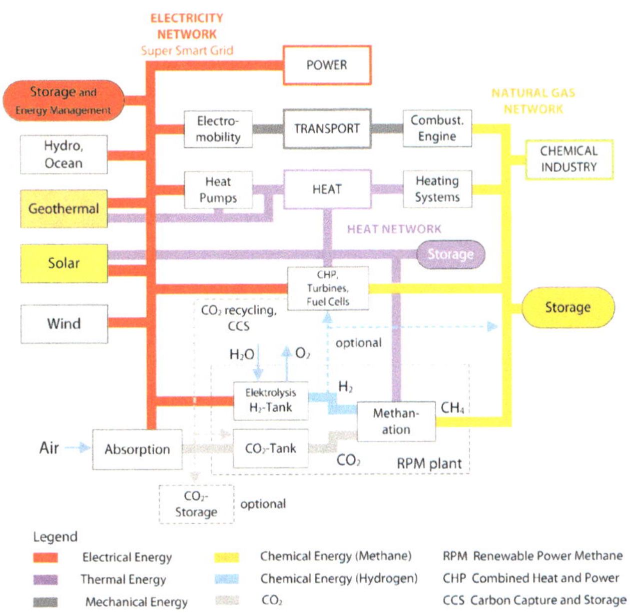

Figure A.11 Hybrid PV-Wind-RPM plant as the integral centrepiece of a future sustainable energy supply system [24]

and with practically unlimited amounts of sustainably provided energy. The RPM approach enables long-term cost stability due to a fully decoupled cost structure from fossil fuels, no net CO_2 emissions and enormous power supply potential offering long-term and sustainable economic growth [24].

This hybrid plant topology might emerge into the role of the key energy supply cornerstone in the world, in particular if mankind intends to economically survive peak oil and physically and economically survive climate change.

A.4 VLS-PV CASE STUDIES

VLS-PV in West Africa

The implementation of VLS-PV systems in West Africa is quite a challenge given the current situation of the power sector in the region. Indeed, the recent development of interconnection systems, the difficult financial situation of the national utilities and the absence of an efficient information management system make it difficult to implement any power transaction in West Africa, let alone new types of transactions such as large scale photovoltaic power plants.

To implement VLS-PV, the strategy should be defined at a regional level, and then spread and managed at a national level, while still being audited and monitored by the regional institutions.

Concretely, these principles could be embodied as shown in Figure A.12.

As an example, a possible approach for Benin was discussed. Benin is an energy-dependent country that relies mainly on expensive imported hydrocarbons for its electricity production. Furthermore, electricity contributes to a very small portion of Beninese primary energy: only 2,2 percent, which represents the third most important source of energy after oil products (38,4 percent) and biomass (59,4 percent).

Given economies of scale in the PV industry, hydrocarbon prices volatility and its upward trends, Benin electricity transport and distribution infrastructure development as well as significant demand growth forecasts and, finally, considerable local solar resource, Benin possesses all assets required for the development of electricity production based on solar technologies.

In this favorable context, we have established a Reference Scenario (see Table A.1) that supposes VLS-PV plants installed capacity representing 10 percent of peak power demand by 2025. According to our experience, the development of such PV capacity in Benin (68 MW) is reasonably achievable from a technical, economic and "project development" perspective.

Helios Energie, a French PV project developing company operating in international markets such as sub-Saharan Africa, has recently conducted a technical and

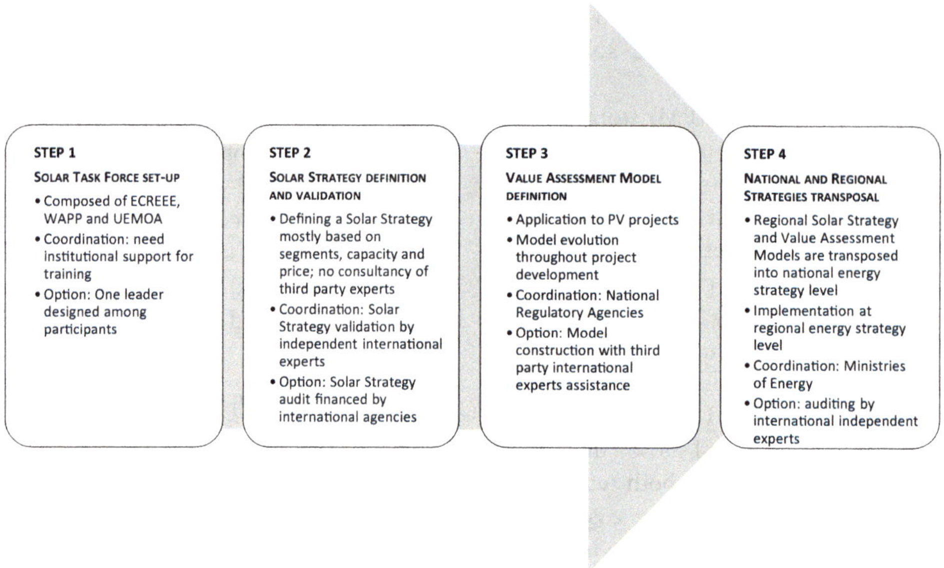

Figure A.12 Steps for implementation strategy

Table A.1 VLS-PV 2025 reference scenario for Benin

Year	2005	2010	2012	2015	2020	2025
Total electricity consumption (GWh)	560	1 002	1 260	1 647	2 484	3 781
Transmission and distribution losses (%)	18	18	17	15	14	14
Total generation (GWh)	661	1 182	1 467	1 894	2 832	4 310
Average annual growth (%)		9	9	8	7	7
Peak power demand (MW)	105	188	233	300	449	683
Average annual growth (%)		9	8	7	7	7
Per capita GDP (×1 000 FCFA)	133	149	168	196	279	408
Average annual growth (%)		2	3.2	5	6	6
VLS-PV installed capacity (MWp)	-	-	6	20	44	68

financial feasibility study for the implementation of a 6 MW photovoltaic plant on the site of Kandi, located in the north of Benin. This study was finalized in April 2011.

The aims of the Kandi project for Benin are: supplying electricity during day time for the northern region of the country, contributing thereafter to the grid stability at this period and off-setting part of the thermal generation that currently supplies the region. Furthermore, this 6 MWp project displays an intrinsic environmental, economic and social value for the region as the first very large scale PV project in the region.

Today, the Kandi project is at an advanced stage of its financial closing process, following the favorable opinion of the French Development Agency (AFD). The selected builder-operator is Siemens Energy France. Construction itself should begin subsequent to the loan agreement by AFD in mid-2012. The electricity produced by this 6 MWp installation, currently expected to be the largest grid connected PV power plant in sub-Saharan Africa, will be amortized over a 25-year period through production sale to SBEE (Société Béninoise d'Energie Électrique).

Once commissioned, this project will represent a significant step forward for the development of VLS-PV power plants in sub-Saharan Africa, showing the feasibility of such a business model in this region of the world.

Figure A.13 Computer image of the 6 MWp PV power plant project in Kandi, north of Benin [25]

VLS-PV in Israel

World electricity production is currently undergoing a quadratic growth with time. This relentless upward trend in electricity production is presumably driven by a combination of population growth and rising industrialization. It is consequently a fact that must be taken into consideration for any long-term planning. For example, by extrapolation through only the coming ten years, electricity generation by 2020 will probably be nearly triple what it was a decade ago in 2000.

On the world energy scale, the state of Israel is of course minuscule, 60 TWh being its estimated 2010 production [26] – i.e. less than 0,3 percent of the world total. However, Israel is a "sunny" country, with the relatively empty Negev desert, and it has the industrial capability of implementing VLS-PV on a relatively massive scale.

In the case of Israel, and its linear electricity growth curve, Figure A.14 indicates that VLS-PV would need to provide 1,92 TWh per year in order to replace the country's present average increase of 366 GW each year of new fossil-fuelled generating capacity. In terms of VLS-PV, taking a nominal annual output of 2 TWh/GWp for a typical desert location [26], this means installing some 960 MW each year.

Naturally, when a bank offers a loan, it expects to be paid back, and there are two classic instruments for paying back the loan on a PV system. One is the so-called feed-in-tariff (FiT) that many governments have adopted. This instrument requires electrical utilities, by law, to purchase PV-generated electricity at a higher rate than it costs them to generate conventional power and a surcharge for all power customers for covering the higher cost. This enables a return on the loan within a time frame that is deemed acceptable to banks. The second method of financing PV is for the government to offer tax incentives to help the purchaser of a PV system cover part of the cost of the initial investment.

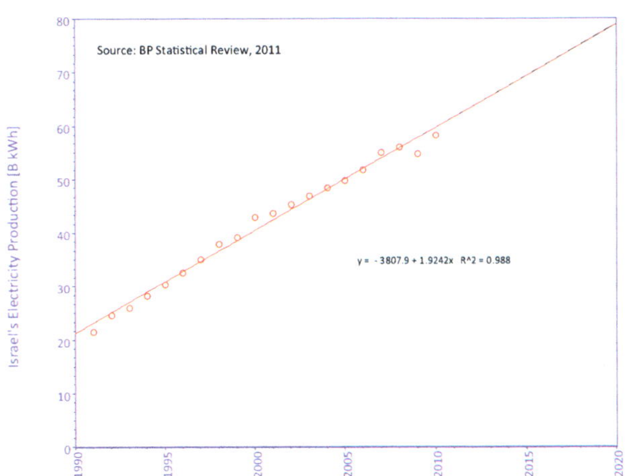

Figure A.14 Israel's electricity production during the past 30 years [27]

In both of the above cases the PV system is treated as an up-front investment that has to return its capital outlay over a period of years. In the case of relatively small PV systems, the amounts of money involved are sufficiently small that many people are prepared to take the risk. But evidently the big lenders – i.e. the banks – have their own ideas about how large the risk may be allowed to become. This may be wisdom on the part of the banks, but if the world is urgently in need of constructing hundreds of GW of VLS-PV plants *each year*, financing of the required investment is unlikely to become available in time to avert catastrophe.

Although there is still some debate in scientific circles as to whether the presently perceived era of global warming is entirely due to anthropogenic causes, there can be little doubt that the increasing use of fossil fuel is definitely a non-negligible contributory factor to this warming. As such, it is in the public good for the world to try and cut down on the burning of fossil fuel. Now many people wrongly assume that this automatically means impeding the development of developing countries. "After all", they argue, "the developed countries are the ones responsible for global warming, so why should the developing countries be made to pay the price of the formers' mistakes?" They clearly have a strong case. However, our claim is that VLS-PV can allow full development of developing countries with a low carbon footprint, while at the same time, enabling the developed countries to clean up their act. The only problem is how to implement VLS-PV on a large enough scale and at a fast enough rate.

Traditionally, the public good is paid for by taxation. In developed societies, such goods as health, education, safer roads, etc. are normally paid for, at least in part, by taxation. Society does not inquire whether a hospital or a school is a good "investment". Neither does it consider how long it will take to return the initial capital cost. It is taken for granted that such capital outlay is for the long-term good of all. And so it should be in the case of "green energy", particularly in a world that is urgently in need in reducing its use of fossil fuel.

In light of the above discussion, we examine the consequences that could emerge if, for example, Israel were to impose an electricity consumption tax (ECT) on all use of electricity in the country. As will be seen, such a tax could not only fund the introduction of VLS-PV on a massive scale, but it would almost certainly lead to a number of additional benefits that could also contribute to lowering the global carbon footprint.

At the present time, the cost of electricity to Israeli consumers is approximately 0,08 EUR/kWh (slightly rounded down for simplicity). Suppose a nominal ECT of 20 percent was to be imposed upon electricity *consumption* (treated here, again for simplicity, as being identical to *production*). From Figure A.14, such a tax during 2010 would have realized 0,96 billion EUR: enough funding – in a single year – for the construction of a 480 MW VLS-PV plant.

Under desert conditions, such a plant would generate, on average, 0,96 TWh of electricity per year. Hence, if no additional fossil-fueled plant were added to the country's existing approximately 10 GW of generating capacity, the production/consumption for 2011 would reach 60,96 TWh. The ECT on this would in turn yield a revenue of 0,975 billion EUR, which could fund a second VLS-PV plant, this one of capacity 488 MW.

A.5 POSSIBLE SCENARIO AND STRATEGY FOR VLS-PV

Expected roles of international cooperation

Significant advances and dramatic cost reductions have been demonstrated in PV technologies over last decade. Further cost reductions of PV are expected due to technological innovations, economies of scale and increased market competition among PV industries worldwide. It is clear that support for high level research and facilitating exchange of information and experiences are crucial if technologies are to be continually expanded, and costs lowered. This can be realized only through the active collaboration of all stakeholders on local, regional and global levels.

Despite the fact that the cost of PV is declining continuously, a VLS-PV system of several hundred MW or GW scale capacity will require increasing amounts of investments. In order to promote VLS-PV systems worldwide, especially in developing countries located in deserts, innovative and favorable financing approaches and active international collaboration are required. Major international collaboration is needed to expand R&D in all areas of VLS-PV systems, especially in policy, regulation, standards and business model development. International cooperation in these areas stimulates the strengthening of international collaboration of existing institutions and activities, as well as the creation of new joint initiatives at regional and global level.

Regional and national programs promoting PV as a mainstream energy source are crucial for further development of large scale PV worldwide. There is an emergence of some of the largest solar promotion programs globally. If such programs are able to achieve the scale envisaged, they have the potential to create a huge wave of demand, which, in turn, will spur increased innovation, efficiency in production, and further cost reduction in PV. The competitiveness of PV against traditional and other alternative energy sources around the world will consequently increase.

In the near future, VLS-PV systems should become an economically viable option to meet the electricity needs of communities in remote desert regions around the world where conventional power plants cannot be built. The rate of deployment of VLS-PV is greatly influenced by the perception of the general public and utilities, local, national and international policies, as well as the availability of suitable standards and codes to govern it. Among a variety of support mechanisms, Feed-in Tariffs (FiTs) would be the main reason behind the strong growth in large scale PV in developed countries.

In the long term, with a solid strategy for implementing of VLS-PV systems, further expansion of large scale PV generation not only can satisfy the increasing energy demand of the region in a sustainable way but also will stimulate the local PV industry and create enough need to justify local manufacturing of PV system components. In addition, diffusion of the various kinds of high-tech knowledge and experience will mean that know-how will be transferred to local renewable energy institutes, utilities and energy companies in order to keep track of worldwide technology developments, technology exchange between universities and scientific institutes.

In January 2012, the International Energy Agency (IEA) and the International Renewable Energy Agency (IRENA) signed a partnership agreement which will strengthen cooperation between the two organizations. There will also be increased collaboration between the two agencies at the Secretariat level, and in energy technology networks. Some themes for international cooperation have been identified. In addition to regional and national programs, this agreement will further advance international collaboration to stimulate the global deployment of large scale PV power generation systems.

Possible contribution of VLS-PV to sustainability

According to the United Nations, the global population will reach 9,3 billion (i.e. approximately 35 percent increase from today's figure) in 2050 and continues to increase until 2100 [28]. Asian Development Bank (ADB) estimates that global GDP will expand to 292 trillion USD in 2050 from 63 trillion USD today [29].

To what extent can PV/VLS-PV contribute to sustainable issues?

In order to answer this question, a simplified model as shown in Figure A.15 was developed. The model looked carefully into the relations between various sustainability factors to retain the consistency among the factors.

Four scenarios were developed, as shown in Table A.2, namely the base scenario, SC-1, SC-2 and SC-3. For all the three SC scenarios, a boost in VLS-PV was assumed. The installation scenario of VLS-PV was identical to the VLS-PV roadmap presented in the previous edition of this book [3] (see Figure A.16).

SC-1 and SC-2 scenarios in comparison with base scenario

With the increase in PV/VLS-PV, consumption of other energy resources decreases considerably. As is shown in

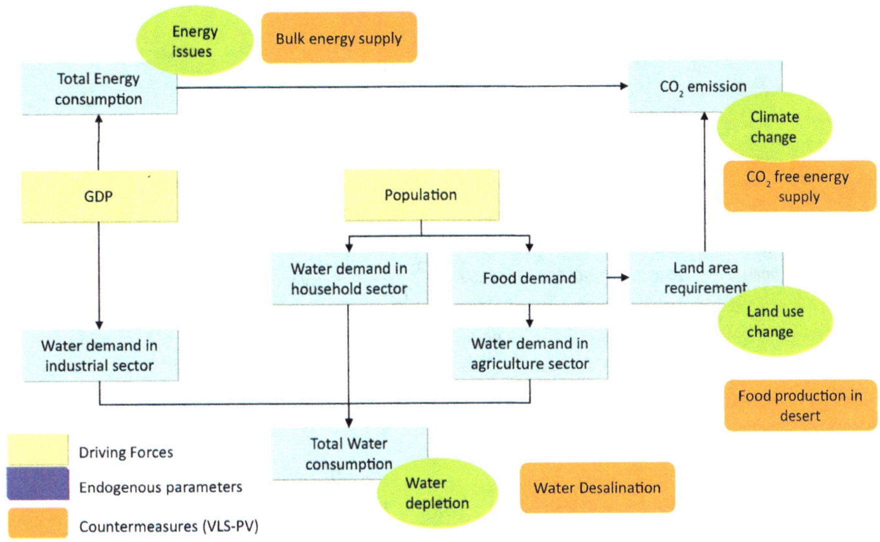

Figure A.15 Model structure

Table A.2 Summary of the scenario descriptions

VLS-PV scenario	Description
BaU	Business as usual. It is assumed that the current trend continues until 2100.
SC–1	Power generated from PV/VLS-PV is used directly in the energy mix. All kinds of energy are replaced by VLS-PV.
SC–2	Power generated from PV is used directly in the energy mix. Only fossil fuels are replaced by PV/VLS-PV power generation.
SC–3	Power generated from PV is used for water production (reverse osmosis). The remaining energy is used for power supply.

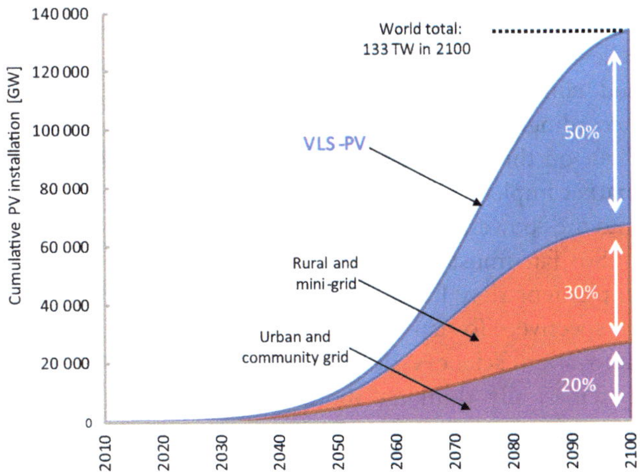

Figure A.16 VLS-PV roadmap: cumulative installation by PV application [3]

Figure A.17, PV/VLS-PV can provide 25 percent of primary energy in 2100. The contributions of PV/VLS-PV in energy saving for coal, oil and natural gas are approximately 2 600 Mtoe, 450 Mtoe, and 2 900 Mtoe in 2100 respectively.

Cumulative CO_2 emission reductions in the projected period reach 531 Gt CO_2-eq in 2100, which is equivalent to 19,5 times today's annual emissions (Figs. A.18 and A.19).

Figure A.20 is the projected temperature increase corresponding to the GHG concentration on the atmosphere. Although the PV/VLS-PV may not be a single and perfect solution for climate change issues, it is clearly shown that the potential contribution is considerable. Combining with other renewable energy technologies or energy efficient measures would be necessary for a sustainable future.

Base scenario versus SC-3 scenario

The SC-3 scenario, on the other hand, is unique. Under this scenario, the power generated from VLS-PV is used for water production. Reverse Osmosis (RO) technologies would be used in order to fulfil the increasing global and water demand. The overall concept of the scenario is to produce water in the most arid and solar resource rich areas, the deserts. The remaining energy is used for power supply. Only VLS-PV in the desert regions are used for desalination applications. Other applications such as roof top PV in urban areas or PV for rural electrification are for conventional electricity use.

Figure A.21 illustrates the additional water demand projected in the base scenario and water production potentials from VLS-PV. Water supply potential from VLS-PV will be enough to fulfil the additional global water demand in 2060 and will be much higher thereafter. The potential will be more than 3,4 times in 2100.

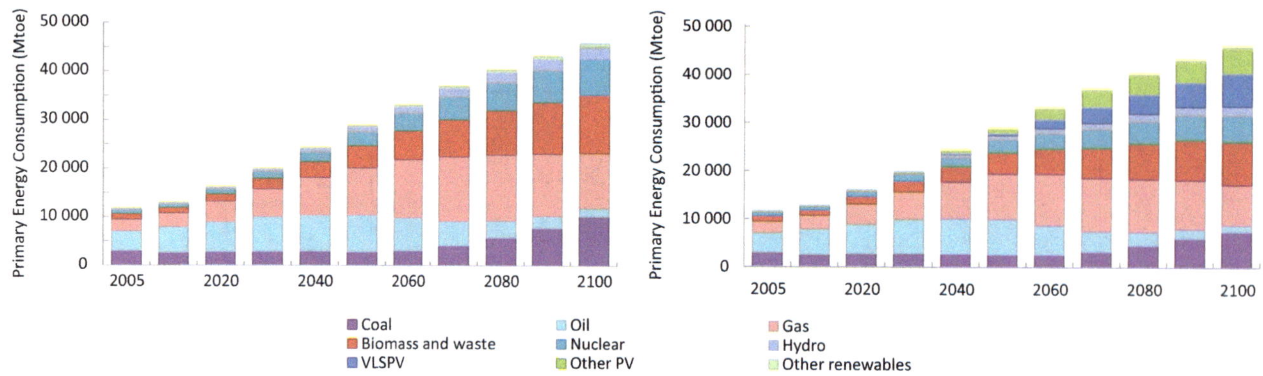

Figure A.17 Primary energy consumption of base scenario (left) and SC-1 scenario (right) [Mtoe]

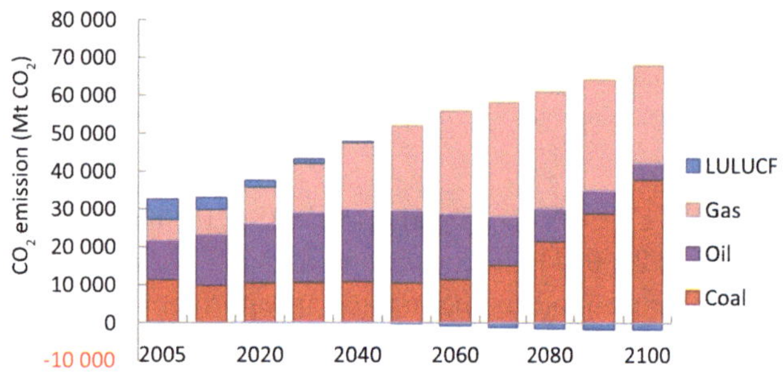

Figure A.18 GHG emissions of base scenario [Mt-CO_2eq]

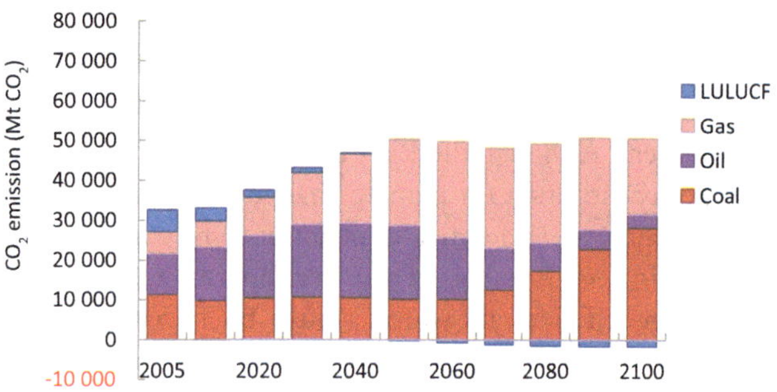

Figure A.19 GHG emissions of SC-1 scenario [Mt-CO_2eq]

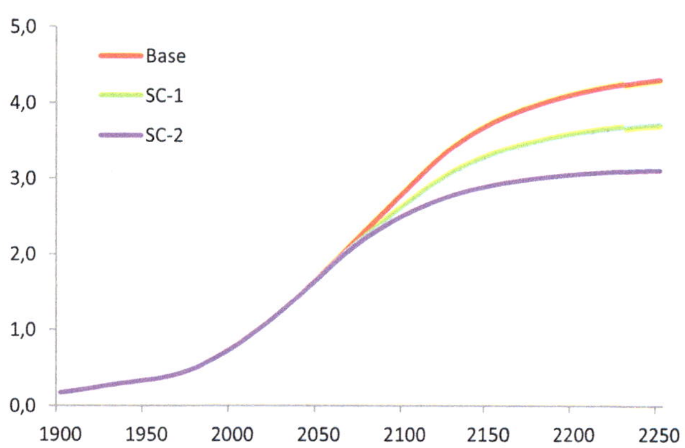

Figure A.20 Temperature change caused in the three scenarios

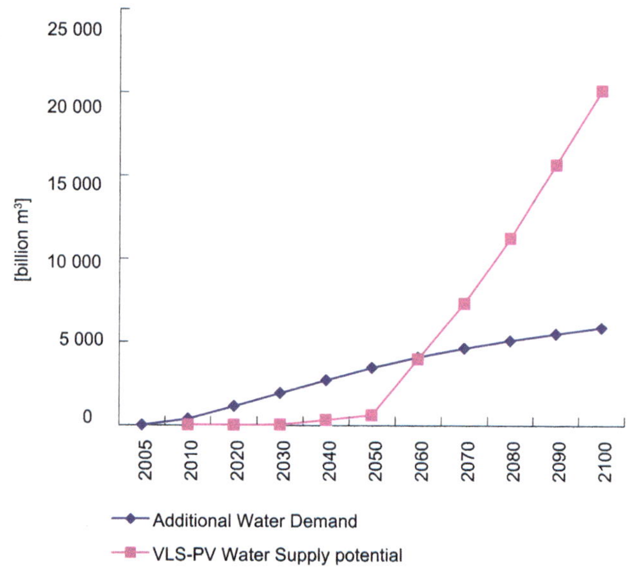

Figure A.21 Water demand and supply capacity (billion m^3)

Strategies for solar power plants in desert regions

Solar power plants in desert regions increasingly count as a realistic energy option. Very large scale ("utility scale") solar power plants have started to appear around the world. From the perspective of the existing global energy system, however, large scale solar power plants remain a promising yet very small niche. They still require financial and political support, entrepreneurs still search for viable business models, and acclaimed benefits such as relieving energy scarcity, job creation, poverty alleviation, and climate change mitigation are still to materialize. Accordingly practitioners and researchers increasingly focus on the challenges of socio-economic feasibility and market implementation to turn desert solar power plants into the "powerhouses of the world".

So far, most studies of socio-economic VLS-PV implementation have taken an informed practitioner's perspective. To tap into the social sciences and the innovation sciences in particular, to suggest ways of systematically thinking about VLS-PV implementation and aggregate existing insights it is necessary to go beyond conventional socioeconomic assessment tools. These normally chart and weigh the pros and cons of a given technological option, and can work well for stable technologies in equally stable societal contexts. However, they fit poorly with the case of solar power plants in deserts. Radical innovations (as opposed to incremental innovations) such as solar power plants in desert regions tend to develop so fast that evaluations at a fixed point in time tend to be instantly outdated and lack serious predictive value. In addition these conventional assessment tools fail to appreciate the intimate relationship between innovation and its societal context. In the case of solar power plants in desert regions this context is in constant flux. For instance, the current financial crises and the political turmoil in a number of North African states (which count as promising solar power plant sites) may greatly affect the odds, forms, implementation trajectory, and benefits of desert solar power plants. Clearly we need to search the innovation sciences for ways to think about VLS-PV implementation concepts that are more dynamic and take changing contexts into consideration. In the following we take our cue from the booming field of sustainability transition studies. This particular field within the innovation sciences asks questions that solar power plants face today: Why do some radical innovations transform from technical possibilities into viable market niches and eventually become part of (or even displace) the dominant technological system, while others do not? What lessons can we draw for the management and implementation of sustainable technologies in uncertain contexts? [30, 31]. We here discuss three such approaches.

Strategic Niche Management (SNM)

SNM is an analysis, policy and practitioner tool that seeks to study and strengthen the dynamics of emergent sustainable technology niches and their transition to full-blown technological systems or "regimes". The SNM approach focuses on the overall level of the niche, in our case solar power plants in desert regions, where individual projects are aggregated. The point of departure is that such radical innovations may be desirable in the future from a sustainability perspective, but face grave competitive disadvantages today. These disadvantages partly stem from the technological and economic uncertainties that accompany most new technology. Other, and arguably more fundamental, disadvantages follow from the fact that radical technologies may not fit well with existing dominant technological practices – precisely because they represent a break with conventional ways of thinking and doing. Consider how national employment, industrialization, and energy security policies support conventional energy players and technologies. So do existing technological capabilities in most countries. Even such mundane matters as insurance rules, planning and construction procedures, energy cost calculation methods (how to account for externalities such as climate change or nuclear risk?), energy user behavior, and so on have been tailored over many years to suit conventional energy stakeholders and technologies. Through these mechanisms and others, existing energy regimes tend to lock-in on conventional power plants and lock-out radical innovations that are at odds with its well-entrenched economic, organizational, legal, technological and other practices. This is why many promising energy innovations eventually succumb or gain only marginal importance.

Alternatively radical energy innovations may develop in niches where they are temporarily shielded from dominant energy regime forces and pressures. SNM seeks to understand and promote sustainable innovation through the deliberate creation, sustenance, and up-scaling of such niche processes. It suggests how proponents can use protected niche spaces to experiment with technical and non-technical components of their novel technology, and thereby find a constellation that fits with opportunities and demand in the outside world. If the learning process works well, lessons learned in individual projects accumulate into a "global niche level". In this perspective, our work on the introduction of VLS-PV strategies in desert regions is an example of a global niche player that accumulates insights on the level of desert solar power niche as a whole. This work includes selection, comparison, framing, contestation, and interpretation of local lessons. Conversely, global niche level insights can be used to frame individual local projects so as to optimize the learning process from an overall niche perspective.

SNM does not merely consider internal niche processes, however. In order to systematically chart the opportunities and barriers for niche technologies in the outside world, it adds the analytical layers of the *sociotechnical regime* and the *sociotechnical landscape* (Figure A.22).

While niches nurture radical energy innovations, incumbent regimes host dominant energy stakeholders with vested interests and stable social networks; regulations, standards, and cognitive routines that may blind actors to developments outside their habits and focus; and conventional energy technologies with their sunk investments and historically grown technical complementarities, such as fits with existing infrastructure. For the niche of desert solar power plants, one may think of the incumbent regime of centralized electricity generation in fossil fuel or nuclear power plants. This regime has long been entrenched within national borders, though in some parts of the world it is rapidly internationalizing [33]. Regime components such as dominant stakeholders, government legislation, and conventional power plants tend to interlock, keep each other in place, and resist radical change. SNM needs to learn about the barriers and opportunities for niche technologies in incumbent regimes and feed these back into niche development processes.

The socio-technical landscape, finally, draws into the analysis exogenous forces on which both niche and regime players have little influence. In some cases such forces seem to constitute additional barriers to change. In the case of desert solar power, think of the political turmoil in desert regions, or financial crises that make financiers wary of risky investment. Other landscape forces and events, however, may put pressure on stable energy regimes and open up opportunities for niche technologies to enter. Examples that may benefit desert solar power include global CO_2 emission reduction agreements, national or international (e.g. European Union) renewable energy targets, or disasters; the Tohuku tsunami of 11 March 2011

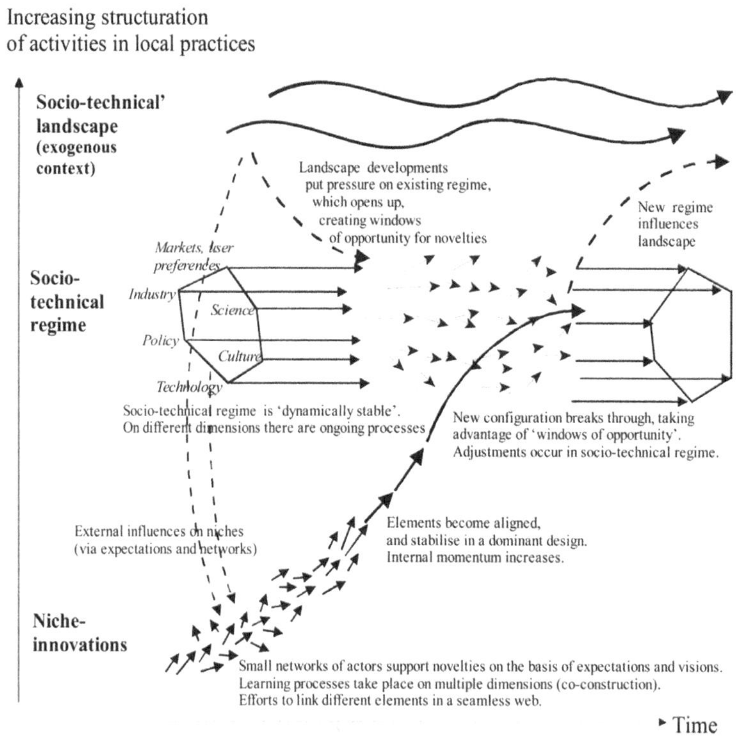

Figure A.22 A dynamic multi-level perspective on radical innovation journeys [32]

and subsequent crisis of the Fukushima nuclear power plants in Japan inspired – at least temporarily – a public and political distrust of nuclear power and renewed interest in renewable energy sources worldwide, as the Chernobyl disaster did decades ago.

In this three-layer analytical framework, the challenge is to propel promising energy technologies from deliberately protected technological niches into viable market niches and eventually latch on to (or even displace) incumbent energy regimes. Landscape pressures on existing regimes provide opportunities for such change. Successful innovation in this perspective is about navigating niche, regime, and landscape dynamics, so as to find a workable innovation journey for the novel technology. Today we can see similar processes emerging in the desert solar power niche. We interpret the recent reframing of desert solar power plants from a utopian very large scale technical possibility to a "utility-scale energy option" as an alignment with the existing electric power regime, and electric power producers in particular. In the EU-MENA region, linking up with forces for a Mediterranean Ring and a European super-grid makes Sahara desert power plants increasingly attractive. Simultaneously the connection of VLS-PV to seawater desalination plants links up the desert solar power niche with quite another regime, that of drinking water supply in desert regions. On the other hand desert solar power plants face particular challenges, especially when they involve international energy systems. Again the EU-MENA region illustrates this: Sahara solar power exports to Europe might prove difficult when European governments provide massive visible and invisible support to domestic conventional power plants that provide domestic employment, turnover, and energy independence. VLS-PV practitioners are already grappling with such opportunities and barriers. SNM is designed to learn about and manage such processes systematically.

For the SNM, ESTEEM (Engage STakeholdErs through a systEmatic toolbox to Manage new energy projects) [34] is about taking seriously and managing social resistance to local projects. Local resistance to novel energy technologies can take various forms, ranging from a simple lack of interest and denial of cooperation to emotional stress and actual sabotage. ESTEEM is a practitioner toolkit to deal with societal acceptance issues in the implementation stage of local projects, when adaptation of the project design is still possible. ESTEEM is typically facilitated by consultants.

New studies of business models

New approaches in the business models literature discuss how entrepreneurs may systematically search and design business models for radical innovations in unstable contexts [34, 35]. Business models are about describing "the rationale of how an organization creates, delivers, and captures value" [36] and are increasingly seen as a source of competitive advantage for a company. Business models are key instruments for bringing new technologies such as VLS-PV to the market [37]. Several scholars are currently working on combining SNM with business model development into one framework, but this work is still in its initial phases. We limit ourselves to outlining selected business model insights and thereby mobilize these for our thinking about the implementation of desert solar power plants. It could be said that the business models approach addresses the project level, but creates a sort of global niche infrastructure to facilitate a joint exchange and learning process [36]. Either way, in terms of content it is used as practitioner experiences to distill the building blocks that a good business model should address. It also defines for instance collaboration with partners and customer involvement. This applies to commercial companies as well as non-commercial organizations such as NGOs or government institutes that aim at creating environmental or social value.

Solar power plants in desert regions are on the brink of finding their way into the world's energy systems. This could mean that preparatory work, done by the International Energy Agency VLS-PV group and others, is currently being "overtaken by market developments" and can be concluded. From the innovation science perspectives, this would be a false conclusion. As policy makers and entrepreneurs experiment with projects and implementation models, now is the time for systematic learning about VLS-PV on a global niche level. For only the confrontation with real life conditions can reveal the technological, business, socio-economic, and geopolitical fits and misfits of solar power plants in desert regions with the demands of the modern energy world. It is now that the main (potential) stakeholders need to learn about expected and unexpected bottlenecks that emerge in individual local projects, accumulate such lessons on an overall niche level, experiment with ways to navigate these real world opportunities and barriers, and invent viable innovation pathways. Importantly, socio-economic or geopolitical bottlenecks might very well feed back into the technology design process: the redesign of solar power concepts for desert regions, originally conceived to serve long-distance international energy flows, to accommodate local economic development and sea water desalination illustrate technology adaptations in the (early) market implementation phase. In sum, the coordination between individual local projects and the overall desert solar power niche community is more important now than ever. The VLS-PV niche requires continued reflection on strategic management on a niche level, and a continued search for social acceptance and business models on a project scale. To facilitate this process we have tapped into innovation sciences work on sustainability transitions. This chapter has outlined the approach of SNM and briefly introduced ESTEEM and business model experimenting. This initial mapping of the territory, however, has merely opened up the topic. A next step would be a more VLS-PV specific and fine-tuned inquiry. A follow-up study should systematically examine existing or ongoing projects for current niche dynamics (in terms of articulating expectations, building social networks, organizing learning processes, shielding and empowering the niche, and emerging business models) and initial encounters with real life contexts (in terms of technological, business, socio-economic, and geopolitical opportunities and barriers, and expected/unexpected bottlenecks that pop up). With the practical tools, we may translate these lessons into a learning-based implementation strategy for policy makers and entrepreneurs.

A.6 CONCLUSIONS

Understandings

Solar energy resources, PV technologies, and renewable energy will help to realize important economic, environmental and social objectives in the twenty-first century, and will be critical elements in achieving sustainable development.

In order to promote renewable energy, the diversity of challenges and resource options, as well as the financing and market conditions among and within regions and countries, implies that different approaches are required. Establishing policies for developing markets, expanding financing options, and developing the required capacity have been discussed in order to initiate and adopt policy changes and to incorporate the goals of sustainable development into their policies.

Since our previous works [1, 2, 3], solar power plants in desert regions increasingly count as a realistic energy option. Very large scale solar power plants have started to appear around the world. In order to accelerate the current transition towards renewable energy, development of VLS-PV power generation systems that can be deployed on a massive scale is needed.

Based on such background, this report presented an overview of relevant issues and the potential of VLS-PV, guidelines for VLS-PV systems, and technical as well as strategic options for implementing VLS-PV.

Our conclusions are summarized as follows.

Projection for the future

Many scenarios for future global energy supply exist, but the roles of PV and other renewable energy options vary greatly between the scenarios. These are heavily dependent on underlying assumptions and projection approaches. In general, the share of PV is expected to increase drastically in most of the scenarios, but it may take a few decades for PV to be a main energy source in the global energy mix.

Our study using a simplified model for sustainability issues suggests that:

- VLS-PV can contribute greatly to global sustainability, but the main contribution will come after 2050.
- The contribution of VLS-PV is heavily dependent on the "business as usual" scenario assumed. Energy efficiency and other renewable energy measures are also of major importance.
- Taking into account energy depletion issues (resulting in increasing energy costs), VLS-PV could be a

cost-effective alternative, although it will still need continuous improvement.

- It should be kept in mind that such innovative changes in the energy supply system require a certain lead time. Also, delay in actions would lead to more serious problems since the driving forces are gaining importance.
- Although the contribution of VLS-PV will be marginal in the next 10 years, immediate action is very important from a sustainability perspective.

PV has already obtained a higher share in the energy mix than some projections presented in the past (for example, IEA: World Energy Outlook), and should continually beat the projections in order to become a main energy source. Meeting the projection by IEA: PV Roadmap [4] and Energy Technology Perspectives [5] until 2050 should be the minimum requirement. Only then can we visualize the VLS-PV roadmap toward 2100 as we proposed in our previous edition [3].

Socio-economic and environmental benefits

Solar electricity generation at utility scale can offer the possibility for socio-economic growth in the region. Because of local availability of energy in desert regions, clean water can be provided, irrigation can be optimized, local food can be produced and local sources of raw materials can be exploited. The generated electricity can directly be used for lighting, communication, education, entertainment and for industrial purposes. Excess electricity can be exported to neighboring regions or beyond. By adapting a regional policy for solar electricity generation at utility scale, a sustainable local market for solar electricity modules and other related components and services will be created and this market can be served by local production. Technology transfer from industrialized countries can provide and improve education and learning, research and development and manufacturing and sales. Local solar electricity generation also provides for security of energy supply in a sustainable way, provides fair access to energy for everybody and reduces the threat of climate change. If managed in the right way, the closed circle between local resources and local needs offers extensive possibilities for further development of the region.

Needless to say, PV technology is environmental friendly. From a life-cycle viewpoint, VLS-PV could contribute to saving and mitigating fossil fuel consumption and greenhouse gas emissions. In addition, applying a VLS-PV policy for decreasing our ecological footprint reveals that VLS-PV has significant potential to implement a sustainable ecological balance.

Engineering and financing approach

Assessment and evaluation of VLS-PV takes place with economic and financial considerations in mind. These are always in coherence with a basic understanding of technical issues.

The quality of execution of the planning is mainly dependent on the qualification and experience of engineers and site managers involved. Beside electrical know-how, basic knowledge of related fields like testing procedures, mechanics and statics as well as soil science and meteorology are required. VLS-PV systems contain a large number of connected conversion devices, a lot of power electronics, measuring systems etc. Depending on the region, more and more VLS-PV installations are located far away from living environments. Therefore the issues of safe, secure and trouble-free operation, maintenance and servicing should increasingly be considered in the course of plant design. From these viewpoints, the technical engineering guidelines we developed in this report facilitate the access to a basic technical understanding of VLS-PV.

The development of a VLS-PV system is a long process where almost all parameters change until the financial closing is reached. The preliminary technical design will continually evolve during the development process, taking into account the economic, financial and legal issues that are raised along the way. In a VLS-PV transaction, the development costs and risks are quite high and the studies are complex. In this respect, the support of governments and development agencies should be directed toward removing the major barriers. Complete studies should point them out and suggest a pluri-disciplinary approach and management of the PV sector support. The study should also anticipate the requirements of the credit and investment committees and the contents of technical and legal due diligences. A high quality level, pluri-disciplinary feasibility study is therefore necessary to secure the financing. It should address, together with the technical aspects, all the key financial and legal issues.

Technical potential

Solar energy resources are available in all populated regions in the world. Looking at the technical variation of PV systems, fixed optimally tilted PV systems are the standard PV power plant system type today. However, depending on local conditions, adapted tracking systems will increase the annual yield per installed unit of capacity and lower the power generation costs.

Considering the technical potential of hybridization between VLS-PV and other renewable energy technology, the weaknesses of such technologies might be solved by engineering ingenuity. For example, due to the complementarity of solar PV and wind power, market expectations for PV may not need to be lowered. Moreover,

PV together with wind power might become the backbone of the global energy supply in the coming decades. Hybrid PV-Wind systems are likely to start to strongly influence the global power system in the next decade by means of sophisticated hybrid concepts, like renewable power methane technology.

In addition, a worldwide grid of ultrahigh voltage DC power lines (or perhaps ambient temperature superconducting cables) fed by the various VLS technologies may be envisaged. Not only conventional PV but also CPV, CSP and wind, each located in climatic regions appropriate to its most favorable economic performance, may be included in the future.

On the other hand, thinking about simple utilization of VLS-PV into the grid, predicting intermittence and energy storage will be important issues. It could be concluded that VLS-PV systems' power output intermittence and prediction can be performed at various time scales, with the use of different instruments, important databases, and mathematical and statistical tools to predict the power output variations. There are some technical options for energy storage. Supporting integration through flexibility enhancement with "storage on grid" can be done with batteries. "Onsite storage" should be conjointly done with batteries for short-term uses and large scale storages, such as pumped-hydro or compressed-air, for power shifting purposes. These may be the best candidates at the moment; further discussion should take place.

International collaboration

In the near future, VLS-PV systems should become an economically viable option to meet the electricity needs of communities in remote desert regions around the world. In the long term, further expansion of large-scale PV generation can not only provide increasing energy supply in sustainable way but also will stimulate a local PV industry and create enough demand to justify local manufacturing of PV system components locally.

As a VLS-PV system of several hundred MW or GW scale capacity will require an increasing amount of investment, innovative and favourable financing approaches and active international collaboration are required, especially in developing countries located in deserts. This international collaboration should include R&D in all areas of the VLS-PV concept, but especially in policy, regulation, standards and business model development, and would stimulate the strengthening of international collaboration of existing institutions and activities, as well as the creation of new joint initiatives at regional and global level. As results, various kinds of high-tech knowledge and experience, know-how will be transferred to local renewable energy institutes, utilities and energy companies in order to keep track of worldwide technology developments, technology exchange between universities and scientific institutes.

The future expansion of VLS-PV depends on successful international cooperation to advance knowledge on development of the VLS-PV concept. In addition, a clear political commitment of governments and a stable regulatory framework towards creating a long-term stable policy is essential for further deployment of the VLS-PV concept worldwide.

Strategic niche management, social acceptance, and business models

Solar power plants in desert regions are on the brink of finding their way into the world's energy systems. This might mean that our preparatory work, is currently being "overtaken by market developments". However, from the innovation science perspective, this would be a false conclusion, and now is the time for systematic learning about VLS-PV on a global niche level. For only the confrontation with real life conditions can reveal the technological, business, socio-economic, and geopolitical fits and misfits of solar power plants in desert regions with the demands of the modern energy world. It is now that the main stakeholders need to learn about expected and unexpected bottlenecks that emerge in individual local projects, accumulate such lessons on an overall niche level, experiment with ways to navigate these real world opportunities and barriers, and invent viable innovation pathways. Importantly, socio-economic or geopolitical bottlenecks might very well feed back into the technology design process: the redesign of solar power concepts for desert regions, originally conceived to serve long-distance international energy flows, to accommodate local economic development and sea water desalination illustrate technology adaptations in the market implementation phase. In sum, the coordination between individual local projects and the overall desert solar power niche community is more important now than ever.

The VLS-PV niche requires continued reflection on strategic management on a niche level, and a continued search for social acceptance and business models on a project scale. To facilitate this process we tapped into innovation sciences work on sustainability transitions, outlined the approach of Strategic Niche Management, and briefly introduced ESTEEM and business model experimenting. This initial mapping of the territory, however, has merely opened up the topic. A next step would be a more VLS-PV specific and fine-tuned inquiry.

Next step and strategy of Task8 activity

Since it was established in 1999, IEA PVPS Task8 has been examining and evaluating the potential of VLS-PV systems from various points of view and with the consideration of different aspects concerning feasibility and practicability. With the present volume and past publications [1, 2, 3], a comprehensive summary of these analyses has been made available. The work of Task8 has shown that VLS-PV can be a viable and competitive building block in the world's future energy supply.

Task8 has observed and analyzed PV, CPV and CSP with the same agnostic interest. The technological development of the three technologies, along with the cost reduction of the energy production especially for PV, leads to the conclusion that PV is or may become the predominant technology to harvest electricity from the power of the Sun.

Throughout the last two decades we have experienced a slight but continuous paradigm shift in economies, which gains more and more momentum as the topic gets more social perception: a more sustainability-oriented energy mix with an important role of renewable energy sources is socially and politically favored in most societies around the globe. With this, as well as with the accompanying maturing of the PV industry, the ideas and proposals presented in the Task8 publications have made their way into common knowledge of all stakeholders – politicians as well as industry players, institutional investors, transnational and non-governmental organizations. The technological aspects in terms of feasibility as well as potential applications worldwide have been and are discussed in depth.

With this in mind, we will conclude that the focus of our work should be refined to add further value to the discussion about VLS-PV. Therefore we decided to widen the scope of our evaluation and to focus on one question: What has to be done besides aspects of feasibility and practicability, to ensure that solar power plants will be a viable and sustainable option for the world's future electrical power supply?

It became clear that besides the public and the governmental/administrative side, the way of thinking and the approaches of the national and transnational electricity utilities have to be understood.

We intend to continue our work to gain further understanding of the requirements for VLS-PV from the utility point of view. Then, we will make a detailed and worldwide analysis of the requirements and expectations of major utilities. The outcome will be a competitive list of suggestions and recommendations for governments as well as for PV manufacturers to fulfill the needs of the utilities as important stakeholders for the future of VLS-PV.

Apart from utilities, we see governments and international organizations as important stakeholders and will make further efforts to understand their requirements and concerns. Eventually, conclusions, suggestions and recommendations of how to overcome hurdles/barriers, from technical and non-technical viewpoints, will be proposed for potential VLS-PV owners/developers.

Also, we have started to learn about the approach of Strategic Niche Management. As our next step, a follow-up study focusing on VLS-PV specific and fine-tuned inquiry will systematically examine existing or ongoing projects for current niche dynamics and initial encounters with real life contexts. With this practical approach, we may translate these lessons into a learning-based implementation strategy for policy makers and entrepreneurs.

As announced in January 2012, the International Energy Agency (IEA) and the International Renewable Energy Agency (IRENA) signed a partnership agreement which will strengthen cooperation between the two organizations. There will also be increased collaboration between the two agencies at the Secretariat level, and in energy technology networks, including the IEA PVPS. Our activity, Task8, will be able to contribute to further advance international collaboration to stimulate the global deployment of large scale PV power generation systems.

NOTE

1 Thermal storage will not be reviewed as it is more in the scope of CSP power plants, which produce higher levels of temperature than PV plants could achieve.

REFERENCES

1 Kurokawa, K. (2003) *Energy from the Desert: Feasibility of Very Large Scale Photovoltaic Power Generation (VLS-PV) Systems*, James & James.

2 Kurokawa, K., Komoto, K., Van Der Vleuten, P., and Faiman, D. (2007) *Energy from the Desert: Practical Proposals for Very Large Scale Photovoltaic Systems*, Earthscan.

3 Komoto, K., Ito, M., Van der Vleuten, P., Faiman, D. and Kurokawa, K. (2009) *Energy from the Desert: Very Large Scale Photovoltaic Systems: Socio-economic, Financial, Technical and Environmental Aspects*, Earthscan.

4 International Energy Agency (2010) Technology Roadmap Solar Photovoltaic Energy

5 International Energy Agency (2010) Energy Technology Perspectives 2010 Scenario & Strategies to 2050.

6 Desertec Industry Initiative (2010) *Energy from deserts – Bringing the Desertec Vision into reality*, Dii, Munich, www.dii-eumena.com/fileadmin/Daten/Downloads/Dii-Brochure_EN.pdf

7 Asia Development Bank (2011) Asia Solar Energy Initiative – A Primer.

8 Stackhouse P.W., Whitlock C.H., (eds) (2008) *Surface meteorology and Solar Energy* (SSE) release 6.0, NASA SSE 6.0, Earth Science Enterprise Program, National Aeronautic and Space Administration (NASA), Langley, http://eosweb.larc.nasa.gov/sse/

9 Breyer Ch., Schmid J. (2010) *Population Density and Area Weighted Solar Irradiation: global overview on solar resource conditions for fixed tilted, 1-axis and 2-axes PV systems*, 25th EU PVSEC/WCPEC-5, Valencia, September 6–10, DOI: 10.4229/25thEUPVSEC2010-4BV.1.91.

10 Breyer Ch. (2011) *The Photovoltaic Reality Ahead: Terawatt Scale Market Potential Powered by Pico to Gigawatt PV Systems and Enabled by High Learning and Growth Rates*, 26th EU PVSEC, Hamburg, September 5–9, DOI: 10.4229/26thEUPVSEC2011-6EP.1.2.

11 Gerlach A.-K., Stetter D., Schmid J., Breyer Ch. (2011) *PV and Wind Power – Complementary Technologies*, 26th EU PVSEC, Hamburg, September 5–9.

12 Breyer Ch., Rieke S., Sterner M., Schmid J. (2011) *Hybrid*

PV-Wind-Renewable Methane Power Plants – A Potential Cornerstone of Global Energy Supply, 26th EU PVSEC, Hamburg, September 5–9.

13 Fthenakis, V.M., Kim, H.C., Alsema, E. (2008) Emissions from photovoltaic life cycles. *Environmental Science & Technology* 42, 2168–2174.

14 Mason, J.E., Fthenakis, V.M., Hansen, T., Kim, H.C. (2006) Energy payback and life-cycle CO_2 emissions of the BOS in an optimized 3.5 MW PV installation. *Progress in Photovoltaics*, 14, 179–90.

15 Awerbuch, S., and Berger, M. (2003) 'Applying portfolio theory to EU electricity planning and policy-making', IEA/EET Working Paper.

16 Solomon, A.A., Faiman, D. and Meron, G. (2010) An energy-based evaluation of the matching possibilities of very large photovoltaic plants to the electricity grid: Israel as a case study. *Energy Policy* 38, 5457–68.

17 Solomon, A.A., Faiman, D. and Meron, G. (2010) The effects on grid matching and ramping requirements, of single and distributed PV systems employing various fixed and sun-tracking technologies. *Energy Policy* 38 5469–81.

18 Solomon, A.A., Faiman, D. and Meron, G. (2010) Grid matching of large-scale wind energy conversion systems, alone and in tandem with large-scale photovoltaic systems: An Israeli case study. *Energy Policy* 38, 7070–81.

19 Solomon, A.A., Faiman, D. and Meron, G. (2010) Properties and uses of storage for enhancing the grid penetration of very large photovoltaic systems. *Energy Policy* 38, 5208–22.

20 Solomon, A.A., Faiman, D. and Meron, G. (2012) Appropriate storage for high-penetration grid-connected photovoltaic plants. *Energy Policy* 40, 335–44.

21 Solomon, A.A., Faiman, D. and Meron, G. (submitted for publication) The role of conventional power plants in a grid fed mainly by PV and storage, and the largest shadow capacity requirement.

22 Perez, R., Kivalov, S., Schlemmer, J., Hemker Jr., K., Hoff, T.E. (2011) Parameterization of site-specific short-term irradiance variability, *Solar Energy* 85, 1343–53.

23 Gerlach A.-K., Stetter D., Schmid J., Breyer Ch. (2011) *PV and Wind Power – Complementary Technologies*, 26th EU PVSEC, Hamburg, September 5–9.

24 Breyer Ch., Rieke S., Sterner M., Schmid J. (2011) *Hybrid PV-Wind-Renewable Methane Power Plants – A Potential Cornerstone of Global Energy Supply*, 26th EU PVSEC, Hamburg, September 5–9.

25 Helios Energie

26 BP (2011) *Statistical Review*. www.bp.com/sectionbodycopy.do?categoryId=7500&contentId=7068481

27 Faiman, D., Raviv, D. and Rosenstreich, R. (2007) Using solar energy to arrest the increasing rate of fossil-fuel consumption: The southwestern states of the USA as case studies. *Energy Policy* 35, 567–76.

28 United Nations (2011) *World Population Prospects – The 2010 Revision*. http://esa.un.org/unpd/wpp/index.htm

29 Asia Development Bank (2011) *ASIA 2050 – realizing the Asian Century*.

30 Grin, J., Rotmans, J. and Schot, J. (eds) 2010 *Transitions to Sustainable Development. New Directions in the Study of Long Term Transformative Change* (London: Routledge).

31 Verbong, G., and Loorbach, D. (eds) (2012) *Governing the Energy Transition. Reality, Illusion, or Necessity?* (London: Routledge).

32 Geels, F. (2002) Technological transitions as evolutionary reconfigurations processes: a multi-level perspective and a case study, *Research Policy* 31, 1257–74.

33 Van der Vleuten, E., and Högselius, P. (2012) "Resisting Change? The Transnational Dynamics of European Energy regimes", in: Verbong and Loorbach, *Governing the Energy Transition* (London: Routledge).

34 Raven, R., Jolivet, E., Mourik, R., and Feenstra, Y. (2009) ESTEEM: Managing societal acceptance in new energy projects: A toolbox method for project managers, *Technological Forecasting and Societal Change* 76, 963–77.

35 E. Jolivet *et al.* (2008) *ESTEEM manual. Create Acceptance deliverable 5. Document ECN-E-08-031* (Energie Centrum Nederland).

36 Osterwalder, A., and Pigneur, Y. (2010) *Business Model Generation. A handbook for visionaries, game changers and challengers* (Hoboken, NJ: John Wiley and Sons).

37 Zott, C, AMit, R., and Massa, L. (2010) The business model: theoretical roots, recent developments and future research. Working paper WP-862, University of Navarra.

PART I
Introduction and overview
Potential of VLS-PV

CHAPTER ONE

Introduction

1.1 OBJECTIVES

The purpose of this study is to examine and evaluate the potential and feasibility of Very Large Scale Photovoltaic Power Generation (VLS-PV) systems, which have capacities ranging from several megawatts to gigawatts, and to develop strategies for implementation of VLS-PV systems in the future.

Our study has comprehensively analyzed the major issues involved in such large scale applications, based on the latest scientific and technological developments and by means of close international cooperation with experts from different countries.

Our first extensive report entitled "Energy from the Desert – Feasibility of Very Large Scale Photovoltaic Power Generation (VLS-PV) Systems" was published in 2003 [1]. The key factors to enable VLS-PV systems feasibility were identified and the benefits of this system's applications for neighboring regions were clarified as well as the potential contribution of system application to global environment protection and renewable energy utilization in the long term was clarified. Mid- and long-term scenario options for making VLS-PV systems feasible in some given areas were also proposed.

As a next step, demonstrative approaches toward the realization of VLS-PV in different deserts around the world, e.g., the Mediterranean region, the Middle East, Asia and Oceania, were studied in depth. The results of these studies were published in 2007, entitled "Energy from the Desert – Practical Proposals for Very Large Scale Photovoltaic Systems" [2]. This report contained state-of-the-art knowledge about how to realize VLS-PV systems in the desert.

For clear understanding of feasibility and impact of VLS-PV projects, we published our third report "Energy from the Desert – Very Large Scale Photovoltaic Systems: Socio-Economic, Financial, Technical and Environmental Aspects" in 2009 [3]. This report focused on various subjects around VLS-PV and proposed a VLS-PV roadmap until 2100.

These publications are often used as a reference for developing VLS-PV strategies in desert regions.

From the perspective of the global energy situation, global warming, and other related issues, it is apparent that VLS-PV systems can:

- contribute substantially to global energy needs;
- become economically and technologically feasible soon;
- contribute significantly to global environment protection;
- contribute significantly to socio-economic development.

It is also concluded that desert regions contain an abundant and inexhaustible source of clean energy and that very large scale solar electricity generation provides economic, social and environmental benefits, security of electricity supply and fair access to affordable and sustainable energy solutions.

During the past 10 years, numerous MW-scale PV systems have been installed around the world and 100 MW-scale PV systems are becoming a reality. Further discussions and plans for over 100 MW-scale and GW-scale PV systems are going on in some regions. However, to accelerate and implement real VLS-PV projects, the feasibility of such projects should be informed to decision-makers in and around desert regions in an appropriate manner.

Based on our previous results and changing market conditions, we have been discussing and evaluating the real feasibility of VLS-PV projects in desert countries, as well as implementation strategies, engineering designs and various technical options for VLS-PV projects.

1.2 CONCEPT OF A VLS-PV SYSTEM

1.2.1 Concept and definition

Solar energy is low-density energy by nature compared with conventional fuels. To utilize it on a large scale, an extensive area is required.

Three approaches are considered to encourage the widespread use of PV systems:

a. Establish small-scale PV systems that are independent of each other. There are two scales for such systems: stand-alone, several-watt to several-hundred-watt-class PV systems for private dwellings; and 2–10 kW-class systems on the roofs of dwellings, as well as 10–100 kW-class systems on office buildings and schools. Both methods are already in use. The former is mainly used to supply electric power in developing countries, so-called Solar Home Systems (SHS), and the latter is mainly used in developed countries.
b. Establish 100–1 000 kW class mid-scale PV systems on unused area on the outskirts of urban areas. Systems of this scale are already in practical use around the world. This category can be easily extended up to multi-megawatt size.
c. Establish PV systems larger than 10 MW on vast, barren, unused areas that enjoy extensive sunlight. In such areas, it would be easy to develop more than 1 GW size of PV systems in total. This approach enables the rapid installation of a large number of PV systems. When the cost of generated electric power comes down sufficiently in the future, many more PV systems will be installed. This will dramatically reduce the cost of electricity, creating a positive cycle between cost and consumption. This may also solve future energy and environmental problems around the globe, so this possibility is worth discussing in more detail.

A third category corresponds to Very Large Scale PV (VLS-PV) systems, which can be defined as follows:

- The size of a VLS-PV system may range from 10 MW to 1 or several GW, consisting of one plant, or an aggregation of multiple units distributed in the same district operating in conjunction with each other.
- The amount of electricity generated by VLS-PV can be considered significant for people in the district, nation, or region.
- VLS-PV systems can be classified by location as follows:
    - land-based (arid to semi-arid, deserts);
    - other concepts (water based, lakes, coastal, or international waters);
    - locality options: developing countries, developed countries.

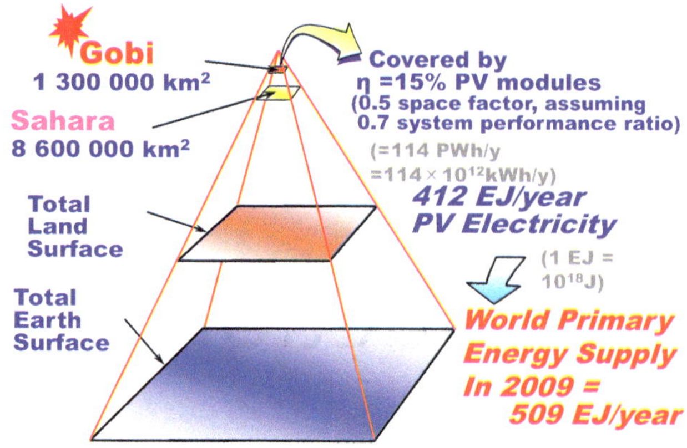

Figure 1.1 Solar pyramid

Figure 1.2 Image of a VLS-PV system in a desert area

Although the concept of VLS-PV includes water-based options, these require much study. As they are just a future possibility, they are outside the scope of this report.

One-third of the land surface of the Earth is covered by very dry deserts, where insolation levels are very high and where there are many unused areas. It is estimated that if a very small part of these areas, approximately 4 percent, were to be used for installing PV systems, the annual energy production would equal the world's energy consumption.

A rough estimation of the potential of VLS-PV can also be drawn as a 'solar pyramid', shown in Figure 1.1, updated from the original version [1]. The 'solar pyramid' shows that the Gobi desert area covering the western part of China and the southern part of Mongolia can generate as much electricity as the present world primary energy supply.

Figure 1.2 shows an image of a VLS-PV system in a desert region.

1.2.2 A synthesized scenario for network evolution

The following scenario for achieving large-scale PV introduction is proposed. As a first step, bulk systems are installed individually in some locations. Then they are interconnected to each other by a power network, and tuned to the growth in demand for regional electricity.

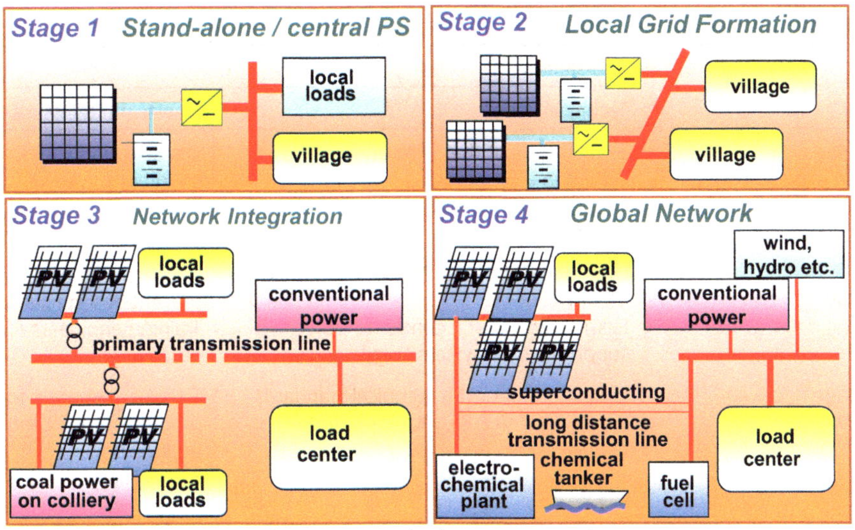

Figure 1.3 A very large-scale PV system deployment scenario

Finally, such a district would become a large power source. This scenario is summarized in Figure 1.3 as follows:

- [Stage 1] A stand-alone, bulk system is introduced to supply electricity to surrounding villages or anti-desertification facilities in the vicinity of deserts.
- [Stage 2] Remote, isolated networks germinate. Multiple systems are connected by a regional grid. This contributes to load levelling and reducing power fluctuations.
- [Stage 3] The regional network is connected to a primary transmission line. Generated energy can be fed into the local HV grid and supplied to e.g. a load centre and industrial zone. Total use combined with other power sources and storage is important for matching the demand pattern and improving the capacity factor of the transmission line.

Furthermore, around the time stage 3 is reached, in the case of a south-to-north tie, seasonal differences between supply and demand can be adjusted. An east-to-west tie can shift peak hours during the day.

- [Stage 4] Finally, a global network is developed. Most of the energy consumed by human beings can be supplied by solar energy.

For the final stage, a breakthrough in advanced energy transportation will be needed on a long-term basis. Therefore superconducting cables, FACTS (flexible AC transmission systems), or chemical media come into account.

1.2.3 A step-by-step approach for VLS-PV development

To realize VLS-PV, it is necessary to identify issues that should be solved and to discuss a practical development scenario. Focusing on the technical development, a basic concept of VLS-PV development consisting of four stages is shown in Table 1.1.

S-0: The R&D stage

To verify the basic characteristics of the PV system in a desert area, the reliability in a desert area and the ability required for grid-connection will be evaluated and investigated as technical issues. Conditions for site selection, planning of cooperation frameworks, including training engineers and funding schemes, will be investigated as non-technical issues. Recently, PV systems with a capacity of more than 10 MW have been constructed in several countries. However, this stage is very important in arranging social conditions for the entire VLS-PV project in regions that have no experience with large-scale PV systems. On the other hand, a lot of companies having know-how and expertise of VLS-PV are already available.

S-1: The pilot stage

A large-scale PV system that has a capacity of tens of MW will be constructed and operated to evaluate and verify the preliminary characteristics of a large-scale PV system. Here, technical issues are of a higher level and shift to concentrated and grid-connected PV systems, for building VLS-PV technical standards. PV module facilities will be constructed and operated, although the PV modules will have to be imported. PV module production will normally be introduced in the next stage. Furthermore, development to prevent desertification such as vegetation and plantations will also be started at this stage.

Table 1.1 Summary of a scenario for VLS-PV development

	Sub-stage	
	Technical issues	Non-technical issues
S-0: R&D stage (may be unnecessary in experienced countries)	Examination of the reliability of a PV system in a desert area. Examination of the required ability of a PV system for grid-connection.	Site selection for VLS-PV based on conditions. Planning project formation including training engineers and funding.
S-1: Pilot stage Scale of system: 10 MW	Development of the O&M methods for VLS-PV. Examination of the control of power supply from PV system to grid-line.	Development of the area around VLS-PV to prevent desertification. Training engineers for on-site PV module production.
S-2: Demonstration stage Scale of system: 100 MW	Development of the technical standards for O&M of VLS-PV, including grid-connection.	Training engineers for on-site mass production of PV modules and for BOS production. Preparation for industrialisation by private investment.
S-3: Deployment stage Scale of system: 1 GW	Building a concept like 'Solar Breeder' in terms of technical and non-technical issues.	

S-2: The demonstration stage

A 100 MW PV system will be constructed and operated according to research methods on grid-connected operation and maintenance when VLS-PV actually takes on a part of the local power supply. The knowledge of grid-connecting VLS-PV to an existing grid-line, which will be obtained at this stage, will form the technical standards for deploying VLS-PV. An involvement of the local utility is always a must. However, non-technical issues will also be advanced for industrialization. Mass production of PV modules will be carried out and on-site BOS production will start moving towards a deployment stage.

S-3: The deployment stage

The deployment stage will be used to verify the capability of VLS-PV as a power source. In this stage, technologies for generating and supplying electricity will be virtually complete. However, for deployment of VLS-PV in the future, some options such as demand control, storage of electricity and recycling components will be required. These will contribute to building the concept of a "solar breeder" and a number of business plans for VLS-PV will be proposed.

From the viewpoint of a VLS-PV scenario, stages S-0 to S-2 shown in Table 1.1 are important for achieving the final stage, and development in each stage should be carried out progressively. If the focus is on technology, then it may be important to start with the R&D stage or the pilot stage when considering overall desert development. If the aim is commercial operation, the initial project to be proposed should be the pilot, or demonstration stage.

Currently there are many countries in the S-1 stage and some of them are likely to progress to the S-2 stage. Such countries may have an advantage for achieving the S-3 stage, as most of them are developed countries and are initiated by strong governmental support schemes. However, other countries and areas will have a potential of realizing VLS-PV projects, too.

1.3 POTENTIAL BENEFITS OF VLS-PV

VLS-PV basically means generation of electricity by means of solar irradiation at utility scale. In our specific case, we have studied utility scale generation of solar electricity in desert regions. By nature, desert regions are rich in sunshine, water is scarce and they are sparsely populated. Living conditions in deserts are harsh and deserts are often considered as unproductive land.

Solar electricity generation at utility scale can change this situation substantially. This will not only provide an infinite, sustainable and inexhaustible source of energy, but also offers the possibility for establishing a source for socio-economic growth in the region.

Because of local availability of energy in desert regions, clean water can be provided when deserts are located close to the sea, irrigation can be optimized, local food can be produced and local sources of raw materials can be exploited. The generated electricity can directly be used for lighting, communication, education, entertainment and for industrial purposes. Excess electricity can be exported to neighboring regions or beyond.

By adapting a regional policy for solar electricity generation at utility scale, a sustainable local market for solar electricity modules and other related components and services might be created and this market could be served by local production. Technology transfer from industrialized countries can provide and improve education and learning, research and development and manufacturing and sales.

Local solar electricity generation also provides for security of energy supply in a sustainable way, provides fair access to energy for everybody and reduces the threat of climate change.

If managed in the right way, such a closed circle between local resources and local needs offers extensive possibilities for further development of the region.

Deserts have the potential to become the future powerhouses of the world, and potential benefits of a VLS-PV policy in desert regions include the following.

Economic benefits

Step-by-step implementation of utility scale PV power plants will reduce and at the end attain independency of imports of fossil fuels for power generation. National budgets will be disburdened. Introducing a solid strategy for the introduction of VLS-PV electricity generation will create a large and sustainable local market for solar panels and related components, maintenance services and materials for a long period of time. Also valuable CDM credits will be generated. The generated electricity can be distributed and used in the local market as well as in the export market and CDM credits can be traded in the international market.

Social benefits

Local exploitation of energy from the sun as well as local production of related components, services and materials will create long-lasting employment in desert regions. For the introduction of state-of-the-art technology, international cooperation and technology transfer will be needed.

Security of energy supply

Because abundant sunshine is and will always be available in desert regions, the sun is the most secure source of energy one could think of. Since solar electricity generation is going to be economically viable with technology evolution, very large scale as well as small scale solar electricity generation will be the most affordable, the cleanest and an inexhaustible source of energy for the near future.

Environmental issues

Solar electricity generation is quiet, has no moving parts and is free from emission of harmful gases. By generation of solar electricity instead of fossil and nuclear electricity, greenhouse gas emissions will be reduced and the effects of climate change will be diminished. Additionally, nuclear waste will be avoided.

Peace/poverty alleviation

The abundant availability of solar energy in desert regions provides affordable and sustainable energy solutions for everybody. This may be an important condition for preventing tensions and political and military conflicts about energy. Moreover, the availability of solar electricity will stimulate economic development in countries with desert regions.

International recognition

By the introduction of a sustainable energy strategy based on utility scale generation of solar electricity, an example will be set that will create international recognition and this example will be followed by many other regions in the world.

Development in related areas

In connection with large scale generation of solar electricity, other important subjects such as irrigation, agriculture, water pumping, desalination, electricity transmission and storage and hydrogen technology will be studied and implemented.

Utility scale solar electricity generation among other solar and renewable electricity generation is a prerequisite for socio-economic development in the region and a basic ingredient in the recipe for human well-being. However, to further deploy such trends and to implement real VLS-PV projects, the main challenge is to define implementation strategies, to make excellent project proposals and to convince local governments, energy companies and financing institutions to become positively involved in realizing ambitious projects for the large-scale generation of solar electricity.

Based on the viewpoints mentioned above, this report will show the overview and potential of VLS-PV, guidelines for VLS-PV systems, technical options and strategic options for implementing VLS-PV, and will propose our next steps into the future.

REFERENCES

1 Kurokawa, K. (2003) *Energy from the Desert: Feasibility of Very Large Scale Photovoltaic Power Generation (VLS-PV) Systems*, James & James.

2 Kurokawa, K., Komoto, K., Van Der Vleuten, P., and Faiman, D. (2007) *Energy from the Desert: Practical Proposals for Very Large Scale Photovoltaic Systems,* Earthscan.

3 Komoto, K., Ito, M., Van der Vleuten, P., Faiman, D. and Kurokawa, K. (2009) *Energy from the Desert: Very Large Scale Photovoltaic Systems: Socio-economic, Financial, Technical and Environmental Aspects*, Earthscan.

CHAPTER TWO

PV and other renewable energy issues

2.1 PROPOSED SCENARIOS AND VISIONS

It has been repeatedly pointed out by scientists that human beings will face several sustainability issues sooner or later if appropriate measures are not taken promptly, especially from the environmental point of view. Their points are clear that the anthropogenic environmental burden will be greater than the capacity of the Earth as the economic activities of the global population expand at the current pace.

However, environmental issues are not the only problems in terms of sustainability. Economic and social issues should also be taken into account, especially for developing countries. According to statistics from IEA, roughly 1,4 billion people, which is about 20 percent of the global population, still have no access to electricity [1]. In developing countries, 3 billion people, equivalent to 56 percent of the population, primarily rely on solid fuels. If those people have access to energy, other sustainability issues including access to water, education, health, and food production, could also be improved.

Since access to energy is one of the key drivers of socioeconomic development, we are facing a kind of dilemma of sustainability.

IEA PVPS Task8 group has been pointing out the importance and the potential of Very Large Scale PV (VLS-PV) as one of the solutions for more than 10 years. The concept has been gradually accepted and is finally coming to fruition. Many VLS-PV projects have been announced and initiated in the commercial base. So, to what extent can PV or solar energy supply contribute to the global energy supply?

In this chapter, the possible contribution of PV in the energy mix or in power generation in future scenarios is reviewed. Various scenarios from different organizations, including international organizations, private companies, and NGOs, are reviewed in order to avoid biased perspectives.

2.1.1 Review of latest global energy scenarios and visions

(1) World Energy Outlook

The World Energy Outlook (WEO), an "official" energy projection from IEA, is a good place to start the review. In the latest outlook in 2011 [2], three scenarios, namely "Current Policies Scenario", "New Policy Scenario", and "450 Scenario" were presented. "Current Policies Scenario" is based on an assumption that no additional policy measures will be adopted after the middle of 2010, while "New Policies Scenario" takes into account declared policy interventions. "450 Scenario" considers the consistency with climate change goals of CO_2 concentration in the atmosphere within 450 ppm to meet the sustainability objectives.

As shown in Figure 2.1, global primary energy supply will reach 16 961 Mtoe in 2035 from 12 132 Mtoe in 2009 in the "New Policy Scenario". The contribution of renewable energy (excluding "hydro" and "biomass and waste") in the scenario is still marginal (approximately 4 percent of overall energy supply) even in 2035. On the other hand, the share of renewable energy expands to 8 percent in the primary energy mix in the "450 scenario".

From the viewpoint of power generation, PV capacity is expected to expand from 22 GW in 2009 to 499 GW (741 TWh) in 2035 for the "New Policies Scenario" and to 901 GW (1 332 TWh) for the "450 Scenario". The increase rates are relatively high compared with other energy sources, but the share of the total energy supply is not considerable for both scenarios (2–4 percent). Considering current installed capacity (67 GW) in 2011 and the current trend, the estimated PV capacity seems underestimated.

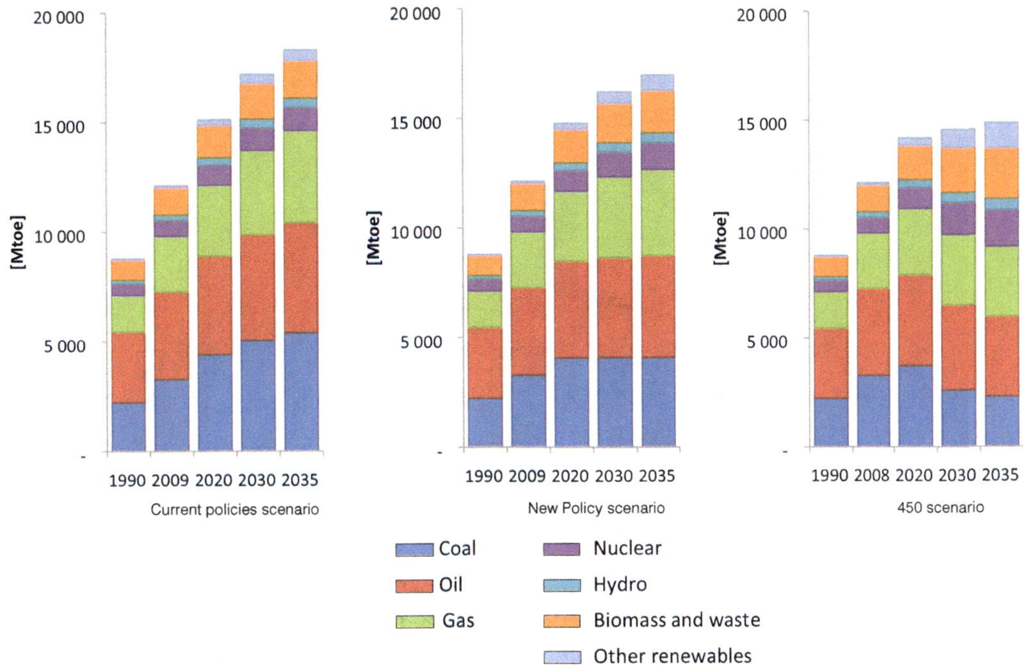

Figure 2.1 Primary Energy Supply (Mtoe) in IEA WEO's three scenarios [2]

(2) Energy Technology Perspective

"Energy Technology Perspective" (ETP) is a very important document from IEA [3]. The report stressed the importance of low-carbon technologies and associated policy measures in order to overcome the threat of climate change. PV is positioned as one of the key technologies for materializing the low-carbon society together with carbon capture and storage, cement sector technologies, CSP, electric and plug-in hybrid vehicles, nuclear energy, and wind energy. For those key technologies, detailed technology roadmaps until 2050 are developed and associated milestones are identified.

The main scenario of ETP is called the "Blue Map Scenario" and a few additional scenarios were developed for sensitivity analysis. The highest contribution of renewable energy is given in the "Blue hi REN" scenario (see Figure 2.2), and the electricity generated from solar (PV+CSP) is over 9 274 TWh (see Figure 2.3). The projection on PV is based on a roadmap developed by IEA [4] (see Figure 2.4).

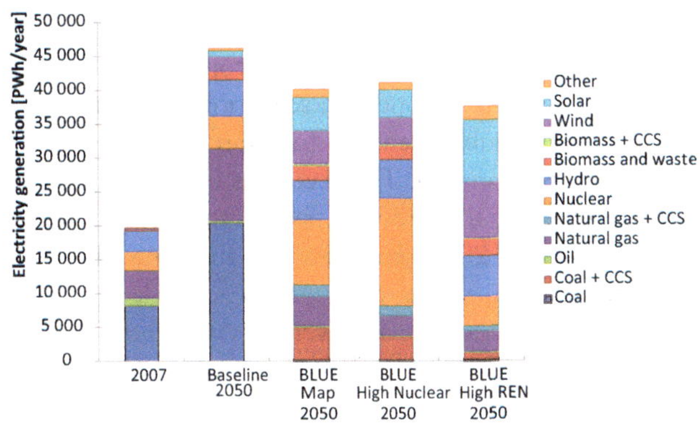

Figure 2.2 Projection of electricity generation in different scenarios in ETP [3]

(3) International Energy Outlook

The "International Energy Outlook", a well-known energy projection from US Energy Information Administration, showed an even more conservative projection than WEO by IEA.

Figure 2.5 shows a projection of renewable power generation. In the Reference scenario, associated solar power generation (both PV and CSP) in 2035 is estimated to be 165 TWh/year and the generation capacity is estimated to be 107 GW. Taking into account the current trends on PV

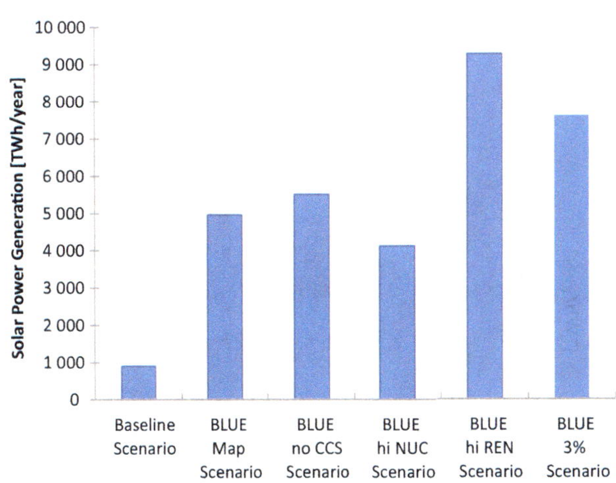

Figure 2.3 Comparison of Solar power generation in different scenarios in ETP [3]

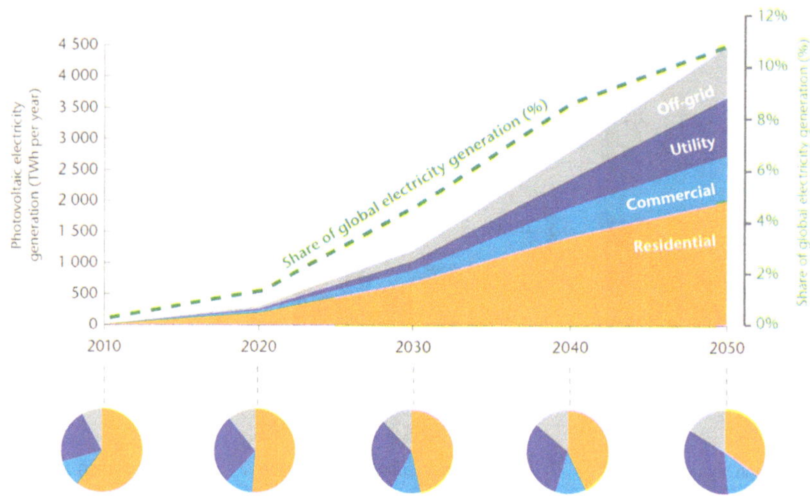

Figure 2.4 PV roadmap presented by IEA [3][4]

and CSP, obviously the potential of solar power generation is underestimated in this scenario. As discussed above, the cumulative installed PV capacity reaches 67 GW in 2011.

(4) Shell Energy Scenario to 2050

Royal Dutch Shell plc, a global oil company commonly known as Shell, is a pioneer in scenario developing. Shell has developed global scenarios for more than 30 years for their strategic planning. The scenarios have been updated every 3 years, and the newest version as shown in Figure 2.6 was disclosed in 2008 [6]. It was the first time that Shell introduced such a long-term vision towards 2050 (the projection period for previous versions was until 2025).

In spite of Shell's position as a global oil company, the aim is not to persuade that the fossil fuel market will continuously expand in the future. A drastic increase is rather seen in renewable energies, while the fossil fuel increase is relatively moderate for both the "Scramble Scenario" and the "Blue Print Scenario".

Although the "Scramble Scenario" assumes less attention of the policy makers to greenhouse gas emissions than the "Blue Scenario", more renewable energy supply is estimated in the "Scramble Scenario".

It is clear that overall primary energy demand is much less in the "Blue Scenario" (Primary Energy Supply: 769 EJ) than in the "Scramble Scenario" (Primary Energy Supply: 880 EJ) due to the effect of energy efficiency measures.

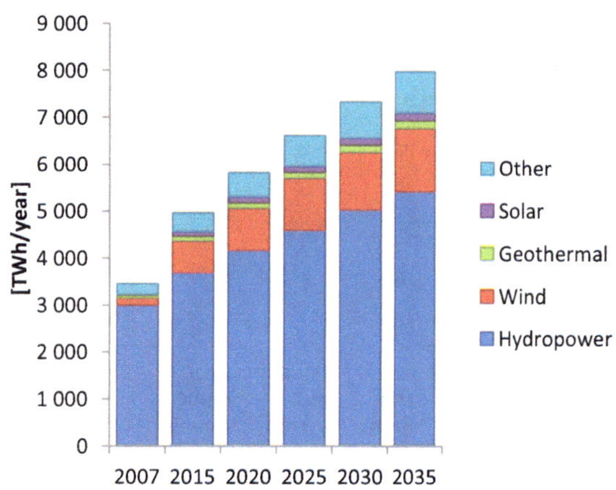

Figure 2.5 Renewable Power generation (TWh) [5]

(5) Energy [R]evolution

Greenpeace and the European Renewable Energy Council (EREC) announced the third edition of the "Energy [R]evolution" scenario in 2011 [7]. In this scenario, renewable energy plays a central role in the global energy supply in 2050. As shown in Figure 2.7, renewable energy contributes approximately 58–80 percent of the global primary energy supply and solar energy is one of the major energy sources among it. It should be noted that the key factor to realize those scenarios are clearly energy efficiency measures as well as renewable energies. Renewable energy technologies alone will not be enough to overcome the sustainability issues. PV capacity in the "Energy [R]evolution scenario" and the "Advanced Energy [R]evolution scenario" are projected to reach to 2 968 GW (4 597 TWh) and 4 318 GW (6 846 TWh) in 2050 respectively.

(6) WWF Energy Report

Another scenario demonstrated by WWF, ECOFYS, and the Office for Metropolitan Architecture (OMA) in 2011 was even more aggressive [8]. The target and message of the paper are "energy can be supplied only by renewable energy".

The transition of the final energy mix is shown in Figure 2.8. As is shown in the figure, solar energy becomes more and more important, especially after 2035.

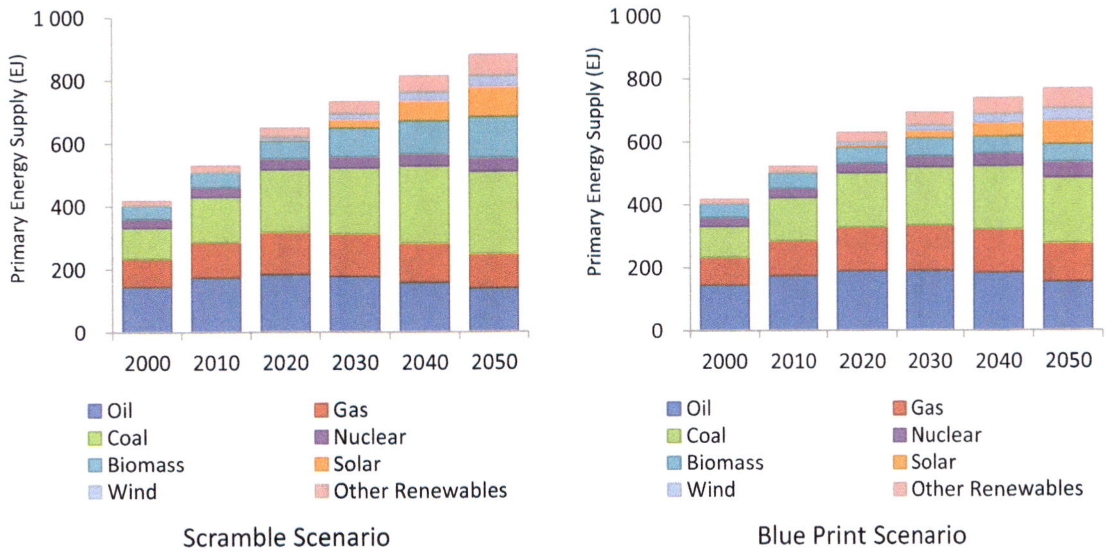

Figure 2.6 Primary energy supply projections in Shell scenarios [6]

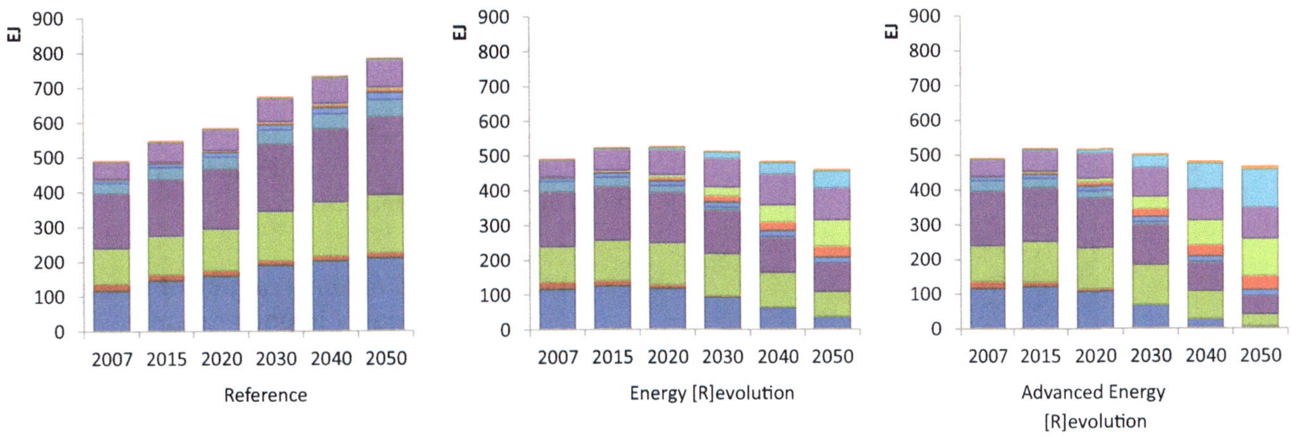

Figure 2.7 Primary energy supply projection in Energy [R]evolution scenarios [7]

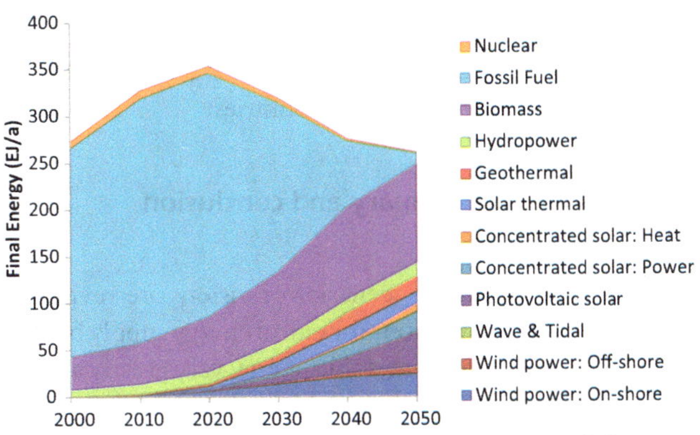

Figure 2.8 Final energy transitions in WWF scenarios [8]

(7) Special Report on Renewable Energy Sources and Climate Change (SREEN)

The Special Report on Renewable Energy Sources and Climate Change Mitigation is prepared by Working group III of the Intergovernmental Panel on Climate Change (IPCC) [9]. The report evaluates and reviews the existing literatures regarding contribution of renewable energies as mitigation measures from science, technology, environment, economy and social perspectives. The report summarized the review that more than half of the scenarios show a contribution of renewable energy in excess of a 17 percent share of primary energy supply in 2030, rising to more than 27 percent in 2050 (see Figure 2.9). Although the range of the renewable energy varies, the highest renewable share was approximately 43 percent in 2030 and 77 percent in 2050. In terms of PV generation, the highest share of PV in global electricity demand is 32,8 percent (74,8 EJ/year).

2.1.2 Comparison of PV power generation among scenarios

The projections of PV power generation vary greatly among the scenarios. In order to make them comparable, the projected power generations from PV in those scenarios are converted into a same units. The summary is shown in Figure 2.10. As presented in the figure, the highest PV power generation among the reviewed scenarios is the "Remind scenario" from IPCC. The generation volume reaches more than 20 000 TWh in 2050. The "100 percent

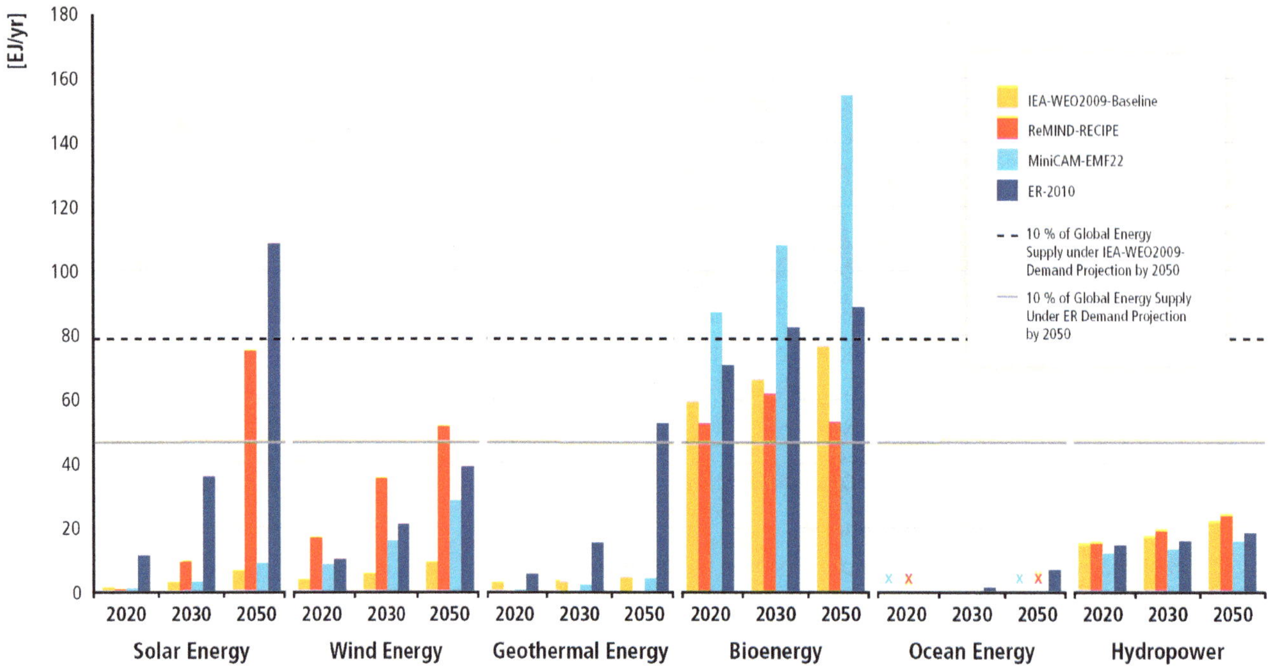

Figure 2.9 Global Renewable Energy development projections by various source reviewed by IPCC [9]

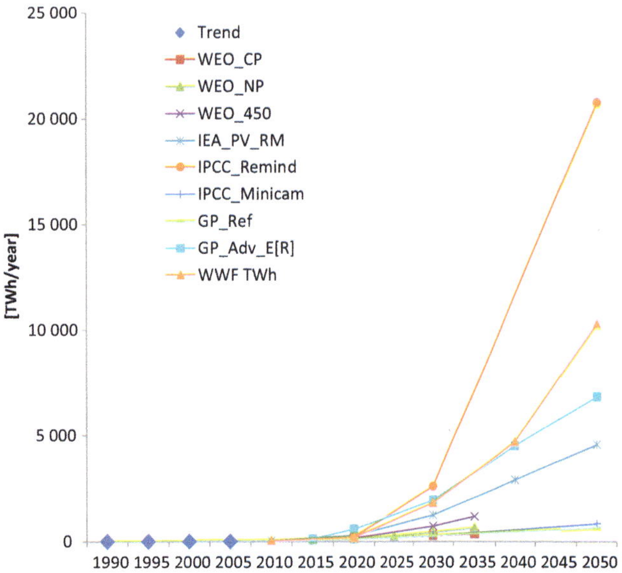

Figure 2.10 Comparison of PV development trajectories in various scenarios in TWh [2]-[3],[5]-[9]

renewable scenario" from WWF and "Energy [R]evolution" also expects relatively high electricity generation from PV.

2.1.3 Historical changes in energy policies/ scenarios and the role of PV

The projected range of future PV potential is so wide that it is very difficult to choose one reliable scenario. In fact, not many analysts or economists expected today's significant growth in PV market 10 years ago. Figure 2.11 shows historical projections of the PV in the World Energy Outlook. In WEO 1998, the PV capacity in 2020 is expected to be 2,8 GW (3,7 TWh), which is far lower than current capacity of 67 GW. As presented in the figure, the PV capacity projections have always been revised upwards. In other words, progress in the PV industry has achieved much higher growth than expected.

2.1.4 National and regional targets of renewable energy and PV in power supply

In order for PV to become a main energy source in the global energy mix in the future, technological development and market penetration speed should be accelerated. One of the key drivers for the acceleration is clearly favorable and stable renewable energy policies. Table 2.1 and Figure 2.12 summarize the renewable energy targets for 2020 and the current progress in various countries.

2.1.5 Summary and conclusion

In this section, existing scenarios or visions are reviewed. There are many projections for global energy supply but the roles of PV and other renewable energy vary greatly among the scenarios. They are heavily dependent on underlying assumptions and projection approaches. In general, PV is expected to increase drastically in most of the scenarios but it might take a few decades to be a main energy source in the global energy mix without innovations. However, PV has achieved more than IEA WEO projections in the past, and should continuously go over the projection in order to become a main energy supply.

Many countries/regions have set ambitious targets for renewable energy and PV. These targets and continuous efforts will accelerate PV penetration in the near future.

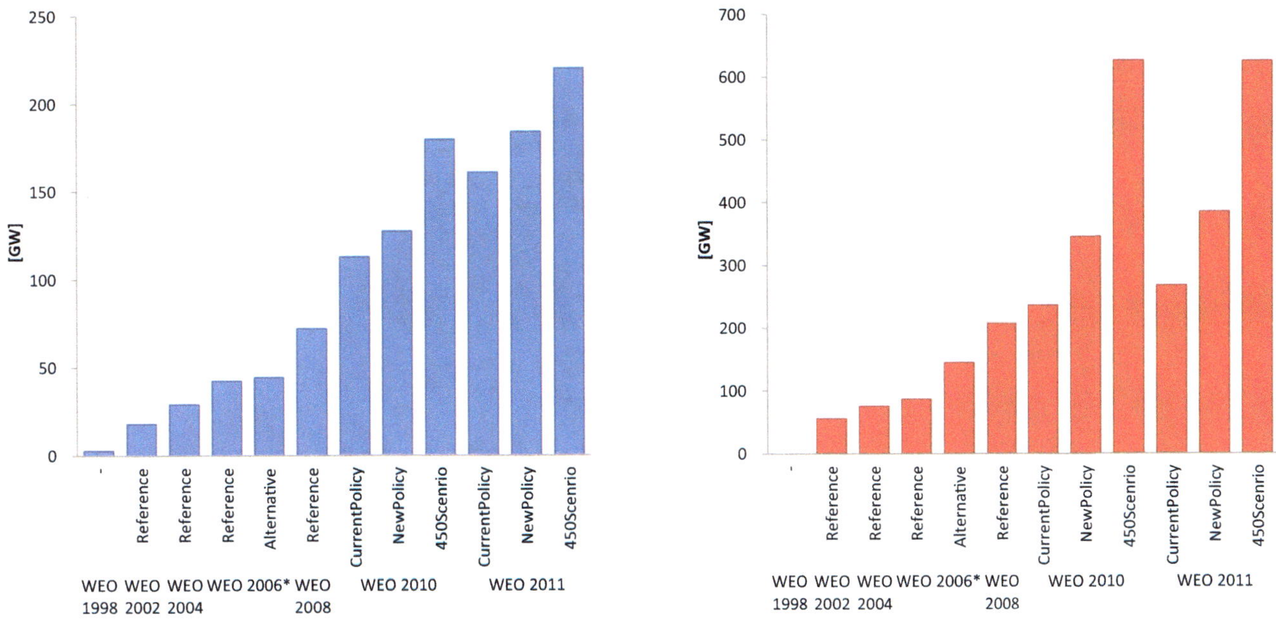

*As for PV capacity projection in 2020 is not presented in WEO 2006. The values presented in the above figure are calculated based on linear increase from 2015 to 2030.

Figure 2.11 Historical changes in PV projections in IEA WEO scenarios for 2020 (right) and 2030 (left) [unit: GW] [2], [10]-[15]

Table 2.1 National targets for PV or CSP in various countries [16]

Country	Energy	Target year	Target
Algeria	PV	2015	5,1 MW
Argentina	Solar	2012	20 MW
China	PV/CSP	2020	1 800 MW
France	PV	2020	4 900 MW
India	PV	2022	12 000 MW
Indonesia	Solar	2025	870 MW
Italy	PV	2016	23 000 MW
Jordan	PV	2015	300 MW
	PV	2020	600 MW
Malaysia	PV	2020	1 250 MW
Morocco	PV/CSP	2020	2 000 MW
Portugal	Solar	2020	1 500 MW
South Africa	PV	2050	14% of generated power
South Korea	PV	2012	1 300 MW
Spain	PV	2020	10 000 MW
Thailand	PV	2011	55 MW
	PV	2016	95 MW
	PV	2022	500 MW
Tunisia	PV	2011	15 MW
	PV	2022	0,5 GW

2.2 PROPOSED SCHEMES AND CONCEPTS

2.2.1 Desertec Industrial Initiative

Several renewable energy sources can be used for a balanced renewable based power supply. An excellent example of how such a fully renewable powered energy system could work has been analyzed for the Desertec project (Figure 2.13) focused on a cooperation between Europe (EU), North Africa and the Middle East (MENA) [17–20]. The historic roots of the Desertec project were laid down in the 1920s, focusing on hydro power [21, 22], and in the 1930s, based on the first PV concepts linked to power lines [23, 24], but were changed in the 1980s to solar hydrogen [25, 26] and in the 2000s to solar thermal electricity generation (STEG), again linked to power lines [17–20]. Desertec is based on all major renewable energy sources and the interconnection of centres of energy supply and centres of energy demand by high voltage direct current (HVDC) power lines. Solar PV, solar thermal and wind power are assumed to be major sources of power, whereas biomass and hydro power might act as a renewable balancing power. The EU-MENA Desertec project gained pace by the Desertec Industrial Initiative (Dii) led by industry giants [27] and might become a blueprint for similar interregional cooperation in other parts of the world, reaching a global power grid.

More than one-quarter of the Earth's land mass is occupied by deserts. They receive solar radiation at excellent irradiation levels (see Section 3.1). For decades it has been a dream of engineers to harvest the solar energy in deserts and supply it to areas of demand. In 1930, Hans Dominik, an engineer of Siemens, outlined the enormous potential of

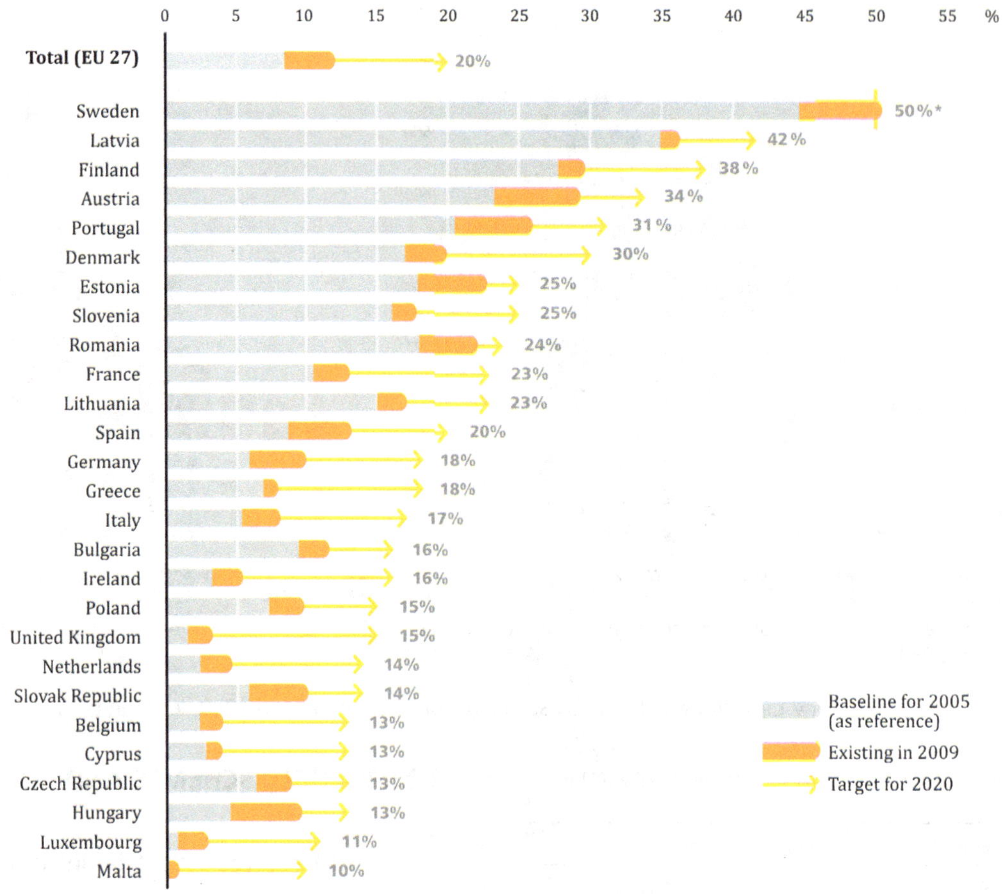

Figure 2.12 Renewable Energy target and its achievement by countries [16]

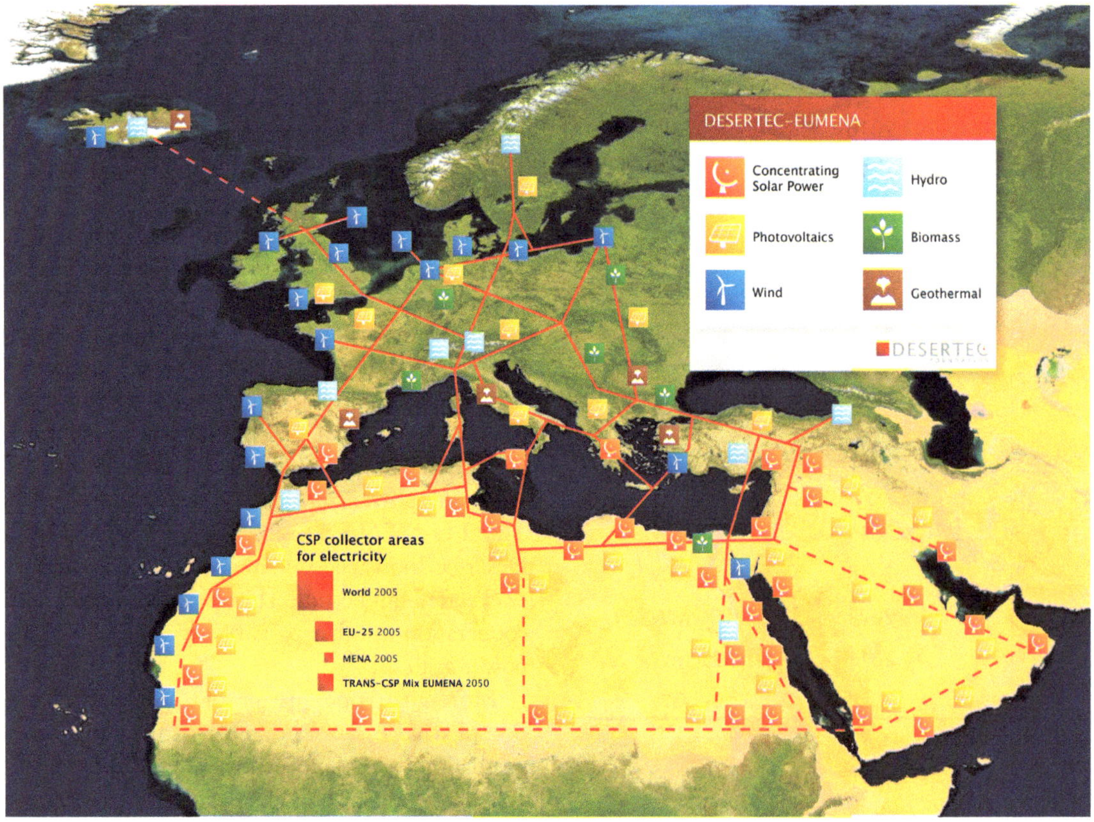

Figure 2.13 Desertec vision for the EU-MENA rim based on renewable energy technologies
The symbols for power sources and lines are only sketching typical locations [17,28].)

solar electricity generated by solar cells in the Sahara desert which was to be transferred to Europe by power lines [24]. In the 1960s, Eduard Justi of the Technical University of Braunschweig in Germany, mentioned that from the point of view of energy deserts will be at least as valuable as coal fields [29].

However, the roots of the Desertec project go back to the 1920s, when Herman Sörgel developed a vision of the future electricity supply for Europe [21, 22]. As a consequence of the Great War, Sörgel became a pacifist and looked for a project which might lead to the broad cooperation of peoples involved. It was planned to lower the mean sea level of the Mediterranean Sea and produce electricity using huge dams for hydro power in the Strait of Gibraltar of a capacity of about 49 GW. The total capacity was planned to exceed 110 GW around the Mediterranean. For comparison, in 1930 Germany's installed power capacity was roughly 15 GW and Germany accounted for about 20 percent of the economy of Europe, Russia and Africa. The Nazi regime was not interested in a cooperation strategy and after World War II the nations thought they had found a new everlasting source of energy: nuclear power.

The vision of an energy cooperation strategy for EU-MENA was regenerated in the 1980s and late 1990s. In the footsteps of Sörgel's, Dominik's and Justi's visions, the Desertec concept for renewable energy, water and climate security became the major agenda of the Trans-Mediterranean Renewable Energy Cooperation (TREC) [30], which was founded by the Club of Rome, the Hamburg Climate Protection Foundation and the National Energy Research Center of Jordan. The international TREC network was focused on the cooperation of professionals in politics, industry, science and the world of finance to make the common concept a reality for both sun-belt nations and technology-belt nations. In early 2009, the Desertec Foundation was established to continue the work of TREC and to globally spread the Desertec vision of an integrated concept which includes energy security and climate protection as well as drinking water production, socio-economic development, security policy and international cooperation – all based on clean power generated in the world's deserts to supply mankind with enough electricity on a sustainable basis.

The Desertec concept is mainly based on three studies conducted by the German Aerospace Center (DLR), which among other topics have analyzed the available resources for renewable energy, the predicted demand for energy and water in EU-MENA till 2050 and the creation of an integrated electricity network between Europe and MENA. The studies are MED-CSP [18] focusing on availability of resources and demand for energy in the Mediterranean area, TRANS-CSP [19] concentrating on the assessment of an interconnection between Europe and MENA and the assessment of solar energy imports and AQUA-CSP which analyzes the demand for water and the possibility of generating freshwater along with electricity by STEGs. These satellite-based studies by the DLR have shown that, using less than 0,3 percent of the entire desert areas of the MENA region, concentrating solar thermal power plants can generate enough electricity to supply current demands in EU-MENA and anticipated increased demands in the future. In addition, this scheme has the potential to alleviate shortages of freshwater in the MENA region.

The global energy supply potential for STEGs alone has already been analyzed [28]. This analysis of global energy supply potential of solar electricity generated only in regions of excellent solar resources, i.e. at least 2 000 kWh/m^2/y direct normal irradiation, clearly shows the true potential of solar power: 90 percent of the world population could be supplied by solar power (solar PV and solar thermal) via HVDC power lines not longer than 3 000 km. The four major backbone technologies, solar PV, solar thermal, wind power and HVDC power lines, are already available and implemented in many regions of the world. The indication is high that based on STEGs a nearly 100 percent renewable power supply could be established. However, the STEG economics are not as favorable as the comparable ones for PV and wind power. Therefore it would be of very high relevance to perform a global energy supply potential for hybrid PV-Wind power plants complemented by a flexible renewable balancing option on the basis of economic competitiveness (Chapter 10).

The Desertec vision discussed so far is largely based on the TRANS-CSP study. This groundbreaking work sketches a possible transformation scenario for EU-MENA from the current unsustainable energy mix towards a sustainable and renewable based power supply (Figure 2.14) [19]. The investment need for the Desertec vision was calculated for the first time in that study and resulted in about 400 billion EUR. In spite of the use of up to 15 percent imported solar energy in the EU, the total ratio of imported energy fuels significantly decreases in the scenario, since no uranium and considerably less coal and natural gas has to be used in the power system.

In October 2009, the Desertec Industrial Initiative (Dii) was founded in Munich as an international consortium, comprising in early 2012 more than 55 companies and institutions, including the Desertec Foundation. By the end of 2012, the major political, legal, regulatory, economic

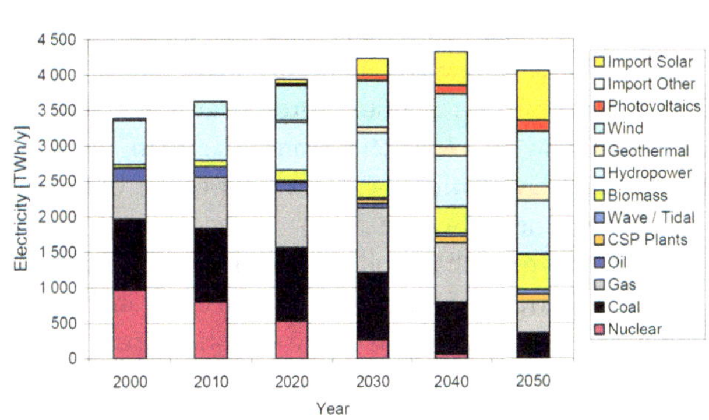

Figure 2.14 TRANS-CSP scenario of the EU electricity mix for the year 2000 to 2050 [19]
(Renewable energies are expected to contribute 80 percent to the electricity mix in the year 2050.)

and technological frameworks to put in place the Desertec vision will have been created. The Dii was founded to bring the Desertec vision into reality [31].

As an industrial initiative, Dii assumes the role of a facilitator. It will ensure that development projects come to fruition, but not make any investments itself either by constructing or operating power plants. The focus of the planning phase is to establish the appropriate framework and create a lasting structure for renewable energies which will make long-term commitments to solar energy plants (solar PV and solar thermal) and wind farms, as well as integrated networks, attractive to public and private investors.

In its pioneering role, Dii has three major objectives:

- Creation of a positive investment climate: to develop the technological, economic, political and regulatory framework, thereby attracting interest as well as enabling investment in renewable energies and interconnected power grids in North Africa and the Middle East.
- Initiation of selected reference projects as a means of demonstrating feasibility and reducing costs.
- Development of a long-term implementation concept (rollout plan) by the year 2050, including guidance on investment and funding. The fastest possible integration of renewable energies into the market should ensue without reliance on subsidies.

The transmission of power from North Africa to the European markets is a key element of the Desertec vision. Dii sees itself as a pioneer in wanting to develop a market, hence it is open to all initiatives which contribute towards such a concept. Dii very much welcomes the foundation of Medgrid [32], an international initiative launched by the French government. Dii and Medgrid complement one another, and to some extent have the same shareholders. Whereas Dii concentrates on the entire value chain through to 2050, from energy production and transfer to development of the energy markets, Medgrid has chosen to focus on selected transmission issues in the Mediterranean region until 2020. In November 2011, the two industrial initiatives signed an agreement to strengthen their cooperation on the development of industrial-scale renewable energy from the deserts and a suitable transmission infrastructure. All of these activities are covered by the Mediterranean Solar Plan (MSP), a political initiative within the framework of the Union for the Mediterranean (Section 2.2.2).

Dii plans to publish the rollout plan in late 2012 indicating the overall attractiveness of the Desertec vision but also emphasizing the necessary framework conditions for long-term stable investment conditions. Dii expects a three-step process for getting the Desertec vision into reality [33]. In the 2010s, the first reference projects need to be established in North Africa for demonstrating the local renewable energy supply contribution and a minor fraction for export into the EU. The scale-up phase is scheduled for the years 2020 to 2035 and includes various renewable power plant projects for local power supply and the establishment of new power transmission routes from the MENA region to Europe. The self-sustained market phase is expected for the mid-2030s and will be characterized by interconnected and integrated energy markets in the EU-MENA region.

The Desertec vision has become a real project for a sustainable energy supply within the EU-MENA region and gained a lot of momentum for harvesting the precious and nearly unlimited energy resources of the deserts of the world. Very large scale PV power plants are expected to become a cornerstone of such an energy system based on a real sustainable fundament.

2.2.2 Mediterranean Solar Plan and relative initiatives for renewable energy in the Mediterranean

In the Mediterranean region, it is recognized that current trends on energy supply and demand are not sustainable and will strain energy security. The following actions should be necessary immediately as alternative countermeasures:

- Increase renewables in power generation and in final use
- Demand-side management policies
- Encourage energy efficiency and energy conservation
- Remove progressively subsidies for fossil fuel
- Invest in research, technology and innovation.

Such actions would require regional cooperation not only within European countries but also including other Mediterranean countries. In particular, investment to the South region would be necessary. From this viewpoint, the Mediterranean Solar Plan and relative initiatives for deploying renewable energy in the Mediterranean region are ongoing, in addition to the Desertec Industrial Initiative described in the previous section.

This section describes outlines of these initiatives.

(1) Mediterranean Solar Plan

The Mediterranean Solar Plan (MSP) is one of the key projects proposed within the Union for the Mediterranean, and intends to increase the use of solar energy and other renewable energy sources for power generation, improve energy efficiency and energy savings, develop electricity grid interconnections and foster and encourage the transfer of know-how and technology towards developing countries.

The final target is the development, by 2020, of 20 GW of new renewable energy installed capacity in the Southern and Eastern Mediterranean countries. It must be noted that 20 GW of new capacity from renewables would represent 11 percent of the total capacity estimated in the south in 2020 and some 6–7 percent of additional electricity generation, according to OME Med Energy Perspective recent scenarios.

The priorities of the MSP can be summarized as below:

- Promote convergence of national energy policies of MSP non-EU countries, including, inter alia, by supporting the setting up of an adequate legal, regulatory, economic, institutional and organizational environment. A viable business climate, in particular, based on stable and transparent investment frameworks, with adequate pricing policies, will be necessary for long-term investments to take place.
- Promote the development of electricity interconnections within the Southern and Eastern Mediterranean, between Southern and Eastern Mediterranean and EU countries, and across the Adriatic Sea in order to establish a viable "green electricity" import and export framework, as well as, where necessary, within the EU in order to allow for renewable electricity trade in the whole region.
- Continue regular dialogues between the stakeholders of the MSP for close coordination and successful implementation of MSP under an appropriate set up.
- Increase efforts aiming at decoupling GDP growth and growth of energy demand, inter alia by improving energy efficiency and energy savings, in particular through the implementation of comprehensive legislative and regulatory frameworks, incentive measures, supported by awareness raising, training, and exchange of experience as a way to progress towards an objective of 20 percent energy savings by 2020. There is a huge potential for energy efficiency improvements in all MSP countries, both EU and non-EU.
- Facilitate extensive cooperation on all technology aspects including, for example, through the establishment of a regional competitive cluster and network of relevant research and scientific institutions active in the field of solar energy.
- Make all EU carbon mechanisms available for the benefit of both sides of the Mediterranean in order to improve the economics of the projects under MSP and fulfill the obligations of the EU developed counties towards GHG emissions reduction under the Kyoto Protocol.

The MSP was launched in 2008, and since then it was slowing down. However, it was re-launched in 2011. The 1st Joint Committee of National Experts for the MSP took place in July 2011 in Barcelona. The meeting was organized by the UfMS Energy Division and attended by representatives of Algeria, Austria, Croatia, Cyprus, Egypt, Finland, France, Germany, Greece, Israel, Italy, Jordan, Malta, Morocco, the Palestinian Authority, Spain, Sweden, Tunisia, Turkey, and the United Kingdom. The European Commission, League of Arab States, European Investment Bank (EIB), Kreditanstalt fuer Wiederaufbau (KFW), and the Paving the Way to the MSP technical assistance project also participated.

The participants endorsed, with some additions, the content structure, time frame, goals and working methods related to the MSP Master Plan, which will be made up of five main building blocks: policy and regulatory frameworks; funding and support schemes; physical infrastructures; renewable energy as an industrial policy tool; and know-how transfer and capacity development

(2) Paving the way for the MSP

The MSP was at a standstill because of the slowing down of the political process of the Union for the Mediterranean, and thus insufficient results were achieved to attract decisive attention from investors.

To accompany the MSP, the EC ENPI (European Neighbourhood and Partnership Instrument) issued the service procurement program "Paving the way", in order to create conditions favorable to the increased use of renewable energy in Southern Mediterranean partner countries. The project will support and coordinate activities among various stakeholders, targeting e.g. ministries in charge of energy, finance and social affairs; industry; the research community; and international financing institutions.

The purpose of the project is that the conditions favourable to the increased use of renewable energy, in general, and solar energy in particular are improved, across all the Southern Mediterranean Partner Countries. The focus is on renewable energy, combined with energy efficiency and savings. The project assists the Mediterranean countries in their efforts to significantly increase the deployment of sustainable energy systems based on solar, wind and other renewable energy sources, combined with energy efficiency and savings.

The following activities are implemented:

- Harmonization of the Legal and Regulatory Framework – benchmarking of the legal and regulatory framework in the nine Mediterranean Partner Countries
- Transfer of Knowledge and Capacity Building – assisting the countries in preparing for the opportunities of industrial development the MSP is offering
- Sustainable Energy Policy – preparing suitable national policies for the promotion of energy efficiency and renewable energy sources
- Support to Investment – considering different methodologies to evaluate the external benefits and how to include them in the overall assessment of MSP projects.

(3) Medgrid

The Medgrid initiative was launched in early July 2010 by France, as it will develop under the umbrella of the MSP.

Figure 2.15 Outline plan of the Medgrid [34]

The aim of Medgrid is to create a network of enterprises in the field of energies with the objective of realizing a large network of underwater electrical connections in the Mediterranean.

Thirteen industrial groups have signed an MoU in order to create a common legal entity devoted to the setting up of a technical and economic conceptual framework, to develop technical and technological cooperation activities with the countries of the South, and to promote an institutional framework favorable for the investments and profitability of the projects proposed in the South.

Figure 2.15 shows their outline plan. The overall objective of the consortium is not to invest in infrastructures but to promote this type of interconnection.

2.2.3 Asia Solar Energy Initiative

In May 2010, the Asian Development Bank (ADB) announced the launch of the Asia Solar Energy Initiative (ASEI) [35], targeted to assist in identifying, developing, and implementing 3 000 megawatts (MW) of solar power in Asia and the Pacific over the next 3 years.

The ASEI aims to create a virtuous cycle of solar energy investments in the region, toward achieving grid parity, so that ADB developing member countries optimally benefit from this clean, inexhaustible energy resource. To meet its target, the ASEI will use an integrated, multipronged approach featuring three interlinked components:

- Knowledge management: The Asia Solar Energy Forum
- Project development: 3 000 MW of solar capacity
- Innovative finance: the Asia Accelerated Solar Energy Development Fund.

The ASEI has the potential to not only drive down costs and to usher in innovation in the solar space, but also to demonstrate replicable models to sustain solar capacity additions in Asia and the Pacific. It is also expected to serve as a model for other regions in the world, such as northern Africa, Latin America, and the Caribbean, which are similarly placed. There has been keen support for the ASEI and ADB leadership in this area.

Beyond the next 3 years, measures of ASEI's success will be the widespread reach of solar energy generation programs across developing member countries (DMCs) in Asia and the Pacific, and the extent of mainstreaming of solar energy as an option for generating capacity addition in all DMCs with significant potential for solar energy in the region. ADB is committed to seeing this success in the future of Asia and the Pacific.

(1) Knowledge management: the Asia Solar Energy Forum

The Asia Solar Energy Forum (ASEF) is envisioned as an independent, nonpartisan, nonpolitical, regional knowledge management platform that links with knowledge hubs globally to facilitate solar energy technology transfer to Asia and the Pacific, provide quality inputs for effective policy and decision making for solar power development, and suggest innovative financing mechanisms to promote solar energy projects. While ADB is committed to supporting the ASEF during its inception, in the long term, the ASEF is conceived to be a self-sustaining, member-supported, nonprofit society, whose members are countries, utilities, financiers, manufacturers, and other stakeholders of solar energy from all over the world.

The ASEF is an inclusive, open, and flexible organization with a light governance structure and optimal delegation of responsibilities to the contributing member organizations for the delivery of agreed priority themes and work areas of high quality and practical value. The private sector is particularly expected to play a prominent role in the management of the ASEF.

The ASEF will have the following primary objectives:

1 Knowledge sharing and consultations
   a. Track and disseminate information to ASEF members on successful solar energy development projects in developed and developing countries, including emerging technologies and lessons from commercial operations of demonstration plants (e.g., molten salt storage for CSP, solar hybrid plants, linear Fresnel plants, and new thin film technologies), and organize major conferences to raise awareness and understanding on issues of topical interest; and facilitate networking.
   b. Explore and discuss financially viable instruments for development in consultation with various stakeholders to facilitate business in solar power generation and smart grid development for grid-connected as well as stand-alone urban and rural solar applications.
   c. Propose to stakeholders innovative financing and risk mitigation mechanisms for long-tenure loans and guarantees that make solar energy projects financially viable.
   d. Conduct matchmaking among the foreign and domestic private sectors, manufacturers, project developers, engineering, procurement, and construction contractors and financiers.
   e. Together with ADB and partner organizations, identify new solar power development proposals in DMCs (developing member countries) and help conduct pre-feasibility studies to attract interest.
2 Policy and regulatory framework development
   a. Together with ADB and partner organizations, analyze best practices of policy, legal, and regulatory frameworks, and share these with DMC governments.
   b. Help enhance institutional capacity of DMCs for developing and implementing policy, legal, and regulatory frameworks for solar energy and smart grid applications.
3 Capacity development for local stakeholders
   a. Assist local utilities in DMCs, cooperating with equipment manufacturers and relevant private sector entities, to localize technology, adapting it to local conditions and parameters, for developing smart grids at transmission and distribution levels (e.g., use of stable current, voltage, and frequency management using storage devices, bill collection mechanisms with two-way meters, and voltage and frequency control systems).
   b. Enhance the capacity of local financing, insurance, and legal firms for appraising solar and smart grid project proposals.
   c. Facilitate training and capacity-building activities pertaining to various aspects of solar energy for different stakeholders as needed or as the opportunity arises.

(2) Project development: 3 000 MW solar capacity

The ASEI is committed to identifying, developing, and assisting implementation of 3 000 MW of solar power over the next 3 years. The phasing is expected to be 500 MW over year 1; 1 000 MW over year 2; and 1 500 MW over year 3.

The targeted projects are expected to demonstrate and model the feasibility and sustainability of various types of public and private solar energy generation projects, including large-scale clustered development of grid-connected projects, building-integrated PV systems, roof-top distributed power generation projects in cities, and off-grid or local-grid rural electrification projects. Pilot demonstration projects to be promoted by DMC governments will also be among the ASEI's focus. Such projects will establish models for financiers and private sector entities that will be investing in projects in ADB DMCs.

The initial projects in year 1 supported under the ASEI are two private sector-led solar PV electricity generation projects in central Thailand: a 73,0 MW project and a 44,5 MW project. As the trustee for the Asia Pacific Carbon Fund, ADB was also selected as a partner for five solar power projects (i.e., two 10,0 MW solar power plants and 10,7 MW of building-integrated PV projects) being developed in China. A 1,5 MW CSP plant has also been implemented near Beijing in China. Other potential projects under consideration over the ASEI's first year included in collaboration with state utilities and private distribution companies are:

1 four demonstration CSP projects in China, 50 MW each in four provinces;
2 a solar power transmission project in Gujarat, India, enabling 500 MW of solar power generation capacity to be connected to the grid;
3 a 5 MW grid-connected solar PV demonstration project and two off-grid, solar-wind hybrid plants (0,5–1,0 MW) in Bangladesh;
4 private large-scale PV and solar thermal projects under concessions in five provinces in China.

For potential projects in year 2 and year 3, the private sector is expected to implement the bulk of projects financed under the ASEI. In this regard, several DMC governments already have ongoing programs to attract investments in solar power generation, notably China, India, and Thailand. A cluster development of solar PV parks in Rajasthan and Gujarat states in India through public-private partnerships is already in the ADB pipeline. Further, the Philippines has in place some manufacturing capacity for solar panels and plans to use this to develop solar energy power generation, which has been identified as a future objective and is envisaged to include 100 MW of solar power generation. Finally, Uzbekistan is also interested in seeking assistance from ADB in rapidly developing its solar energy resources.

The size of PV projects tends to be small, as developers benefit from the modular nature of the technology. With better understanding of the CSP technology, developers are expected to also pursue large grid-connected projects that would allow more capacity to be added in later years through fewer transactions.

The ASEI also proposes to make considerable investments in systems for generating and gathering insolation data in DMCs, e.g. solar resource mapping for portfolio development, through current or proposed programs. The ASEI will focus on using satellite-based global data to identify about a dozen large potential sites across a country for detailed ground-level measurements. Subsequently, based on identification of probable high potential sites and third-party resource modeling results (considering other factors such as land availability, transmission accessibility, and tariff and regulatory regimes), assistance would be provided for inviting bids from the private sector to develop bankable solar energy projects. Together with targeted incentive programs that draw on international experience, early resource mapping is expected to provide a portfolio of projects in a systematic manner.

To implement the ASEI component of developing 3 000 MW of solar power in 3 years, ADB plans to provide direct technical assistance to DMC governments for conducting pre-feasibility studies, preparing projects, and structuring solar generation projects as public-private partnerships. Moreover, ADB intends to provide assistance in terms of financing and expects to leverage additional funds for solar energy investment. Efforts are already under way to find innovative ways and means of attracting large amounts of private investment. ADB will leverage instruments, such as London interbank offered rate (LIBOR) based loans, donor contributions, grant funds, innovative risk mitigation mechanisms, carbon market support measures, and direct support, which will contribute to the improved financial viability of solar projects in the region.

(3) Innovative finance: the Asia Accelerated Solar Energy Development Fund

To supplement its direct financing, ADB devised the Asia Accelerated Solar Energy Development Fund (AASEDF). The AASEDF is a solar technology-specific fund intended to support solar project and program preparation in DMCs by mitigating related risks, thereby attracting investment at a lower cost, and providing power generation incentives so that consumers are not required to bear the high cost of the technology before it becomes fully commercial in a particular market.

The AASEDF is intended to catalyze solar capacity in Asia and the Pacific. It will encourage visible private sector participation by keeping transaction and opportunity costs low for the sector and by offering risk mitigation and incentive generation products while solar technology remains at the pre-commercial stage, ultimately unburdening end-consumers of the initial high cost of solar power.

More specifically, AASEDF resources are proposed to be used for the following activities:

1 Project and program preparation
2 Risk mitigation, such as exploration risks, specification risks, offtake and payment risks
3 Generation incentives.

The AASEDF is structured as a multidonor fund under the Clean Energy Financing Partnership Facility administered by ADB. While all DMCs will be eligible for support, the AASEDF will prioritize projects in large markets that accelerate the deployment of significant solar energy generation capacity and cost reduction that would contribute largely to lowering costs, consequently making solar technology affordable to many more markets.

There will be two key conditions for AASEDF support: (i) the host country government shares part of the cost of transition projects by way of fiscal or tax incentives with project developers, or by announcing adequate FITs; and (ii) the procurement process (for selection of project developers or purchase of solar equipment) is open to all ADB member countries in a fair and equitable manner.

REFERENCES

1 International Energy Agency. (n.d.). *WEO-2010 new Electricity access Database.* Retrieved August 1, 2011, from http://www.iea.org/weo/electricity.asp
2 International Energy Agency. (2011). *World Energy Outlook 2011.* Paris.
3 International Energy Agency. (2010). *Energy Technology Perspectives 2010 Scenario & Strategies to 2050.* Paris.
4 International Energy Agency. (2010). *Technology Roadmap Solar Photovoltaic Energy.* Paris.
5 U.S. Energy Information Administration. (2010). *International Energy Outlook 2010.*
6 Shell International BV. (2008). *Shell Energy Scenarios to 2050.*
7 Greenpeace International. European Renewable Energy Council (EREC). (2010). *Energy Revolution.*
8 WWF. (2011). *The Energy Report: 100% renewable energy by 2050.*
9 IPCC. (2011). *Special Report on Renewable Energy Sources and Climate Change Mitigation.*
10 International Energy Agency. (1998). *World Energy Outlook 1998.* Paris.
11 International Energy Agency. (2002). *World Energy Outlook 2002.* Paris.
12 International Energy Agency. (2004). *World Energy Outlook 2004.* Paris.
13 International Energy Agency. (2006). *World Energy Outlook 2006.* Paris.
14 International Energy Agency. (2008). *World Energy Outlook 2008.* Paris.
15 International Energy Agency. (2010). *World Energy Outlook 2010.* Paris.
16 REN21. (2011). *Renewables 2011 Global Status Report.*

17 Knies G. (ed.) (2009). *Clean Power from Deserts – The Desertec Concept for Energy, Water and Climate Security*, Whitebook 4th Edn, Desertec Foundation, Hamburg, www.desertec.org

18 Trieb F., Schillings C., Kronshage S., Klann U., Viebahn P., May N., Wilde R., Paul C., Kabariti M., Bennouna A., Nokraschy El H., Hassan S., Yussef L.G., Hasni T., and Bassam El N. (2005). *Concentrating Solar Power for the Mediterranean Region (MED-CSP)*, German Aerospace Center (DLR) by order of Federal Ministry for the Environment (BMU), Berlin, www.dlr.de/tt/med-csp

19 Trieb F., Schillings C., Kronshage S., Viebahn P., May N., and Paul C., (eds) (2006). *Trans-Mediterranean Interconnection for Concentrating Solar Power (TRANS-CSP)*, Research of German Aerospace Center (DLR) by order of Federal Ministry for the Environment, Nature Conservation and Nuclear Safety, Berlin, www.dlr.de/tt/trans-csp

20 Czisch G. (2005). *Szenarien zur zukünftigen Stromversorgung – Kostenoptimierte Variationen zur Versorgung Europas und seiner Nachbarn mit Strom aus erneuerbaren Energien*, Dissertation, University of Kassel, https://kobra.bibliothek.uni-kassel.de/bitstream/urn:nbn:de:hebis:34-200604119596/1/DissVersion0502.pdf

21 Sörgel H. (1932). *Atlantropa*, Fretz & Wasmuth, Zürich.

22 Voigt W. (2007). *Atlantropa. Weltbauen am Mittelmeer. Ein Architektentraum der Moderne*, Grosser + Stein, Pforzheim.

23 Schottky W., 1929. Photoelectric Generator, US Patent 2,040,632, filed in May 1.

24 Dominik H. (1930). *Wunschträume beim Jahreswechsel*, Beilage: Technik der Zeit, Berliner Tageblatt, Berlin, December 31.

25 Winter C.J. (1981). *Wasserstoff als Sekundärenergieträger*, Deutsche Forschungs- und Versuchsanstalt für Luft- und Raumfahrt (DFVLR), Stuttgart.

26 Sprengel U., and Hoyer W. (1989). *Solarer Wassersoff – Energieträger der Zukunft*, Deutsche Forschungsanstalt für Luft- und Raumfahrt (DLR), Stuttgart.

27 Dii – Desertec Industry Initiative (2010). *Energy from deserts – Bringing the Desertec Vision into reality*, Dii, Munich, www.dii-eumena.com/fileadmin/Daten/Downloads/Dii-Brochure_EN.pdf

28 Breyer Ch., and Knies G. (2009). *Global Energy Supply Potential of Concentrating Solar Power*, Proceedings SolarPACES 2009, Berlin, September 15–18.

29 Justi E. (1960). *Die zukünftige Energieversorgung der Menschheit*, Vandenhoeck & Ruprecht, Göttingen.

30 Knies G. (2006). *Deserts as sustainable powerhouses and inexhaustible waterworks for the world*, Trans-Mediterranean Renewable Energy Cooperation (TREC), Global Conference on Renewable Energy Approaches for Desert Regions, Amman, September 18–22.

31 Dii – Desertec Industrial Initiative (2012). *Bringing the Desertec Vision into Reality*, Dii, Munich, www.dii-eumena.com

32 Medgrid (2012). *Opening new lines for sustainable electricity*, Medgrid, Paris, www.medgrid-psm.com

33 Son van P. (2011). *Dii's longterm vision for a EU-MENA market*, Desertec Industrial Initiative, 2nd Dii Desert Energy Conference, Cairo, November 2–3.

34 http://www.medgrid-psm.com/en/project/the-vision/

35 Asia Development Bank (2011). *Asia Solar Energy Initiative – A Primer.*

CHAPTER THREE

Global potential of solar energy

3.1 SOLAR RESOURCES

Energy resource availability is decisive for all power technologies. The basis for all solar energy technologies is the amount and the composition of global horizontal irradiation. Different solar system concepts have been developed for optimizing the energy yield. In this chapter, solar resources adapted to systems' requirements are derived and compared among each other.

3.1.1 Global horizontal irradiation as basis for all solar resources

Measurement stations all around the world record solar irradiation on the ground. The irradiance consists of the direct irradiance from the sun, and diffuse irradiance from different sources, e.g. the rest of the hemisphere or ground reflection. Another valuable source of solar irradiation data are meteorological satellites in space. Combining satellite data and ground measurements as reference enables global solar irradiation data to be recorded. The National Aeronautics and Space Administration (NASA) compiled several global datasets for different meteorological variables, so for global horizontal irradiation (GHI) and its diffuse fraction on global scale for a 1° × 1° mesh, i.e. average data are available for every 1° step of latitude and longitude, respectively. In this chapter the NASA Surface meteorology and Solar Energy (SSE) 6.0 dataset is used for all calculations [1, 2].

Core solar resource data are the beam and diffuse fraction of global horizontal irradiation in this chapter. All further solar resources discussed in this chapter are based on global horizontal irradiation (GHI) depicted in Figure 3.1. Here, GHI input data are provided by NASA SSE 6.0 [2] and used in a 1° × 1° mesh of latitude and longitude. The key driver for beneficial yield results of tracking systems is the diffuse fraction of GHI besides the geographic location of the site and overall solar irradiation.

Global horizontal irradiation is more or less homogenously distributed on Earth. Variations of GHI for most populated regions in the world are typically within a range of a factor of two, e.g. Central Europe (~ 1 100 $kWh/m^2/y$) and Arabian Peninsula (~ 2 200 $kWh/m^2/y$). GHI is best in the Atacama desert and in the Sahara desert. Excellent and very good solar conditions of more than 1 700 $kWh/m^2/y$ are available all around the world. Tropical regions near to the Equator show good but not excellent solar conditions despite their best geographic location for solar exposure. Local climate conditions, mainly daily or seasonal cloud patterns, can have enormous impact on GHI. Besides tropical regions near the Equator, this can be found for monsoon effects in India and South-East China as a consequence of regional topography and Pacific clouds.

Absolute and relative diffuse irradiation is strongly coupled to regional climatic conditions, in particular to daily and seasonal occurrence of clouds. Tropical regions near the Equator show these effects, e.g. Amazonas rim, Central Africa and South-East Asia, but also India. Most deserts on Earth are located between 15° and 40° N/S. Diffuse irradiation in deserts is quite low and ranges typically between 15% and 30%. Diffuse irradiation is lost for solar power technologies only capable of converting direct solar irradiation, thus sites of low diffuse irradiation are most appropriate for these technologies. Solar system technologies increasing insolation on an active surface by tracking the sun's position are most beneficial for lower diffuse irradiation.

3.1.2 Overview of solar system concepts

A major advantage of solar PV technology is the need for no moving parts; this is still the case for fixed optimally tilted systems. To give up such a great advantage very good

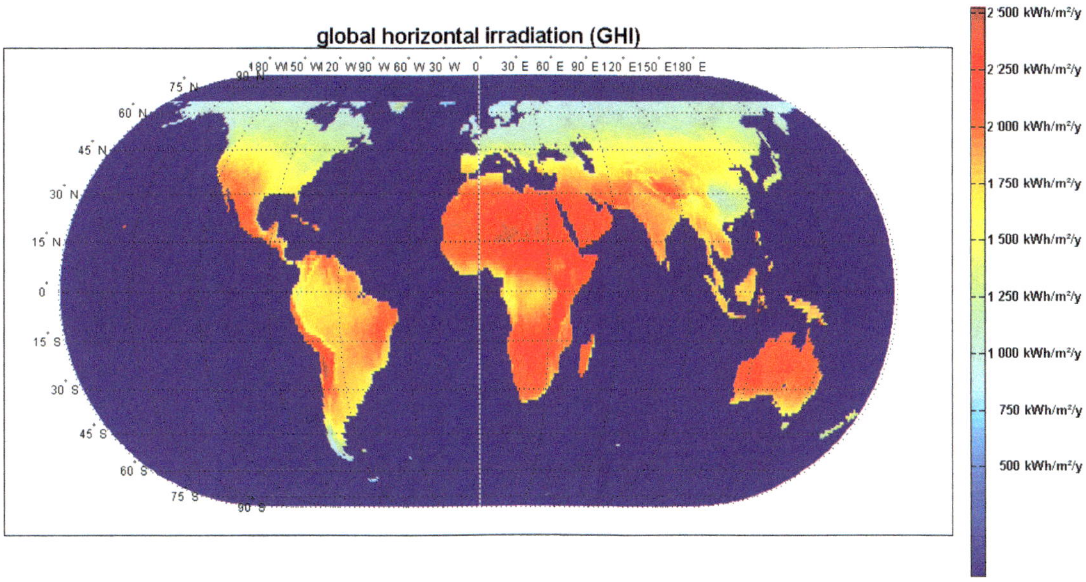

Figure 3.1 Annual global horizontal irradiation (GHI)

reasons are needed; these might be given for specific tracking systems. 1-axis east-west horizontal tracking systems are best adapted for regions of large seasonal deviation of daily zenith angles. An advantage of horizontal tracking and fixed tilted systems is shading in only one direction, which significantly reduces the necessary minimum site area thus the balance of system-related costs is kept at the lowest possible level. 1-axis north-south horizontal continuous tracking systems follow the sun over the day. 1-axis north-south optimally tilted tracking systems are better adapted to respective site latitude but need a larger plant area than fixed optimally tilted or 1-axis north-south horizontal continuous tracking due to two-directional shading. 1-axis vertical optimally tilted systems also need a larger plant area as a consequence of two-directional shading. The PV system of highest annual irradiation is the 2-axes non-concentrating tracking system. Large plant area is needed due to two-directional shading effects and, moreover, a higher mechanical stress and further investment costs have to be tackled due to more moving parts and engines. 2-axes concentrating tracking variation is only needed for high concentrating PV (HCPV) systems and solar thermal electricity generation (STEG). A brief discussion of selected systems in respect to their respective solar resource is given in the following.

GHI needs to be transferred to irradiation data received by a tilted PV module. For well adapted calculation of direct and diffuse irradiation, Hay and Davis introduced a model peer-review published by Hay and McKay [3] which takes into account the circumsolar diffuse and horizon-brightening diffuse components of diffuse irradiation on a tilted surface. Reindl *et al.* [4] practically improved the horizon-brightening part after a suggestion of Klucher [5]. This standard model for describing direct and all relevant diffuse parts of solar irradiance on a tilted surface is called the Hay–Davis–Klucher–Reindl (HDKR) model, and is comprehensively discussed in Duffie and Beckmann [6]. A great advantage of the HDKR model are good output results using only very limited input data, i.e. monthly data for global horizontal irradiation, diffuse horizontal irradiation, geographic coordinates and system geometry.

3.1.3 Fixed optimally tilted systems

The fixed optimally tilted PV system is a typical PV power plant system and is used as a reference system for tracking systems.

The major cost components of PV power plants are PV modules, inverters, further so-called balance of system (BOS) components and the site for placing the plant. A large fraction of BOS cost is dependent on the total area needed for the plant, which is a driver for placing rows of modules very close to each other. However, the solar energy source, the sun, shines only for a short proportion of annual sunshine hours perpendicular to the ground. Therefore, modules are tilted for optimized annual irradiation on the modules' surface. Standard textbook knowledge suggests the tilt angle should be equal to the latitude of the specific site of a PV system for optimizing annual irradiation on modules. Nevertheless, optimized tilt angles and lowest possible area demand for a PV plant form an inherent trade-off.

The annual incident irradiation on tilted module surface has been evaluated as a function of module tilt angle. The tilt angle of highest annual irradiation is defined as the optimum tilt angle, which optimizes the irradiance on module surface over the entire year. An optimized tilt angle increases annual irradiation on module surface up to 500 kWh/m²/y and 35%. Irradiation at latitudes near the Equator cannot be significantly improved by tilt angles, whereas for latitudes lower than 30°S and higher than 30°N a significant increase in irradiation on module surface can be expected. Usually, one assumes the optimum tilt angle

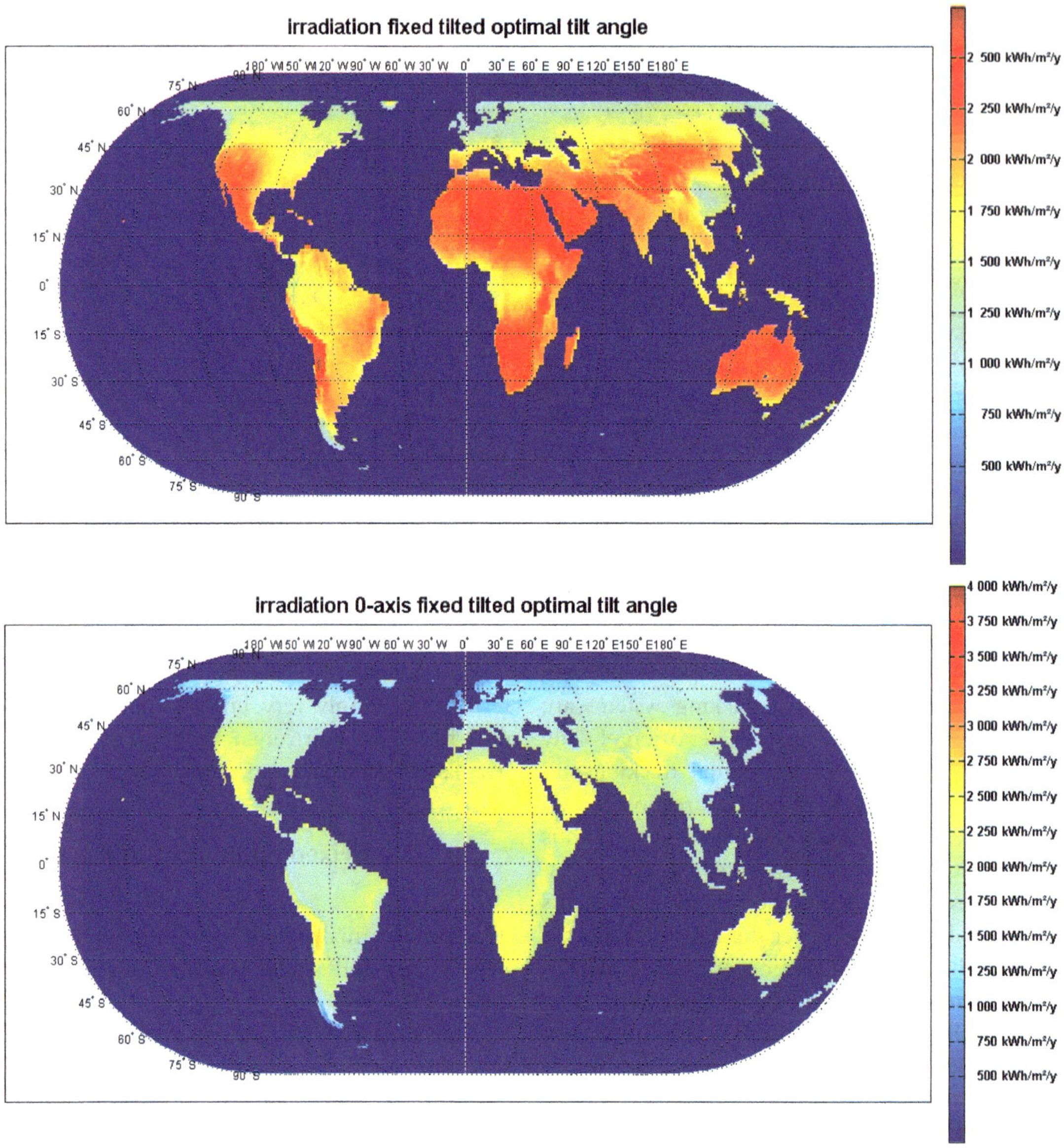

Figure 3.2 Annual irradiation on module surface of fixed optimally tilted PV systems at tilt angle for optimized irradiation [8]
Top and bottom visualizations are identical except colour coding for subsequent comparison reasons.)

to be the latitude of the site of interest as a rule of thumb. Some parts in the world, e.g. Central and North Europe, Tibet and Thailand, deviate by about 10° or more from this educated guess, mainly because of different occurrence of clouds [7].

Figure 3.2 shows the annual irradiation on module surface of fixed optimally tilted PV systems at tilt angle for optimized irradiation. Every coordinate of a 1° × 1° mesh of latitude and longitude within 65°S and 65°N is separately optimized for maximum annual irradiation on module. The HDKR model is performed by calculating each month of the year and selecting the tilt angle of maximum annual irradiation.

Notably, solar resource is homogenously available for nearly all people in the world (Figure 3.2). This is a significant contrast to all other sources of energy. There is only a maximal factor of about 2,5 between the highest and lowest irradiation sites in the world, in particular where people live. More or less even availability of solar resource triggers a very important social impact of PV as the most democratic energy technology.

Solar irradiation is mainly determined by latitude and weather conditions (Figure 3.2), which can be observed for tropical regions in South America, Africa and South Asia, but also for special weather conditions like in India (monsoon) or South East China.

3.1.4 1-axis horizontal north–south continuous tracking system

Identical horizontal continuous tracking PV systems can be built either in east-west or in north-south direction of axis. East-west axis systems optimize the module's orientation to the zenith angle of the sun, thus they are adapted to seasonal variations of the position of the sun. On the

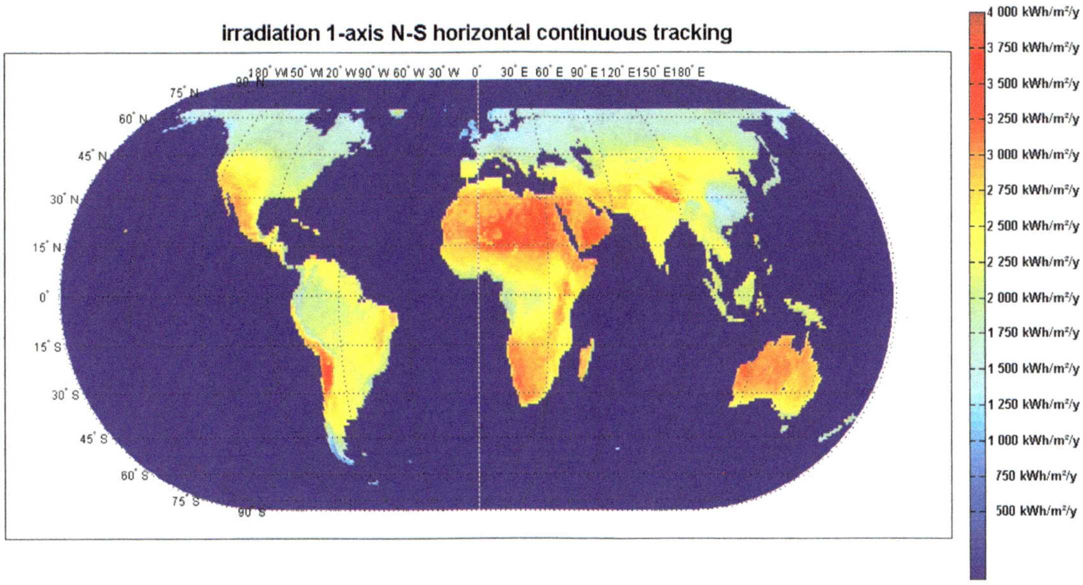

Figure 3.3 Annual irradiation on module surface of 1-axis horizontal north-south continuous tracking PV systems [8]

contrary, north-south axis systems follow the position of the sun over the day and therefore optimize irradiation on a daily basis.

Figure 3.3 shows the annual irradiation on module surface of 1-axis horizontal north-south continuous tracking PV systems. As well, every coordinate of a 1° × 1° mesh of latitude and longitude within 65°S and 65°N is separately optimized for maximum annual irradiation on module. The HDKR model is performed by calculating each month of the year and selecting the tilt angle of maximum annual irradiation.

As shown, for all regions in the world irradiation on module surface is increased by 1-axis horizontal north-south continuously tracking PV systems in reference to fixed optimally tilted systems. The absolute irradiation increase reaches 500–800 kWh/m²/y for regions of good and excellent solar conditions, whereas regions of moderate solar conditions achieve higher irradiation of 100–300 kWh/m²/y. This translates to a relative increase of about 20–30% for the sunny regions and 10–20% for the regions of moderate GHI. Absolute and relative increase are highest for sites of excellent solar conditions and low level of diffuse insolation, hence the irradiation increasing effect of tracking is maximized.

Summarizing the results, 1-axis horizontal north-south continuous tracking PV systems might be competitive with fixed optimally tilted and 1-axis horizontal east-west continuous tracking PV systems. An increase in irradiation of 20–30% for sunny and very sunny regions might justify higher capital and operational expenditures, especially for plant operators achieving further advantages of more full load hours of the PV plant.

3.1.5 2-axes non-concentrating continuous tracking system: global normal irradiation

Highest possible irradiation on modules can be achieved by 2-axes non-concentrating tracking, hence module surface is always perpendicular to direct radiant beam of the sun, but for non-concentrating systems additional diffuse irradiance can be converted. This highest possible solar resource is defined as global normal irradiation (GNI).

Figure 3.4 shows the annual irradiation on module surface of 2-axes non-concentrating continuous tracking PV systems. As well, every coordinate of a 1° × 1° mesh of latitude and longitude within 65°S and 65°N is separately optimized for maximum annual irradiation on module. The HDKR model is performed by calculating each month of the year and selecting the tilt angle of maximum annual irradiation.

For all regions in the world irradiation on module surface is increased by 2-axes non-concentrating continuous tracking (GNI) PV systems in reference to fixed optimally tilted systems. Absolute irradiation increase reaches 600–1 000 kWh/m²/y for most regions in the world, whereas regions of high absolute and relative diffuse horizontal irradiation achieve higher irradiation of only 200–400 kWh/m²/y. This translates to a relative increase of about 30–45% for nearly all regions in the world and 15–25% for regions of high absolute diffuse horizontal irradiation. Absolute and relative increase is highest for sites of excellent solar conditions but still high for regions of moderate GHI except those of high absolute diffuse horizontal irradiation.

Comparing 2-axes non-concentrating to 1-axis horizontal north-south continuous tracking PV systems, only regions higher than 30°N and 30°S receive significantly more irradiation on module surface and reach 10–20% more irradiation which is similar to fixed optimally tilted PV systems versus GHI and also a consequence of the tilt angle

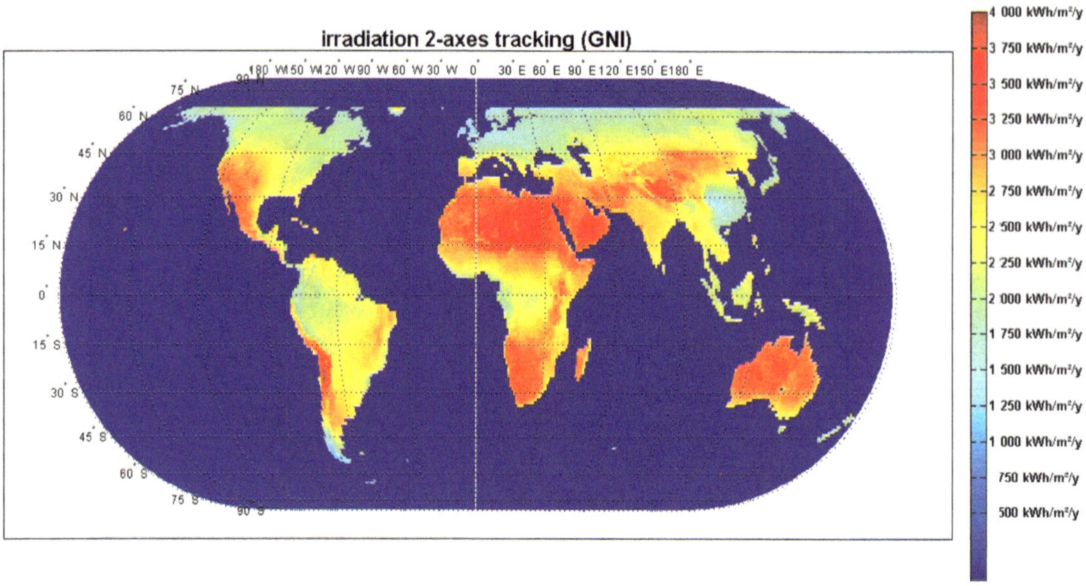

Figure 3.4 Annual irradiation on module surface of 2-axes non-concentrating continuous tracking PV systems [8]

advantage of higher absolute latitude sites. Furthermore 2-axes non-concentrating continuous tracking PV systems receive more irradiation on module surface than 1-axis vertical optimally tilted PV systems. However, this increase is limited to only 100–200 kWh/m²/y for absolute latitudes within 45°N/S. 2-axes non-concentrating continuous tracking also improves 1-axis optimally tilted north-south continuous tracking PV systems, but this effect is rather low, as only 70–130 kWh/m²/y more irradiation can be realized.

Summarizing the results, 2-axes non-concentrating continuous tracking (GNI) is the best solar resource available on Earth in terms of irradiation as a consequence of optimum tracking modes and maximal irradiation, in particular including diffuse irradiation. Nevertheless, 2-axes non-concentrating continuous tracking (GNI) PV systems might not be competitive with 1-axis horizontal north-south, 1-axis vertical optimally tilted and 1-axis optimally tilted north-south continuous tracking PV systems as a consequence of very limited additional irradiation on module surface. 2-axes tracking is more complex in its technical realization due to one additional axis resulting in more capital and operational expenditures.

3.1.6 2-axes concentrating continuous tracking system: direct normal irradiation

Concentrating solar power technologies are specially adapted for converting direct solar irradiation. For optimizing insolation 2-axes continuous tracking systems are applied, or at least quasi 2-axes systems, e.g. long rows of 1-axis horizontal north-south continuous tracking systems using a one-dimensional focal line. As a consequence of 2-axes tracking and direct solar irradiation, the solar resource for concentrating solar power technologies is called direct normal irradiation (DNI).

Figure 3.5 shows annual irradiation on module surface of 2-axes concentrating continuous tracking PV systems. As well, every coordinate of a 1° × 1° mesh of latitude and longitude within 65°S and 65°N is separately optimized for maximum annual irradiation on module. The HDKR model is performed by calculating each month of the year and selecting the tilt angle of maximum annual irradiation.

For all regions in the world irradiation on module surface is decreased by 2-axes concentrating continuously tracking (DNI) PV systems in reference to 2-axes non-concentrating and 1-axis horizontal north-south continuous tracking PV systems. The absolute irradiation decrease reaches 550–750 kWh/m²/y for most sunny and very sunny regions in the world, whereas regions of low absolute and relative diffuse horizontal irradiation achieve the lowest irradiation losses. This translates to a relative decrease of about 10–30% for regions of good and very good solar conditions.

2-axes concentrating solar power technologies need to overcome the aforementioned irradiation loss and the higher capital and operational expenditures for active sun tracking and the necessity for 2-axes tracking to be additionally compensated. Both factors are best addressed by higher power conversion efficiency of the respective solar power systems. High concentrating PV (HCPV) currently achieve maximum DC-efficiencies of up to 26% [9], compared with measured string efficiencies of about 15–18% of respective low- and non-concentrating PV systems [10]. These numbers are best illustrated for an exemplary site in the Sahara desert. GNI and DNI are 3 400 kWh/m²/y and 2 650 kWh/m²/y, respectively, hence GNI is 750 kWh/m²/y higher than DNI, i.e. slightly more than 20% irradiation is lost for concentrating solar power systems. Currently the highest power conversion efficiencies of non-concentrating and concentrating PV technologies are about 16% and 26% for entire PV systems accounted for string power. Assuming a low ground cover ratio of 20%, the 2-axes non-concentrating and the 2-axes

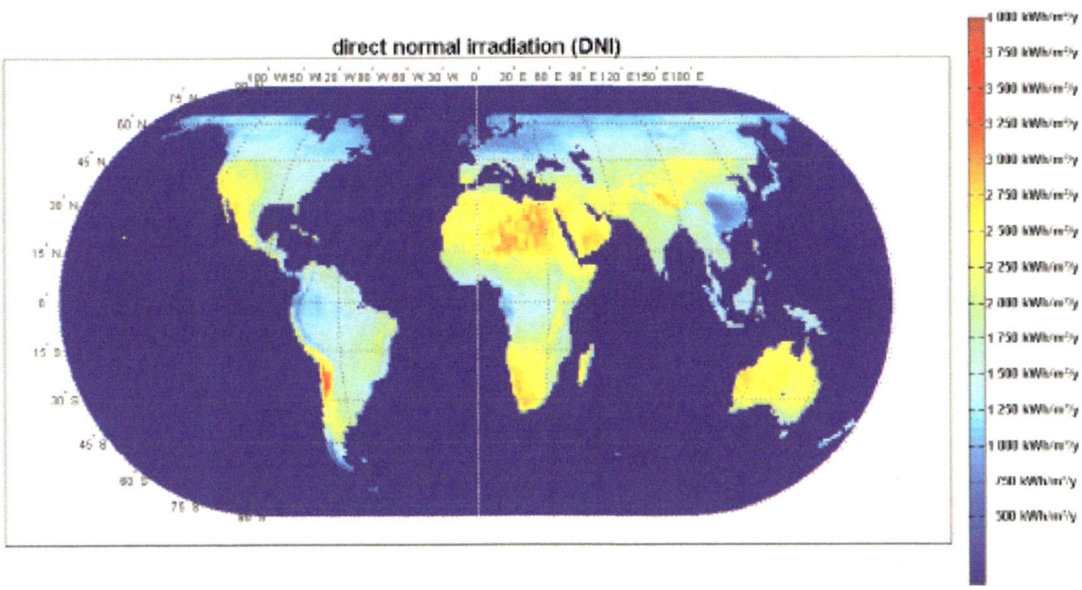

Figure 3.5 Annual irradiation on module surface of 2-axes concentrating continuous tracking PV systems [8]

concentrating PV systems are able to generate about 110 $GWh_{el}/km^2/y$ and 150 $GWh_{el}/km^2/y$, i.e. HCPV systems generate about 35% more electricity than conventional PV systems. Therefore concentrating PV systems might be competitive with non-concentrating PV systems if capital and operational expenditures increases were less than 35%.

In theory the irradiance at standard test conditions is assumed to be 1 000 W/m² for non-concentrating and 850 W/m² for concentrating PV systems, as a consequence of lost diffuse irradiance [11]. However, in practice concentrating irradiance (DNI) reaches up to 94% of non-concentrating irradiance (GNI) and 1 000 W/m² direct normal irradiance is the standard operation condition [12]. Due to that fact, the first manufacturers of HCPV modules, e.g. Soitec CPV (former Concentrix Solar), label their panels 1 000 W/m². Up to now the deviation of practical measurements and theory cannot be explained properly, but it is a further push for HCPV systems.

Summarizing the results, the solar resource for 2-axes concentrating continuous tracking (DNI) PV systems is about 10–30% lower than for comparable 2-axes non-concentrating continuous tracking PV systems, which can use GNI. However, significantly higher power conversion efficiencies of HCPV in reference to non-concentrating PV systems result in about 35% higher electricity generation potential per area unit. The higher cost of HCPV systems due to more complex technology and the necessity of using 2-axes tracking have to be below 35% to be competitive in very sunny regions of low diffuse irradiation.

In total, the solar resource is available in all populated regions in the world. Fixed optimally tilted PV systems are the standard PV power plant system type, however depending on local conditions adapted tracking systems improve the annual yield per installed capacity and the power generation cost.

3.2 ECONOMIC MARKET POTENTIAL OF SOLAR PV

Installations of photovoltaic (PV) power plants have shown high growth rates around the world [13]. As a consequence of this growth, PV electricity generation costs continually decrease. The opposite trend is shown by power generation costs due to increasing fossil fuel prices and by electricity prices for end-users. The intersection of these two trends is commonly referred to as grid-parity and fuel-parity, respectively, where grid-parity allows for cost neutral residential and commercial end-user PV installations and fuel-parity indicates cost neutral PV power plant investments. Moreover, small PV applications are very competitive for a large majority of the 1,4 billion people without access to electricity and are complemented by renewables-based mini-grids in rural regions of higher income.

Here, a comprehensive analysis of the total economic PV market potential until the year 2020 is shown. The analysis is based on the economic PV market potential including all major global PV market segments in an integrated manner. Furthermore, the contribution of PV to the global energy supply is critically discussed, since the PV contribution potential is assumed to be of minor relevance only in almost all important energy scenarios.

3.2.1 Overview of PV scenarios

Energy scenarios are a very helpful tool for guiding different stakeholder groups to what might happen in future in dependence on various policy options. Such scenarios show long-term consequences of potential policy decisions and enable an assessment of whether respective developments would be acceptable in reference to relevant

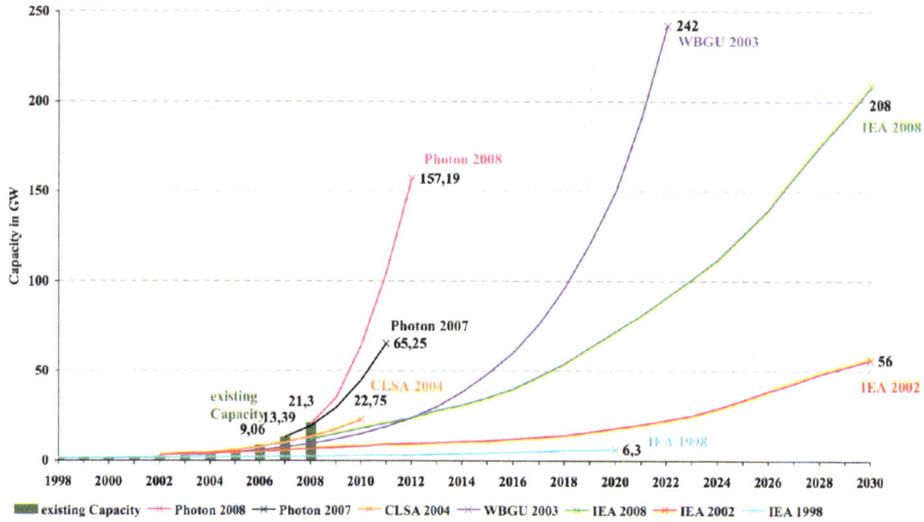

Figure 3.6 Forecasts of PV development and real cumulated installed PV capacity for 2008 The references are: 'Photon 2008' [15], 'Photon 2007' [16], 'CLSA 2004' [17], 'WBGU 2003' [18] and 'IEA 2008' [19].)

constraints, e.g. diminishing fuel resources, greenhouse gas emissions or social cost of energy supply.

Figure 3.6 shows an example of the comparison of recent development of PV production with forecasts of various institutions [14]. Here, the production numbers for PV are shown in the graph and the cumulated installed PV capacity as of end of year would be 7,0 GW (2006), 9,5 GW (2007) and 15,7 GW (2008) according to the European Photovoltaic Industry Association (EPIA) [13].

Results indicate that the projections of the International Energy Agency (IEA) structurally underestimate the growth rate of PV and even those of the German Advisory Council on Global Change remain significantly below the real development. Market research performed with deep insights and a broad coverage of the PV industry, such as studies compiled under the leadership of Michael Rogol, show the necessity for upward adaption, as well, but implying substantial higher growth rates.

In Figure 3.7, PV market projections from various sources are visualized in an integrated view in order to display the substantial variance in market expectations in the current decade until the year 2020 for the cumulated installed PV capacity and annual installations. To be included in this integrated view, the studies need to cover at least the time span from the year 2010 to 2020, therefore all short-term reports have been excluded, i.e. those by PV industry research companies like Photon Consulting, EuPD Research and IHS Research but also the reports of all financial analysts.

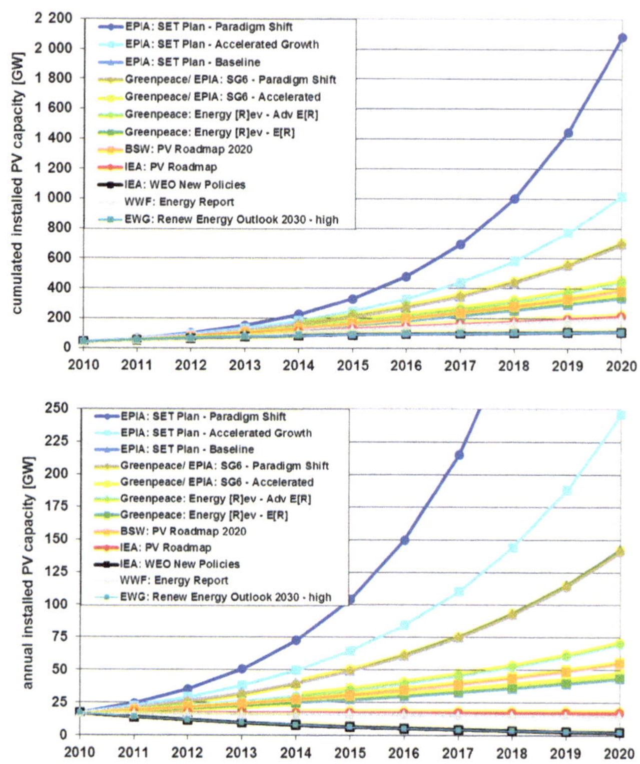

Figure 3.7 Projections of cumulated (top) and annual (bottom) installed PV capacity of various institutions.
(EPIA figures are only provided for Europe and have been extended on a global basis according to total installed power plant capacity [20]. Data are taken from EPIA [21], Greenpeace [22, 23], BSW [24], IEA [25, 26], WWF [27] and EWG [28].)

The most obvious result of the integrated view of the relevant studies covering the PV market development in the 2010s is the enormous range of market projections. There are roughly three fundamentally different projections: firstly, PV will become a major source of global energy supply (EPIA and partly Greenpeace); secondly, PV will grow but the growth rate will decline to a fraction of the mean of the past 15 years (BSW and partly Greenpeace); thirdly, the growth rates of annual PV installations will become negative, i.e. a decline in further market development is expected (IEA, WWF and EWG). The projected annual PV market growth rates for these three major groups amount to roughly 25–45% (EPIA and partly Greenpeace), 10–12% (BSW and partly Greenpeace) and 0–17% (IEA, WWF and EWG). The IEA assumes a cumulated installed PV capacity of 110 GW in the year 2020 in its most progressive 'New Policies' scenario in the World Energy Outlook [26]. This is in vast contrast to the average annual growth rate of PV for all major diffusion phases which has never been below 30% and for the last 15 years was about 45% (Figure 3.8).

The annual installations at the end of the 2010s again show three fundamentally different expectations: the very conservative group projects average installations at the end of the 2010s below the numbers of 2010, the moderate group expects annual installations between 40 GW and 70 GW and the progressive group predicts that annual installations might be 140 GW or higher. This can be compared with overall power plant installation numbers having doubled from about 80 GW in the 1990s to about 160 GW in the 2000s [31]. Moreover, power plants around the world are operating at about 4 300 annual full load hours in average, whereas PV power plants will range between 1 000 and 2 500 full load hours. Either PV complements other sources of electricity or PV plus storage solutions fulfil all requirements for full power supply. In any case, new PV installations in the order annually of about 150 GW are not technically limited at all.

Comparing the market projections with the realized growth in the past, the reasonable expectation might be the progressive assessment by EPIA and some Greenpeace scenarios. It remains unclear why some institutions assume a decline in market growth or even a decline in annual installations, i.e. negative growth rates, strongly violating the long-term growth trend of PV (Figure 3.8). No arguments have been found in these studies that could explain this sudden deviation from the trend that has been ongoing for more than five decades. This is very surprising, as nearly all of these institutions have a drastic view of the business-as-usual scenario and some of them claim a reduction in greenhouse gas emissions as soon as possible, e.g. WWF and EWG.

3.2.2 Major PV diffusion phases – consequence of high growth rates and learning rates

The average annual growth rate of global PV production has increased from about 33% in the space age and during off-grid diffusion to about 45% for the last 15 years during on-grid diffusion (Figure 3.8) [29]. In the history of PV, three major inventions led to new and sustainable markets for PV systems. In the 1950s, the introduction of PV power supply in space as a least cost option started the first PV market diffusion phase. The second PV diffusion phase was driven by off-grid PV applications and started the fast growth in PV production in the 1970s. The third PV market diffusion phase was enabled by the political invention of roof-top programs and feed-in tariff laws in the 1990s [30, 31]. The fourth diffusion phase comprises commercial utility-scale PV power plants and is starting right now.

Figure 3.9 shows total installed PV capacity (top), growth rate classification of cumulated installed PV

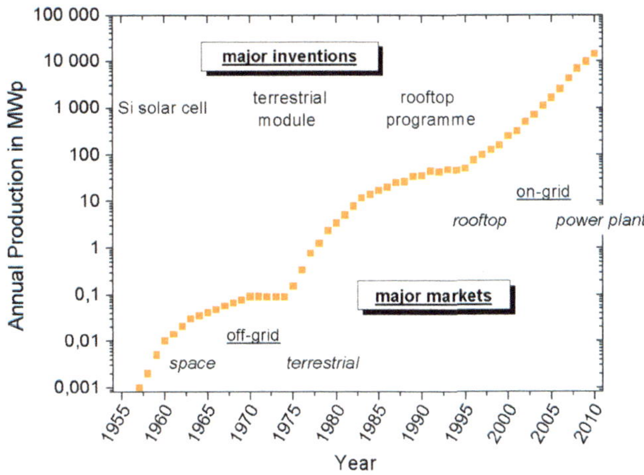

Figure 3.8 Historic PV production depending on major inventions and market segments [29]

capacity (center) and ratio of PV installations to total power plant capacity per country by the end of 2010 (bottom).

By end of 2010 about 40 GW of cumulated PV power capacity had been installed globally and, most interestingly, PV products found their markets in all countries in the world. From the year 2009 to 2010, the cumulated installed PV capacities grew by more than 50% in the majority of the countries in the world, and more than 50 countries achieved growth rates of more than 100%. The future PV market potential and growth ahead is illustrated by the ratio of already installed PV capacity to total installed power plant capacities per country. Some countries have already broken the 1% threshold. An important threshold of market saturation is the ratio of about 50% PV to total installed capacity, however still without significant storage capacities. Including storage this level would be substantially higher. Such high PV penetration ratios are possible due to the relatively low full load hours of PV systems and the good match of PV power feed-in to the load curve as a consequence of daily sunshine characteristics.

The sustainable PV market growth over more than five decades has been possible due to the favorable fundamental economics of PV technology. The basis for this development is the modular and scalable nature of PV applications and production. Modular PV products can be found in the market from the sub-watt class (e.g. solar calculators), to the watt range (e.g. pico systems and solar home systems) [33], to the kilowatt size (e.g. residential roof-top systems) [30] up to the multi-megawatt dimension (utility-scale power plants) [31]. The industrial value chain of PV is highly scalable and characterized by nearly continuous production flows for all production steps from metallurgical silicon (Si), to Si refinery, ingoting, wafering, cell and module manufacturing (or integrated PV thin film module production), inverter production and even system assembly, in particular of large scale power plants. Most industries based on modular and continuous production flows are characterized by an enormous cost reduction as a consequence of historic industrial production. Accordingly, PV technology shows a stable long trend of reducing PV

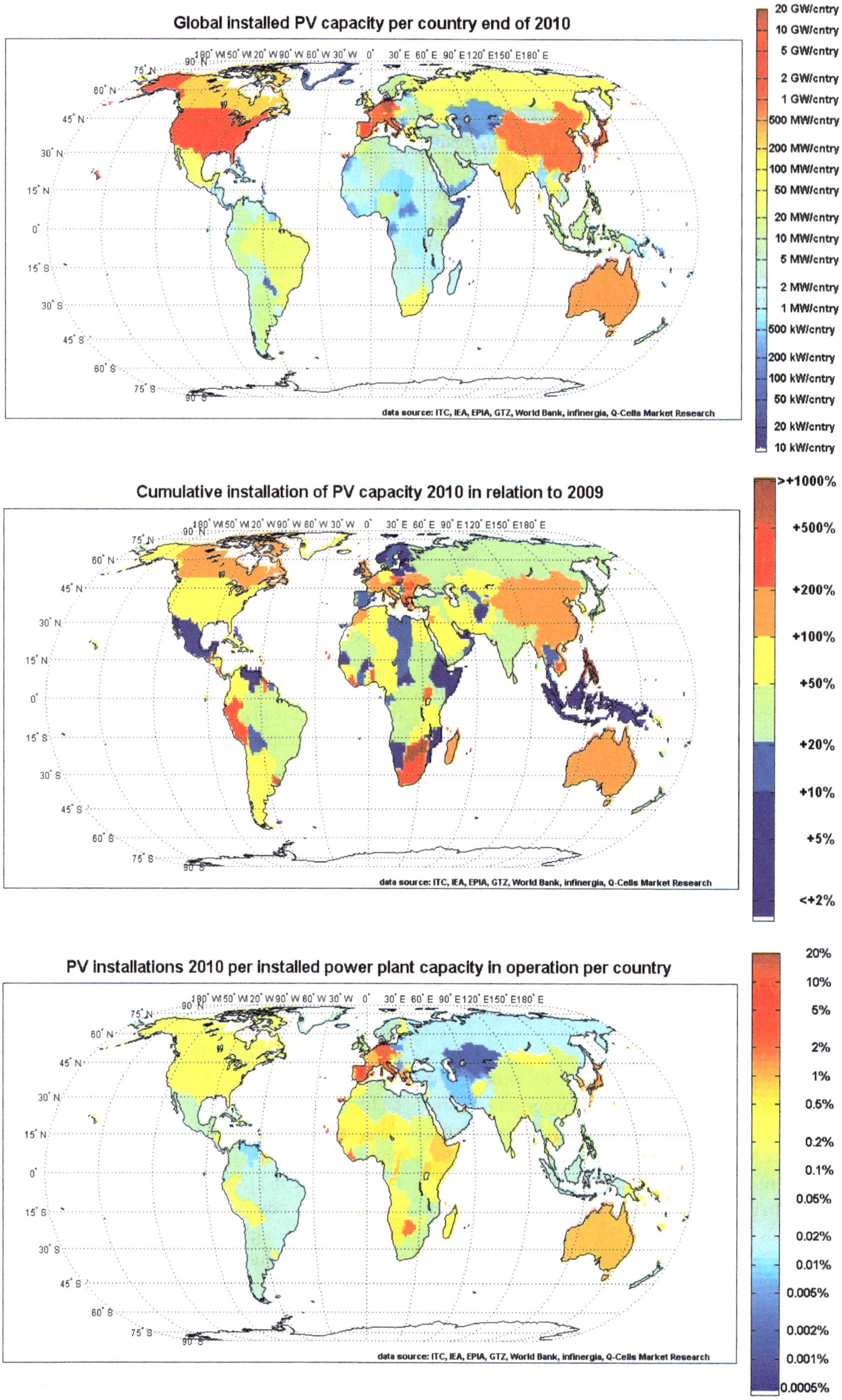

Figure 3.9 Total installed PV capacity (top), growth rate classification of cumulated installed PV capacity (center) and ratio of PV installations to total power plant capacity per country by the end of 2010 (bottom) [32] Data for total power plant capacity have been taken from Platts [20].)

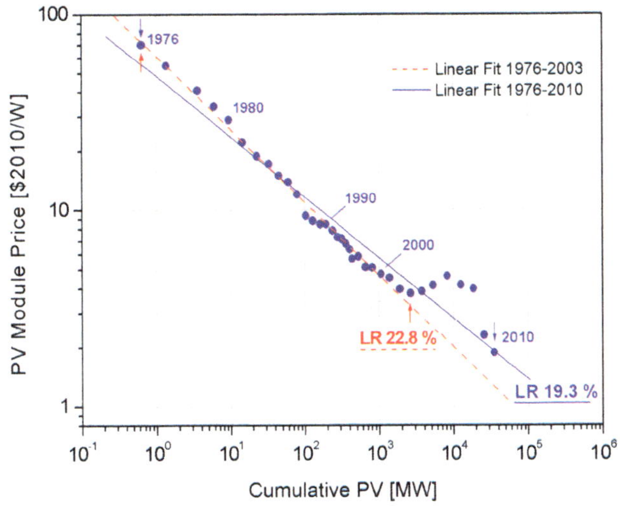

Figure 3.10 Learning curve for PV modules for the mid-1970s to 2010 [29]

module cost per doubling of cumulated production of about 20% for the entire period from the mid-1970s until 2010, as shown in Figure 3.10.

The best approximation for the cost is the price curve as information rated in Wp. Oscillations around the trend are mainly caused by varying PV industry market dynamics and therefore profit margins, documented by applying different learning rates of 22.8% and 19.3% for the periods 1976–2003 and 1976–2010, respectively [29].

Stable learning rates (LR) can be expected in the years to come due to fast increasing corporate PV R&D investments [29]. A broader discussion of the PV learning curve can be found elsewhere [29, 34]. For a proper discussion, the PV learning curve on the system level has to be taken into account, however data are very limited.

3.2.3 Grid-parity of PV systems

Grid-parity for end-users is defined by the parity of PV electricity generation cost and cost of electricity supply [30]. The most appropriate method for cost calculation is the levelized cost of electricity (LCOE) approach [35] summarized and adapted to PV in Equation 3.1.

$$LCOE = \frac{Capex}{Y_{ref} \cdot PerfR} \cdot \left(\frac{WACC \cdot (1+WACC)^N}{(1+WACC)^N - 1} + k \right) \quad \text{(Eq. 3.1a)}$$

$$WACC = \frac{E}{E+D} \cdot k_E + \frac{D}{E+D} \cdot k_D \quad \text{(Eq. 3.1b)}$$

$$k = k_{ins} + k_{O\&M} \quad \text{(Eq. 3.1c)}$$

Equation 3.1 Levelized cost of electricity (*LCOE*) for PV systems [30]

Abbreviations stand for: capital expenditures (*Capex*), reference yield for specific PV system at specific site (Y_{ref}), performance ratio (*PerfR*), weighted average cost of capital (*WACC*), lifetime of PV system (*N*), annual operation and maintenance expenditures (*Opex*), annual cost of *Opex* in percentage of *Capex* (*k*), equity (*E*), debt (*D*), return on equity (k_E), cost of debt (k_D), annual insurance cost in percentage of *Capex* (k_{ins}) and annual *Opex* in percentage of *Capex* ($k_{O\&M}$).

For analyzing the grid-parity dynamics in time, the critical input parameters are the progress ratio of PV, the growth rate of the global PV industry, both key drivers of the experience curve, and the electricity price trends.

The experience curve approach is an empirical law of cost reduction in industries. It was observed that for each doubling of cumulated output the specific cost decreases by a nearly stable percentage (Equation 3.2). This stable cost reduction is defined as the learning or experience rate. For use in calculations, the progress ratio is introduced, which is defined as unity minus learning rate.

$$c_x = c_0 \cdot \left(\frac{P_x}{P_0} \right)^{\frac{\log progress\ ratio}{\log 2}} \quad \text{(Eq. 3.2a)}$$

$$P_x = \sum_{t=0}^{T} P_t \quad \text{(Eq. 3.2b)}$$

$$P_t = P_{t-1} \cdot (1 + GR_t) \ \text{ for } t \geq 1 \quad \text{(Eq. 3.2c)}$$

$$P_x = P_0 \cdot \prod_{t=0}^{T} (1 + GR_t) \quad \text{(Eq. 3.2d)}$$

Equation 3.2 Empirical law of experience curves [30]

Abbreviations stand for: cost at historically cumulated output level of P_x (c_x), cost at initial output level P_0 (c_0), historically cumulated output level (P_x), initial output level (P_0), unity minus learning rate defined as (progress ratio), annual production of a specific year (P_t), and growth rate of a specific year (GR_t).

Here, Eqs. 3.2b and 3.2d are equivalent. The variables *Capex* and c_x are identical and describe the specific investment cost in a PV system in cost/*Wp*. Also, combination of Eqs. 3.2a and 3.2d effectively transforms the cost function from production volume dependence to time dependence, which is often more convenient for scenario analyses.

The empirical law of experience curves (Eq. 3.2a) drives the levelized cost of electricity of PV systems (Eq. 3.1a) via the *Capex* parameter which has to compete against local electricity prices in respective markets.

The dynamics of grid-parity have been analyzed for the residential and industrial market segment for more than 150 countries representing more than 98% of world population and more than 99% of global gross domestic product (Figure 3.11) [30].

The key parameters for the model are set to *Capex* for residential and industrial systems in 2010 of 2,70 EUR/W and 2,40 EUR/W, PV system learning rate of 20%, PV installation growth rate of 30%, WACC of 6,4% and *Opex*

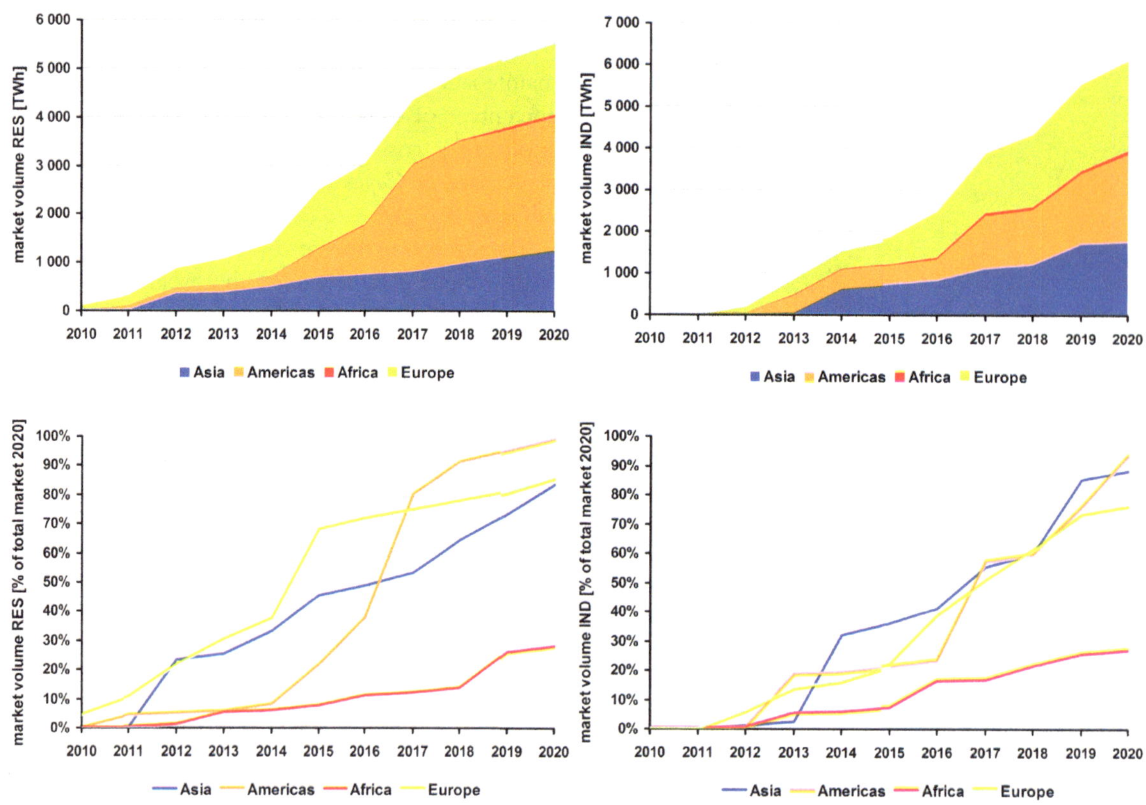

Figure 3.11 Grid-parity market volume for residential (left) and industrial (right) segments in absolute (top) and relative (bottom) numbers for all regions in the world in the years 2010 to 2020 [30]
Numbers for the market volume include a 1–4% growth rate in electricity depending on the market maturity.)

including insurance of 1,5% of *Capex*. The lifetime is assumed to steadily increase from 25 years in 2010 to 30 years in 2015 and stay constant afterwards. Solar irradiation on modules at optimal fixed-tilted angle for each location [36] is averaged by population distribution for respective countries and aggregated regions [37]. The PV system performance ratio is assumed to constantly increase for residential and industrial systems from 75% and 78% in 2010 to 80% and 82% in 2015, respectively, and remain constant afterwards. Electricity prices are assumed to increase in the same manner as in recent years by 5%, 3% and 1% per year for electricity price levels of 0–0,15 EUR/kWh, 0,15–0,30 EUR/kWh and more than 0,30 EUR/kWh, respectively. The data and method are described in more detail elsewhere [30].

Based on this model, the first residential grid-parity events occur today in all regions in the world and continue throughout the entire decade. Cyprus, Italy, the Caribbean and West Africa are markets where grid-parity is reached first. At the end of this decade more than 80% of market segments in Europe, the Americas and Asia are beyond residential grid-parity. Due to energy subsidies in South Africa and Egypt, which represent more than 60% of African electricity generation, Africa is an exception in this point. Residential grid-parity is complemented by highly economic off-grid PV in rural regions of developing countries (see Section 3.2.5). This is the case for about 1,4 billion people in the world, mostly living in Africa and South Asia. PV systems are in operation in all countries in the world (Figure 3.9) [32]. Therefore sustainable markets might grow very fast after grid-parity, in particular in case of low technical and legal obstacles.

Similarily, the first industrial grid-parity events occur today in all regions in the world and often on islands. They continue throughout the entire decade (Figure 3.11). Very early market segments are Cyprus, West Africa, the Seychelles, Caribbean, Cambodia and Fiji. Europe, the Americas and Asia-Pacific show quite similar characteristics of industrial grid-parity events throughout the entire decade. At the end of the decade more than 75% of market segments in Europe, the Americas and Asia are beyond industrial grid-parity. Again, Africa is an exception, due to extensive energy subsidies in South Africa and Egypt. Further exceptions are mainly oil producing countries used to substantially subsidizing their energy markets, e.g. Russia, Saudi Arabia, Libya, Venezuela, Iran, Iraq, Kuwait, Qatar, Oman and Angola. However, most of these countries can generate significant benefits by applying PV systems on the power plant level [31], as analyzed in Section 3.2.4.

Based on the global grid-parity analysis, the market potential for on-grid end-user PV electricity is estimated. On the 2010 electricity consumption basis, the grid-parity market volume of the entire power market by end of the 2010s is about 13 000 TWh in total composed of about 3 900 TWh (residential), about 3 600 TWh (small medium enterprises) and about 5 500 TWh (industrial).

To translate the aforementioned market volume into a cumulated installed PV capacity potential, the following assumptions are made: a population weighted average annual irradiation on fixed optimally tilted PV systems on a per country level [37], a global performance ratio of 80% (residential) and 82% (industrial), a global maximum PV supply contribution between 10% (without storage) and 80% (including storage) and a global market penetration of this supply contribution of not more than 50%, i.e. maximum PV share is assumed to not exceed 40% of total power supply. The 10% PV supply contribution threshold requires a flexible power plant portfolio but enables the full utilization of the PV capacity even without storage. Based on these assumptions, an upper and a lower limit of the sustainable economic market potential for PV systems is estimated for the year 2020. The upper limit for the grid-parity economic market potential would be the net metering grid-parity, i.e. parity of PV LCOE and the electricity price for end-users. A more appropriate upper limit would be the cost of PV plus storage solutions compared with the electricity price. First work in progress insights indicate roughly a PV plus storage grid-parity about four years after the net metering grid-parity. The lowest limit for the economic grid-parity market potential assumes the case of no storage availability, thus an effective 10% energy contribution based on the maximal fully utilizable PV power in the grid.

In summary, these assumptions lead to a total fully economically sustainable installed end-user PV capacity potential of about 980 GW without any storage solutions (lower limit) and up to 3 930 GW (upper limit) including storage in case of the financial net metering approach. However, the appropriate economic PV plus storage grid-parity comprises the storage cost resulting in 2 070 GW (upper limit) full economic market potential by 2020, if advanced economic storage systems for PV electricity are available.

3.2.4 Fuel-parity of PV power plants

As a consequence of fast decreasing PV LCOE, PV power plants become more cost competitive than fossil fuel fired power plants. Beyond fuel-parity further cost reduction in power generation can be realized by combining PV and fossil power plants to hybrid PV-Fossil power plants, i.e. for periods of sunshine the conventional power plant can be reduced in power output or completely shut down.

The most appropriate method for cost calculation is the LCOE approach (Equation 3.1) summarized and adapted to fossil fuel fired power plants in Equation 3.3:

$$LCOE_i = \frac{Capex_i \cdot crf + Opex_{i,fix}}{FLh_{i,el}} + Opex_{i,var} + \frac{fuel_i}{PE_{th,i} \cdot \eta_{i,el}} + \frac{carbon \cdot GHG_i}{\eta_{i,el}} \qquad \text{(Eq. 3.3a)}$$

$$crf = \frac{WACC \cdot (1+WACC)^N}{(1+WACC)^N - 1} \qquad \text{(Eq. 3.3b)}$$

$$fuel_i = fuel_{crude\ oil} \cdot cf_i \qquad \text{(Eq. 3.3c)}$$

$$fuel_{crude\ oil} = fuel_{crude\ oil,2010} \cdot \left(1 + r_{crude\ oil}\right)^{y-2010} \qquad \text{(Eq. 3.3d)}$$

Equation 3.3 Levelized cost of electricity (*LCOE*) for PV and fossil fuel fired power plants [31]

Abbreviations stand for: capital expenditures (*Capex*), annuity factor (*crf*), annual operation and maintenance expenditures (*Opex*), oil/natural gas and coal fossil plants as index (*i*), annual fixed *Opex* of fossil plants ($Opex_{i,fix}$), variable *Opex* of fossil plants ($Opex_{i,var}$), annual full load hours of fossil plants ($FLh_{i,el}$), fuel cost for fossil plants ($fuel_i$), thermal energy conversion factor of fossil plants ($PE_{th,i}$), primary to electric energy conversion efficiency of fossil plants ($\eta_{i,el}$), carbon emission cost (*carbon*), carbon emission intensity per thermal energy of fossil plants (GHG_i), weighted average cost of capital (*WACC*) defined in Eq. 3.1b, lifetime of plants (*N*), fuel cost ($fuel_i$), fuel cost of crude oil ($fuel_{crude\ oil}$), ratio of fossil fuel to crude oil as coupling factor (cf_i), fuel cost of crude oil in the year 2010 ($fuel_{crude\ oil,2010}$), annual escalation rate of crude oil price ($r_{crude\ oil}$) and year (*y*).

From an end-user perspective grid-parity is a good definition for sustainable PV economics. This must be regarded differently from the utility point of view. Large power generation companies are mainly used to operate large power plants, which is also possible for PV by operating several large scale multi 10–100 MW power plants. PV power plants can be built in the 10 MW scale but also for a power capacity of more than 1 GW [38]. Large scale PV power plants become attractive for utilities in case of favorable economics. Consequently, PV power plants are competing with fossil fuel fired power plants, in particular oil, natural gas and coal fired power plants. Competitiveness is best measured by calculating and comparing LCOE for all power plants. Fuel-parity is therefore defined by the parity of PV LCOE to the LCOE of respective fossil fuel fired power plants at the respective geographical location plus the cost of reduced full load hours (FLh) of fossil power plants. Moreover, being beyond fuel-parity automatically implies economic benefits of CO_2 reduction, as fossil fuel fired power plants emit large quantities of greenhouse gases (GHG) in contrast to PV power plants contributing only 2% to 5% of specific GHG per kWh compared with fossil power plants on basis of total life cycle analysis [39]. The first fuel-parity events have already started in very sunny regions supplied by oil fired power plants, like on the Arabian Peninsula [31].

The total global installed fossil fuel power plant capacity amounts to about 3 180 GW, being about 67% of total global installed power plant capacity, and by the end of 2008 it consisted of about 440 GW (oil), 1 230 GW (gas) and 1 510 GW (coal). These fossil power plants generated 13 683

TWh, being about 68% of total global power generation in 2008. The contribution by fuel was 1 104 TWh (oil), 4 303 TWh (gas) and 8 273 TWh (coal). Comparing the installed capacity and the generated electricity makes it possible to characterize the power technologies by their FLh, being 2 520 FLh (oil), 3 500 FLh (gas) and 5 460 FLh (coal) [26].

Several requirements need to be fulfilled for a successful hybridization of PV and fossil fuel fired power plants, in particular applicability of the hybridization concept, flexible fossil power plant operation modes and sophisticated energy meteorology. These prerequisites are fulfilled for hybrid PV-Fossil power plants and discussed in more detail elsewhere [31].

Necessary inputs for the evaluation of the global market potential of hybrid PV-Fossil power plants are globally distributed and georeferenced solar resource data for fixed optimally tilted PV systems [36] and the coordinates and capacities of all fossil fuel fired power plants in the world [20]. The georeferenced power plants are sorted by the respective local solar irradiation of fixed optimally tilted PV modules.

There are thousands of oil, gas and coal fired power plants of more than 2 000 kWh/m^2/y located in very sunny regions. The total fossil fuel fired power plant capacity of more than 2 000 kWh/m^2/y in the world located at very sunny sites is about 150 GW (oil), 250 GW (gas) and 290 GW (coal) [31]. Both, oil and gas fired power plants are able to adjust their power generation on a minute scale, i.e. by using state-of-the-art energy meteorology able to forecast 24 hours ahead. Thus there is no fundamental problem in combining PV power plants with oil and gas power plants to give hybrid power plants. In the case of coal power plants excellent energy meteorology has to be applied, because otherwise new plants will have to be built such as integrated gasification combined cycle (IGCC) coal plants, since they are as flexible as oil and gas fired power plants.

For practical reasons only power plants are considered in the following in case of at least 2 000 FLh of all power plants in one country per fuel type. This limit reflects a high probability that the respective power plants are also in operation during daytime when the PV power plants feed in their power [31].

For a simplified economic description, a price coupling of natural gas and coal to the crude oil price is used in this chapter. Fossil fuel prices for crude oil, natural gas and steam coal considerably deviate in different markets in the world, but the overall price trend is similar and relative price differences have decreased in the last decade [31]. Long-term price escalation as a consequence of increase in demand and degrading and diminishing resources is reflected in fossil fuel prices. Dependence of natural gas and coal price on crude oil price can be found within market fluctuations over recent decades. The fossil fuel prices can be compared on the basis of harmonized thermal energy units, like USD per barrel (of oil equivalent), showing a long-term price ratio of natural gas to crude oil of about 0,7–0,9, whereas the ratio of coal to crude oil is about 0,2–0,4. The coupling of natural gas and coal to crude oil is sensible because both are used for their thermal energy content but factors such as relative availability, local energy logistics and respective power plant efficiencies create price offsets. Long-term price coupling of natural gas on crude oil is expected by the IEA to be about 0,9 for the US and 0,8 for Europe and Japan [40].

The applied economic scenario for fossil fuel power plants is defined as follows: Capex of 800 EUR/kW (oil), 750 EUR/kW (gas) and 1 500 EUR/kW (coal), annual fixed Opex of 17 EUR/kW (oil), 15 EUR/kW (gas) and 20 EUR/kW (coal), annual variable Opex of 1 EUR/MWh$_{el}$ for all fossil power plants, power plant lifetime of 30 years (oil and gas) and 40 years (coal), average power plant efficiency in the year 2010 of 40% (oil), 50% (gas) and 35% (coal), annual increase of the absolute average power plant efficiency of 1% (oil and coal) and 0,5% (gas), coupling factor of 0,8 (gas) and 0,3 (coal), WACC of 6,8%, exchange rate USD/EUR of 1,40, crude oil fuel price of 80 USD/barrel in 2010, annual crude oil price escalation rate of 3% in real terms, no cost for existing CO_2 emissions of the entire scenario period and a thermal energy conversion of 1,680 6 MWh$_{th}$/barrel. In general the scenario covers a business-as-usual approach and reflects a realistic estimate of all major economic drivers, except the price for fossil fuels being very likely too conservative. The range of the most fundamental price, crude oil, is 80 to 107 USD/barrel from 2010 to 2020. If the depletion and degradation rate of fossil fuels stays at the rate of the 2000s, the real price could be twice as high at the end of the 2010s as assumed in the applied scenario. Numbers mentioned are for fossil fuel power plants of multi-100 MW. Data are taken from various sources described elsewhere [31].

The PV scenario assumptions are: PV power plant Capex of 1,80 to 2,00 EUR/Wp (depending on local least cost conditions), Opex of 1,5% of Capex, performance ratio of 80%, local irradiation of fixed optimally tilted systems [36], weighted average cost of capital of 6,8%, plant lifetime of 25 years, annual power degradation of 0,4%, learning rate for modules and inverters of 20% (2010 to 2012) and 15% (2013 to 2020) and for remaining BOS components no further learning to be conservative, global PV growth rate of 40% (2010 to 2012) and 30% (2013 to 2020).

The PV scenario setting can be considered as realistic, but may be slightly too conservative. The PV growth rates have been higher for the last 15 years (Figure 3.8), hence the cost reduction in time might be faster. The most competitive utility-scale market segments in the world are China and Germany, which show average fully-loaded PV system Capex of about 1,9–2,1 EUR/Wp in the year 2010 [41] being in line with the realistic scenario assumptions. Most competitive PV industry leaders achieve an even better cost level. The true costs of PV power plants in Germany equipped with CdTe modules from First Solar are found to be slightly below 1,6 EUR/Wp. The two c-Si module cost leaders achieve fully-loaded module cost of on average about 1,02 EUR/Wp. The fully-loaded average non-module cost in China and Germany for c-Si PV power plants are 0,67 EUR/Wp and 0,72 EUR/Wp, respectively. As a consequence the fully-loaded system cost for very competitive c-Si PV power plants has been about 1,7 EUR/Wp, composed of c-Si module cost leaders and the two most cost efficient PV markets. Accordingly, leading PV industry players have been able to offer PV power plants for 1,6–1,7 EUR/Wp for cost efficient PV markets [41].

Comprehensive hybridization economics of PV and fossil fuel power plants can be derived on the basis of the scenario assumptions for PV power plants and fossil fuel power plants, the LCOE modelling (Equations 3.1 and 3.3), the experience curve approach (Equation 3.2 and Figure 3.10) and the georeferenced dataset of all fossil fuel power plants. A key assumption is the close physical location of PV and fossil power plants. Therefore no additional storage is needed, no substantial grid constraints have to be feared and electricity supply security is provided.

Upgrading existing fossil fuel power plants by PV power plants to hybrid PV-Oil, PV-Gas or PV-Coal power plants is economically favorable for PV LCOE lower than respective oil, gas or coal LCOE. The precise calculations need to include the slightly higher capital cost of fossil power plants by reducing their FLh in order of the PV FLh. This effect can be calculated by Equation 3.3a and has to be generated additionally by the PV component, thus lowering the PV LCOE of the hybrid PV-Fossil power plant. All breakeven analyses, i.e. fuel-parity analyses in this chapter, take this effect into account. A fair comparison of PV and fossil fuel power plants would be on a total plant LCOE basis, i.e. including all cost components. However, for estimating the competitiveness of the PV hybridization approach a fuel-only LCOE calculation is helpful, since only the marginal costs of the fossil fuel of the power plants are taken into account, i.e. in case of lower PV LCOE than fuel LCOE a non-hybridization strategy of respective power plant owners would definitely cause higher power generation costs and lead to higher prices for the end-users, thus losing competitiveness either to competitors or to other regions on a macro-economic level.

When applying LCOE data for fossil fuel fired power plants and PV power plants for all coordinates in the world, a detailed world map of local fuel-parity events dynamics can be derived [31] and based thereupon a global demand curve for hybrid PV-Fossil power plants. Since all fossil fuel power plants are georeferenced, the year of financially beneficial hybridization for the different fuel types for all existing fossil fuel power plants can be derived and plotted in an integrated manner. The global hybrid PV-Fossil power plant demand curve based on fuel-parity is depicted in Figure 3.12 for the case of total plant and fuel-only LCOE parity for fossil power plants operated in countries of an average of at least 2 000 FLh and for plants firing coal of at least bituminous coal quality.

In Figure 3.12, fossil power plant capacity, i.e. oil, natural gas and coal, is counted only in case of PV LCOE (plus higher fossil capital cost due to reduced FLh) lower

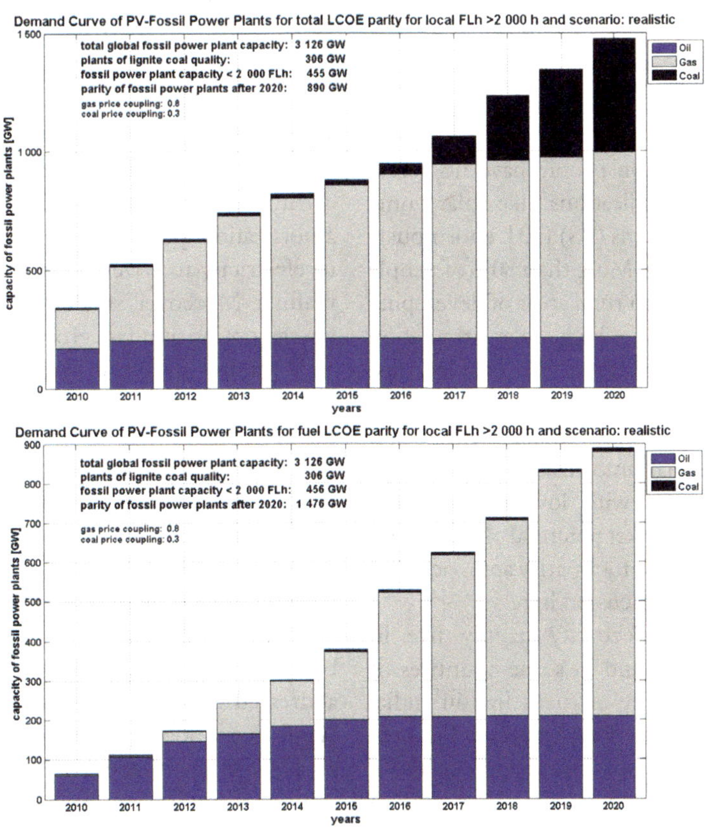

Figure 3.12 Global hybrid PV-Fossil power plant demand curve in the 2010s on total plant (top) and fuel-only (bottom) LCOE parity for fixed optimally tilted PV power plants in the 2010s [31]

than total and fuel-only fossil LCOE. All coordinates are aggregated to the fuel categories. Power plant scenario assumptions are applied as defined in this section. Countries operating respective fossil fuel power plants on average of at least 2 000 FLh and coal plants firing coal of at least bituminous coal quality are included in the analysis.

The global hybrid PV-Fossil power plant market potential is structured as follows: Total plant LCOE parity is already given for about 350 GW (oil and gas) in 2010, rising to 750 GW (oil, gas and start of coal) in the middle of the 2010s and reaching about 1 500 GW (oil, gas and coal) by the end of the 2010s. Fuel-only LCOE parity is already given for about 60 GW (oil) in 2010, rising to about 380 GW (oil, gas and start of coal) in the middle of the 2010s and reaching about 900 GW (oil, gas and very little coal) by the end of the 2010s. Global total fossil power plant capacity was about 3 130 GW by early 2009. About 460 GW of that capacity is identified as being operated at less than 2 000 FLh and therefore excluded from the analysis. Further about 310 GW coal power capacity is excluded due to the use of low quality coal not tradable on the world market. The remaining about 2 370 GW fossil fuel power plant capacity can be economically upgraded by PV power plants by 2020 to approximately 63% and 38% for total plant and fuel-only LCOE parity, respectively.

In the year 2020, fuel-parity of PV power plants and conventional fossil fuel fired power plants might be in the order of 1 500 GW (upper limit), whereas a capacity of approximately 900 GW fulfils the most aggressive criteria of PV LCOE parity to fuel-only LCOE of fossil power plants (lower limit).

3.2.5 Rural off-grid market potential of PV systems

Throughout the world 1,4 billion people have no access to electricity [26]. Small PV applications, like solar home systems (SHS) and PV pico systems (PS) [33], enormously improve local standards of living. More than 80% of people without access to electricity live in rural areas of developing countries [26] typically showing high solar irradiation [33]. Thus attractively priced SHS and PS enable a fast financial amortization compared with conventional energy costs. Increasing demand for small PV applications and constant reduction of financial amortization create large market potentials in countries with low electrification rates. The off-grid economic market potential of PV cannot be estimated by the grid-parity or fuel-parity approach, but by an amortization calculation discussed here.

People without access to electricity mainly live in sub-Saharan Africa, South Asia and in some countries of the Pacific Rim. A large majority of them live in India, Bangladesh, Indonesia, Nigeria, Pakistan and Ethiopia [33]. Georeferenced analysis of solar irradiation of those regions where the 1,4 billion people without access to electricity live confirms that a majority of them have access to very good to excellent solar resources of about 1 800–2 200 kWh/m^2/y irradiation on module surfaces of fixed optimally tilted PV systems. Most people without access to electricity live in developing countries of gross domestic product (GDP) per capita of less than 2 500 USD. Only a few of those countries have PV installations on a promising level (Figure 3.9) [33].

People without access to electricity spend money on kerosene for lamps, dry cell batteries for radios and tape recorders and charging mobile phones. All these basic energy needs are able to be substituted by SHS and PS. The conventional energy costs are paid week by week, or month by month, whereas PV systems have to be bought in total but can be used for years. The attractiveness of PV applications is distinguished by their financial amortization period. For calculating the financial amortization of SHS and PS their specific cost including replacement cost of key components need to be taken into account and compared with the substituted energy cost, mainly driven by expenditures for kerosene and dry cell batteries. Details on the assumptions for the amortization calculation are given elsewhere [33].

Amortization of complete PV systems ranges between 6–18 months (PS) and 12–36 months (SHS) but strongly depends on energy consumption patterns, as shown in Figure 3.13. Economics are based on the following assumptions: system cost between 8 EUR/Wp (SHS) and 23 EUR/Wp (PS), monthly kerosene consumption between 5 l (PS) and 20 l (SHS), monthly cost of batteries 5,2 EUR (SHS), local diesel prices, inflation of 10% plus a real interest rate of 3% [33]. Thus, small PV applications are the least energy cost option for people without access to electricity in rural areas. Analysis shows that upfront investment seems to be less of a financial burden than anticipated. The capitalized value of the various small PV systems range between 10–45 (PS) and 5–20 (SHS) times the original capital expenditure in most countries [33].

Attractive prices for PS and SHS lead to short financial amortization periods and enable people without access to electricity to cover their energy needs in a sustainable manner. Moreover, small PV applications are the least cost energy option and fast growing global on-grid PV markets and scaling effects in local distribution channels will further reduce system costs. Excellent economics of small PV applications might significantly accelerate the growth rate of off-grid PV markets in the years to come.

A local market evaluation for Ethiopia, one of the poorest countries in the world, has led to a total market potential of about 280 MW for roughly 70 million people without electricity access. These numbers correspond to about 30 Wp per small PV system. The market for the 1,4 billion people without access to electricity might be evenly addressed by small PV systems of the average size of 30 Wp for very poor regions, larger scale small PV systems of an average size of about 100 Wp for less poor regions and mini-grid applications providing annually about 300 kWh/user. The derived global market potential for PS and SHS in very poor rural regions for residential purposes might

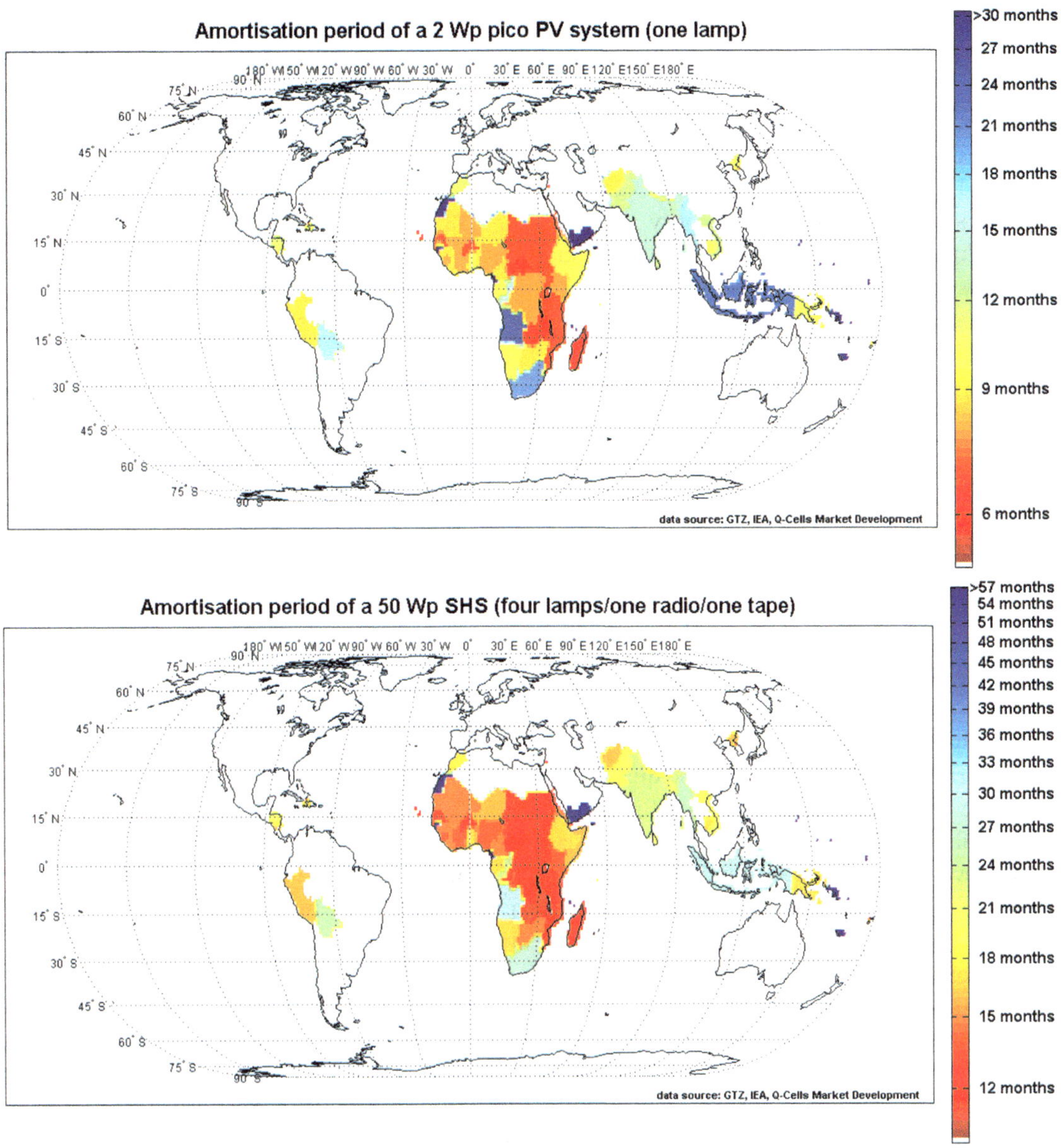

Figure 3.13 Amortization period for 2 Wp PS (top) and 50 Wp SHS (bottom) [33] (Coloured countries show electrification rates of less than 80%.)

be about 2 GW, whereas the market potential for less poor rural regions might represent about 6 GW. The market potential for residential electricity in rural PV based mini-grids might be in the order of 16 GW [33]. PV based mini-grids are higher in LCOE than on-grid power solutions, but lower in LCOE than comparable diesel powered mini-grids as a result of fast decreasing PV LCOE [33].

However, advanced commercial applications and public services lead to an even higher market potential. Most well-emerging and developed countries in the world show a residential fraction of total final electricity consumption of roughly-one third [33], i.e. commercial, industrial and public consumption is two times higher than the residential one. Therefore, the commercial and public (schools, health centres, lanterns, water pumping, telecommunication, etc.) small PV off-grid market might be two times larger than the market for residential applications.

The total addressable economic market potential for rural off-grid PV systems can be estimated at roughly 70 GW being composed of about 24 GW residential and about 46 GW commercial and public demand or about 24 GW PS and SHS (residential, commercial and public) and about 46 GW mini-grids. Roughly two-thirds of the off-grid PV market potential is due to mini-grid applications, but two-thirds of the not electrified people will get access to electricity by small PV solutions like PV pico systems or solar home systems.

3.2.6 Total sustainable economic PV market potential

As presented in Sections 3.2.3 to 3.2.5, there are three major classes of PV markets and hence two different parity concepts for deriving the sustainable on-grid economic market potential and an additional amortization approach for the off-grid segment. Some countries reach grid-parity first, e.g. Spain, Germany and Japan, and others initially achieve fuel-parity, e.g. Saudi Arabia, UAE, India and Mexico [30, 31]. In the mid- to long-term all countries reach grid-parity and fuel-parity, therefore the overlap capacity of the two approaches needs to be quantified on a per country basis in order to avoid double counting. The highly profitable PV off-grid markets (Section 3.2.5) are characterized by no overlap with the grid-parity and fuel-parity approach, thus they are not included in the following overlap considerations.

Based on the grid-parity and fuel-parity concept, an upper and a lower limit of the sustainable economic market potential for PV systems can be estimated. The upper limit for the grid-parity would be the net metering grid-parity, i.e. parity of PV LCOE and the electricity price for the end-users. The lower and more appropriate limit would be the cost of PV plus storage solutions compared with the electricity price. The upper limit for the fuel-parity would be the comparison to the total plant LCOE of fossil fuel fired power plants reduced by increasing capital cost due to lower FLh as a consequence of hybridization with PV systems. The lower limit for the fuel-parity refers to the marginal cost of fossil fuel fired power plants, i.e. the fuel-only LCOE. To avoid double counting of capacities the overlap of grid-parity and fuel-parity has to be quantified. Therefore, in case of reaching grid-parity and fuel-parity the smaller accessible market potential for PV systems has been counted as overlap and excluded from the integrated total parity capacity. This consideration has been applied on an annual and per country basis and aggregated afterwards on a worldwide scale. This does not imply that it is more favorable to install the PV systems in a more decentralized end-user structure than in a more centralized utility-scale manner. The global economic market potential for PV systems analyzed in this work is based on a global grid-parity analysis for more than 99% of global electricity consumption (Section 3.2.3) and a comparable global fuel-parity analysis for all fossil fuel fired power plants in the world on a per power plant basis (Section 3.2.4). The outcome is depicted in Figure 3.14 for an integrated sustainable economic market potential estimate for PV systems on the basis of a learning rate of 20% for modules and inverters and 15% for the other balance of system (BOS) components.

The key assumptions are: grid-parity and fuel-parity approach, learning rate of 20% for modules and inverters and 15% for other BOS components, growth rate of 30% in 2010s, substitution of 80% of the single end-user power demand and 50% maximal addressable market potential in case of grid-parity and hybrid PV-Fossil power plants without any storage capacities in case of fuel-parity. As a result, the total economic market potential can be estimated as about 2 700 GW to 4 200 GW for the year 2020.

The economic market potential for PV systems is comparably low in the very early 2010s but growing fast during the entire decade, particularly for the lower limit of PV plus storage grid-parity and marginal cost fuel-parity. In the beginning, the majority of the global economic market

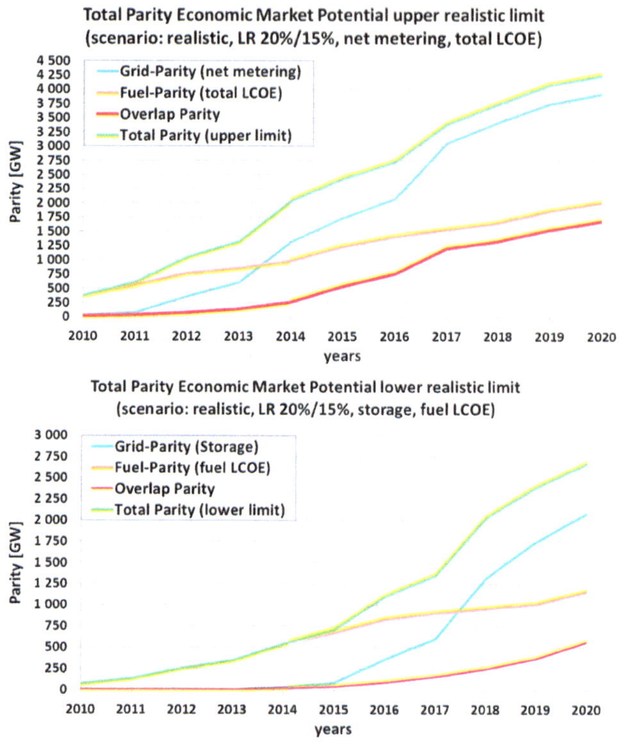

Figure 3.14 Total economic market potential for PV systems in the 2010s in the upper (top) and lower limit (bottom)

potential is driven by fuel-parity (typically PV power plants) and later by grid-parity (typically decentralized roof-top applications).

A lower learning rate reduces the market growth but would be no show-stopper at all. Significantly lower learning rates of 10% for modules and inverters and 5% for the additional BOS components would reduce the potential accessible in the 2010s, but PV market diffusion would be delayed only by a few years, typically less than four years. A more detailed sensitivity analysis on the impact of learning rates on the economic PV market potential is discussed elsewhere [31, 34].

As a result, the economic market potential for total installed on-grid PV capacity in year 2020 can be estimated as about 2 700 GW to 4 200 GW based on grid-parity and fuel-parity analyses. The additional market potential for rural off-grid PV applications, however, is by at least one order of magnitude smaller and estimated to be about 70 GW (Section 3.2.5). Nevertheless, this capacity would be equivalent to a quantum jump for the 1,4 billion people currently without access to electricity. The total overall economic PV market potential can thus be estimated to roughly 2 800 GW to 4 300 GW based on grid-parity, fuel-parity and off-grid amortization analyses.

3.2.7 Discussion of the economic PV market potential and expectations by PV scenarios

The fully economic market potential for installed PV capacity in year 2020 based on grid-parity, fuel-parity and rural off-grid economic analyses can be estimated to about 2 800 GW to 4 300 GW on a global scale (Section 3.2.6).

It is obvious that only a fraction of the economic market potential derived in Section 3.2.6 will be realized in the same time. In this section a realization of 20% (pessimistic), 35% (realistic) and 50% (optimistic) of the full PV market potential is assumed. When applying these scenario assumptions one can expect a cumulated installed PV capacity in the year 2020 for a pessimistic case of about 560–860 GW, for a realistic case about 980–1 500 GW and for an optimistic case about 1 400–2 150 GW. To be more on the safe side the final expectation is about 600 GW (pessimistic), about 1 000 GW (realistic) and about 1 600 GW (optimistic) and summarized in Table 3.1. These three cases are roughly equivalent to average annual growth rates of new installations of 20%, 30% and 40%, i.e. fully in line with the long-term growth of global PV installations (Figure 3.8). Technological restrictions for these installation numbers need not to be feared, since the fuel-parity approach requires only grid access of PV power plants near to existing power plants, many grid-parity market segments become profitable even including local storage options and small PV applications are already highly profitable including storage.

These PV market expectations are in drastic contrast to market projections of major institutions in the field of PV scenarios, since only the progressive group (Section 3.2.1) fulfils the pessimistic case derived in this chapter, i.e. only EPIA and partly Greenpeace can really imagine that PV can show such fast progress towards the long-term target of becoming a major source of energy supply. The "SET FOR 2020" study from EPIA [21] is the only publication expecting a cumulated installed PV capacity in the order of the realistic case derived here.

It is not surprising that the two global leading organizations in the fields of PV association (EPIA) and campaigning for protecting and conserving the environment (Greenpeace) possess the know-how and visionary leadership for realistic energy scenarios in the field of PV [21–23].

On the contrary, it has to be stated that the global leading organizations in the fields of energy (IEA) and conservation of the environment (WWF) assume a substantially lower diffusion of PV, and thus cannot imagine the true potential of PV as a near-term game changer in the global energy business [25–27]. Even an organization like the EWG does not recognize the true near-term potential of PV despite being focused on the short-term shortage in conventional energy fuels and strategies for a long-term secure energy supply based on cost competitive renewable energy sources [28]. The three organizations expect a cumulated installed PV capacity in the year 2020 of 110 GW (IEA in its most progressive "New Policies" scenario), 111 GW (EWG in the "High" scenario) and 194 GW (WWF in the "100% RE by 2050" scenario). Even the result of 600 GW cumulated installed capacity for the pessimistic case assumptions derived here are more than three (WWF) and five (IEA and EWG) times higher than the most progressive assumptions of these organizations, despite taking into account only the fully economic sustainable PV market potential in the decade of the 2010s.

One of the most prominent characteristics of PV is the faster than expected progress mainly based on the long-term stable high learning rate and growth rate. The learning rate is driven by modular, highly scalable but not very complex production processes and still retains enormous room for further progress. The growth rate is driven by the highly modular character of the PV technology ranging from 2 Wp PV pico systems to 1 GW very large scale PV power plants generating sustainable economic benefits in outer space and on Earth, off-grid and on-grid, in residential and power plant applications. Additionally, the PV power technology is the only established energy technology based on the fundamental cost structure of semiconductor electronics without any moving parts and the potential for a further substantial increase in lifetime, decrease in energy pay back time and hence increase in energy return on energy invested. In every market segment captured by PV during the major diffusion phases in recent decades, the PV applications set standards in establishing a least cost energy option: firstly in powering space applications, secondly in the off-grid field and currently in parallel for on-grid decentralized residential roof-top and on-grid more centralized power plants.

In consequence, it has to be considered as highly probable that the technological basis of PV coupled to the economic characteristics will drive sustainable and high PV market growth rates in the decade of the 2010s which will result in a structural change of the global energy system in the following one to two decades due to a historically very fast increase in market share.

In a broader perspective, the three major future power technologies in relation to minimized fully loaded social cost must be considered to be solar PV, wind power and hydro power. But only solar PV and wind power have access to nearly abundant resources, whereas the solar resource is

Table 3.1 Cumulative global installed PV capacity by end of the year 2020 depending on scenario assumptions and resulting average annual growth rates of new installations

Scenario	Cumulative global installed PV capacity by end of the year 2020	Average annual market growth rate
Pessimistic	600 GWp	20%
Realistic	1 000 GWp	30%
Optimistic	1 600 GWp	40%

the most homogeneously distributed energy source in the world. It is a godsend that the two least cost energy options for the twenty-first century are fully complementary to each other [42].

A first global analysis of the complementary characteristics of PV and wind power plants gives plenty of indications that these two major renewable power technologies complement each other to a very high extent due to the fundamental underlying solar and wind resources [42]. Consequently, PV and wind power plants are finally no competition to each other and the findings for the global hybrid PV-Fossil power plant demand curve (Section 3.2.4) need not be lowered. However, it seems likely that hybrid PV-Wind-Fossil power plants can capture a significant market share in the 2010s and further reduce the remaining full load hours of the fossil fuel power plant component of the hybrid power plant.

A very promising option in the mid- to long-term is renewable power methane (RPM) [43]. RPM would enable hybrid PV-Wind-RPM power plants establishing a fully dispatchable power supply based on fluctuating wind and solar resources. In the concept of RPM the excess power is converted in a first step by electrolysis into hydrogen and in a second step by methanation into methane. Besides electricity only water and carbon dioxide are needed. A major advantage of RPM is the step-by-step switch from current fossil methane (natural gas) to the future renewable power methane, since the entire downstream infrastructure can be used, i.e. transmission pipelines, distribution networks and the methane (gas) power plants. A first economic analysis on the global impact potential of RPM based on hybrid PV-Wind-RPM power plants finds indications that this approach will become competitive in the early 2020s. The complementarity of PV and wind power enormously reduces further investments in the energy system and will enable the power sector to offer highly competitive solutions for the heat and transportation sector and maybe even for the chemical industry via renewable power generated methane [44].

REFERENCES

1 Stackhouse P.W. and Whitlock C.H., (eds) (2008) *Surface meteorology and Solar Energy (SSE)* release 6.0, NASA SSE 6.0, Earth Science Enterprise Program, National Aeronautic and Space Administration (NASA), Langley, http://eosweb.larc.nasa.gov/sse/

2 Stackhouse P.W. and Whitlock C.H., (eds) (2009) *Surface meteorology and Solar Energy (SSE)* release 6.0 Methodology, NASA SSE 6.0, Earth Science Enterprise Program, National Aeronautic and Space Administration (NASA), Langley, http://eosweb.larc.nasa.gov/sse/documents/SSE6Methodology.pdf

3 Hay J.E. and McKay D.C. (1985) Estimating Solar Irradiance on Inclined Surfaces: A Review and Assessment of Methodologies, *International Journal of Solar Energy*, 3, 203–240.

4 Reindl D.T., Beckmann W.A. and Duffie J.A. (1990) Evaluation of Hourly Tilted Surface Radiation Models, *Solar Energy*, 45, 9.

5 Klucher T.M. (1979) Evaluation of Models to predict Insolation on tilted Surfaces, *Solar Energy*, 23, 111–114.

6 Duffie J.A. and Beckmann W.A. (2006) *Solar Engineering of Thermal Processes*, John Wiley & Sons., Hoboken.

7 Breyer Ch. and Schmid J. (2010) Global Distribution of optimal Tilt Angles for fixed tilted PV Systems, *Proceedings 25th European Photovoltaic Solar Energy Conference*, Valencia, September 6–10.

8 Breyer Ch. and Schmid J. (2010) Population Density and Area weighted Solar Irradiation: global Overview on Solar Resource Conditions for fixed tilted, 1-axis and 2-axes PV Systems, *Proceedings 25th European Photovoltaic Solar Energy Conference*, Valencia, September 6–10, DOI: 10.4229/25thEUPVSEC2010-4BV.1.91.

9 Gombert A., Gerstmaier T., Heile I., Riesen van S., Röttger M. and Wüllner J. (2010) Analysis of Long Term Operation Data of Flatcon CPV Systems, *Proceedings 25th European Photovoltaic Solar Energy Conference*, Valencia, September 6–10.

10 Judkins Z.S., Johnston K.W., Almy C., Linderman R.J., Wares, B., Barton N.A., Dawson M. and Peurach J. (2010) Performance Results of a Low-concentration Photovoltaic System Based on High Efficiency Back Contact Cells, *Proceedings 25th European Photovoltaic Solar Energy Conference*, Valencia, September 6–10.

11 Gueymard C.A., Myers D. and Emery K. (2002) Proposed Reference Irradiance Spectra for Solar Energy Systems Testing, *Solar Energy*, 73, 443–67.

12 Lerchenmüller H. (2011) DNI Values in good CPV Sites, private communication, April 19.

13 EPIA (European Photovoltaic Industry Association) (2011) Global Market Outlook for Photovoltaics until 2015, EPIA, Brussels, www.epia.org/publications/photovoltaic-publications-global-market-outlook/global-market-outlook-for-photovoltaics-until-2015.html

14 Gredler C. (2008) Das Wachstumspotenzial der Photovoltaik und der Windkraft – divergierende Wahrnehmungen zentraler Akteure, Diploma thesis, University of Salzburg.

15 Rogol M., Farber M., Flynn H., Meyers M., Paap S., Porter C., Rogol J. and Song J. (2008) Solar Annual 2008: Four Peaks, PHOTON Consulting, Boston, Autumn.

16 Rogol M., Flynn H., Porter C., Rogol J. and Song J. (2007) Solar Annual 2007: Big Things in a Small Package, Photon Consulting, Boston, Autumn.

17 Rogol M., Doi S. and Wilkinson A. (2004) Sun screen: Investment opportunities in solar power, CLSA Asia-Pacific Markets, Hong Kong, July.

18 WBGU – German Advisory Council on Global Change (2003) *World in Transition: Towards Sustainable Energy Systems*, WBGU, Berlin, Earthscan, London, www.wbgu.de/fileadmin/templates/dateien/veroeffentlichungen/hauptgutachten/jg2003/wbgu_jg2003_engl.pdf

19 IEA (International Energy Agency) (2008) *World Energy Outlook 2008*, IEA, Paris, www.iea.org/textbase/nppdf/free/2008/weo2008.pdf

20 Platts (2009) UDI World Electric Power Plants data base, Platts – A Division of McGraw-Hill, Washington, version of March 31.

21 EPIA (European Photovoltaic Industry Association) (2009) *Set for 2020: Solar Photovoltaic Electricity – A mainstream power source in Europe by 2020*, EPIA, Brussels.

22 EPIA (European Photovoltaic Industry Association) (2011) *Solar Generation 6 – Solar Photovoltaic Electricity Empowering the World*, EPIA and Greenpeace International, Brussels and Amsterdam, www.epia.org/index.php?eID=tx_

nawsecuredl&u=0&file=fileadmin/EPIA_docs/documents/SG6/Solar_Generation_6__2011_Full_report_Final.pdf

23 Teske S. (ed.) (2010) *Energy revolution: A Sustainable World Energy Outlook*, Greenpeace International and EREC, Amsterdam and Brussels, www.greenpeace.org/international/Global/international/publications/climate/2010/fullreport.pdf

24 BSW (Bundesverband Solarwirtschaft) (2010) *Wegweiser Solarwirtschaft: PV-Roadmap 2020*, study from Roland Berger Strategy Consultants and Prognos on behalf of BSW, Berlin, www.solarwirtschaft.de/fileadmin/content_files/wegweiser_sw_pvrm-lang.pdf

25 IEA (International Energy Agency) (2010) *Technology Roadmap: Solar photovoltaic energy*, IEA, Paris, www.iea.org/papers/2009/PV_roadmap_targets_printing.pdf

26 IEA (International Energy Agency) (2010) *World Energy Outlook 2010*, IEA, Paris.

27 WWF (World Wild Fund for Nature International) (2011) *The Energy Report: 100% Renewable Energy by 2050*, WWF, Ecofys and Office for Metropolitan Architecture, Gland, http://assets.panda.org/downloads/101223_energy_report_final_print_2.pdf

28 Peter S. and Lehmann H. (2008) *Renewable Energy Outlook 2030: Energy Watch Group Global Renewable Energy Scenarios*, Energy Watch Group, Berlin, www.energywatchgroup.org/fileadmin/global/pdf/2008-11-07_EWG_REO_2030_E.pdf

29 Breyer Ch., Birkner Ch., Kersten F., Gerlach A., Stryi-Hipp G., Goldschmidt J.Ch., Montoro D.F. and Riede M. (2010) Research and Development Investments in PV – A limiting Factor for a fast PV Diffusion?, *Proceedings 25th European Photovoltaic Solar Energy Conference*, Valencia, September 6–10.

30 Breyer Ch. and Gerlach A. (2010) Global Overview on Grid-Parity Event Dynamics, *Proceedings 25th European Photovoltaic Solar Energy Conference*, Valencia, September 6–10, DOI: 10.4229/25thEUPVSEC2010-6CV.4.11.

31 Breyer Ch., Görig M., Gerlach A.-K. and Schmid J. (2011) Economics of Hybrid PV-Fossil Power Plants, *Proceedings 26th European Photovoltaic Solar Energy Conference*, Hamburg, September 5–9.

32 Werner C., Gerlach A., Adelmann P. and Breyer Ch. (2011) Global Cumulative Installed Photovoltaic Capacity and Respective International Trade Flows, *Proceedings 26th European Photovoltaic Solar Energy Conference*, Hamburg, September 5–9.

33 Breyer Ch., Werner C., Rolland S., and Adelmann P. (2011) Off-Grid Photovoltaic Applications in Regions of Low Electrification: High Demand, Fast Financial Amortization and Large Market Potential, *Proceedings 26th European Photovoltaic Solar Energy Conference*, Hamburg, September 5–9.

34 Kersten F., Doll R., Huljić D.M., Görig M.A., Breyer Ch., Müller J. and Wawer P. (2011) Learning from the sun – PV experience curve and the levers of cost reduction, *Proceedings 26th European Photovoltaic Solar Energy Conference*, September 5–9.

35 Short W., Packey D.J. and Holt T. (1995) *A Manual for the Economic Evaluation of Energy Efficiency and Renewable Energy Technologies*, NREL, NREL/TP-462-5173, Golden.

36 Breyer Ch. and Schmid J. (2010) Global Distribution of optimal Tilt Angles for fixed tilted PV Systems, *Proceedings 25th European Photovoltaic Solar Energy Conference*, Valencia, September 6–10.

37 Breyer Ch. and Schmid J. (2010) Population Density and Area weighted Solar Irradiation: global Overview on Solar Resource Conditions for fixed tilted, 1-axis and 2-axes PV Systems, *Proceedings 25th European Photovoltaic Solar Energy Conference*, Valencia, September 6–10, DOI: 10.4229/25thEUPVSEC2010-4BV.1.91.

38 Komoto K., Ito M., Vleuten van der P., Faiman D. and Kurokawa K. (eds) (2009) *Energy from the Desert – Very Large Scale Photovoltaic Systems: Socio-economic, Financial, Technical and Environmental Aspects*, Earthscan, London.

39 Fthenakis V.M. and Kim H.C. (2011) Photovoltaics: Life-cycle analyses, *Solar Energy*, 85 (1609)1628.

40 IEA (International Energy Agency) (2004) *World Energy Outlook 2004*, IEA, Paris, www.iea.org/textbase/nppdf/free/2004/weo2004.pdf

41 Bolman C., Boas R., Farber M., Meyers M., Porter C., Rogol M., Song J., Tracy P., Trangucci R. and Zuboff G. (2011) *Solar Annual 2010–2011: Cash In*, Photon Consulting, Boston.

42 Gerlach A.-K., Stetter D., Schmid J. and Breyer Ch. (2011) PV and Wind Power – Complementary Technologies, *Proceedings 26th European Photovoltaic Solar Energy Conference*, Hamburg, September 5–9.

43 Sterner M. (2009) Bioenergy and renewable power methane in integrated 100% renewable energy systems, Dissertation, University of Kassel.

CHAPTER FOUR

Overview of large-scale solar power generation

4.1 CURRENT STATUS OF LARGE-SCALE SOLAR POWER GENERATION

4.1.1 Trends in large-scale PV systems

Installation of large-scale (MW-scale) PV systems has been rising substantially year by year throughout the world. The starting point and milestones for this development have been the feed-in-tariff schemes in European countries. These schemes created an attractive market for developers, investors, etc. and a new business model of MW-scale PV systems was established.

At the beginning of the 2000s, the number of large-scale PV systems with a capacity of more than 1 MW operated in the world was around 10, and the largest capacity among them was 3,3 MW (Serre, Italy), as surveyed in the first edition of Task8 report [1]. In 2011, the number of MW-scale PV systems is over 1 500, and the capacity of such MW-scale PV systems is expanding year by year, as shown in Figure 4.1.

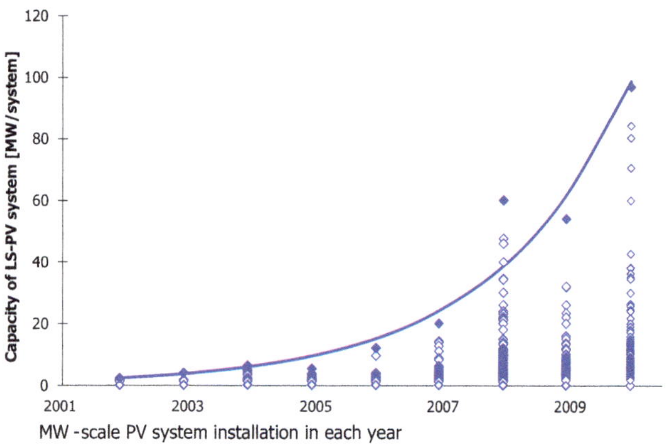

Figure 4.1 Trends in MW-scale PV systems installation

Table 4.1 shows examples of MW-scale PV systems, which have a capacity of more than 20 MW [2, 3, 4]. The number of PV systems over 20 MW in 2011 is more than eighty. Among them, the largest PV system, Perovo I-V PV power plant in Ukraine, is 100 MW, constructed in December 2011. The second largest PV system, Sarnia PV power plant in Ontario, Canada, is 97 MW_{AC} (80 MW_{DC}).

Tables 4.2, 4.3 and 4.4 show specific information on Perovo 100 MW PV plant in Ukraine, Sernia 97 MW PV plant in Canada and Ohotnikovo 80 MW PV plant in Ukraine. Figure 4.2 shows the trends in annual installed capacity of the MW-scale in each year compared with that of all PV applications installed in the world [5, 6]. While the ratio of MW-scale PV systems was only 3%, i.e. negligibly small in 2002, it can be shown that the ratio achieved about 15% of all PV applications installed in 2010.

Figure 4.3 shows the trends in annual installed capacity of the MW-scale by two categories, e.g. 10 MW-scale PV (>10 MW) and other MW-scale PV (<10 MW). 10 MW-scale PV has appeared since 2006, and in 2010 about 60% was installed as 10 MW-scale PV.

From the current point of view, it can be concluded that 10 MW-scale PV is very common. Indeed, 50 MW-scale PV (>50 MW) has appeared since 2008. Although 50 MW-scale PV is included in the 10 MW-scale PV in Figure 4.3, the number of such PV systems in the world is around 20, and the cumulative capacity of those is estimated as more than 1,2 GW.

Figure 4.4 shows the trends in average capacity of MW-scale PV systems. The average capacity of 10 MW-scale PV systems was almost 18,5 MW in 2010, while the average of all MW-scale PV systems was about 8,3 MW. In contrast, the average capacity of PV systems less than 10 MW was about 5 MW.

Figure 4.5 shows the trends in cumulative installed capacity of MW-scale compared with that of all PV applications installed in the world [5, 6]. The ratio of MW-scale PV systems has achieved 17–20% of all PV applications.

Table 4.1 Examples of MW-scale PV systems around the world

Name	Country	Location	PV capacity [MW]	Installed or operated in
Pervo I-IV PV power plant	Ukraine	Perovo	100	2011
Sarnia PV power plant	Canada	Sarnia (Ontario)	97 (80AC)	2010
Montalto di Castro PV power plant	Italy	Montalto di Castro (Lazio)	84,2	2010
Solarpark Senftenberg II, III	Germany	Senftenberg	82	2011
Solarpark Finsterwalde I, II, III	Germany	Finsterwalde	80,2	2010
Ohotnikovo PV power plant	Ukraine	Ohotnikovo	80	2011
Lobpuri PV power plant	Thailand	Lobpuri	73	2011
Solarpark Lieberose	Germany	Turnow-Preilack	71	2011
San Bellino PV power plant	Italy	San Bellino	70,6	2010
Solarpark Alt Daber	Germany	Wittrock	70	2011
Starwood SSM I, II, III	Canada	Sault Ste, Marie	68	2011
Parc Solaire Gabarden	France	Losse	67,2	2011
Parque Fotovoltaico Olmedilla de Alarcón	Spain	Olmedilla (Castilla-La Mancha)	60	2008
Copper Mountain Solar Facility	USA	Boulder City, NV	60 (48AC)	2010
Avenal Solar Facility	USA	Kettleman Hills, CA	57 (45AC)	2011
Solarpark Strasskirchen	Germany	Strasskirchen	54	2009
Solarpark Tutow I,II	Germany	Tutow (Mecklenburg- Vorpommern)	52	2011
Serenissima PV power plant	Italy	Canaro	48	2011
Parque Fotovoltaico Puertollano	Spain	Puertollano (Castila-La Mancha)	47,6	2008
Solarpark Zerbst	Germany	Zerbst	46	2011
Moura photovoltaic power plant	Portugal	Moura (Alentejo)	46	2008
Solarpark Köthen	Germany	Köthen	45	2011
Cellino San Marco PV power plant	Italy	Cellino San Marco	42,7	2010
Masquite Solar I	USA	Sonoran desert, AZ	42	2011
Solarpark Fürstenwalde	Germany	Früstenwalde	40	2011
Solarpark Waldpolenz	Germany	Brandis	40	2008
FVE Ralsko	Czech Republic	Ralsko	38,3	2010
Solarpark Reckahn I, II, III	Germany	Reckahn	38	2011
Cimarron Solar Facility	USA	Cimarron, NM	38 (30AC)	2010
Long Island Solar Farm	USA	Long Island, NY	37 (32AC)	2011
Parco Fotovoltaico Alfonsine	Italy	Alfonsine	36,2	2010
San Luis Valley Solar Ranch	USA	Mosca, CO	35	2011
Solarpark Veprek	Czech Republic	Veprek	35	2010
Parco Fotovoltaico Sant' Alberto	Italy	Sant' Alberto	34,6	2010
Planta Solar La Magascona & La Magasquila	Spain	Trujillo (Cáceres)	34,5	2008
Solarpark Ernsthof	Germany	Dörlesberg	34,4	2010
Planta Solar Arnedo	Spain	Arnedo (La Rioja)	34,2	2008
Webberville Solar Facility	USA	Webbervillem, TX	34 (30AC)	2011
Parc Solaire Curbans	France	Curbans	33	2011
Huerta Solar Dulcinea	Spain	Dulcinea (Cuenca)	31,8	2009
Parc Solaire Les Mees	France	Les Mees	31	2011

continued...

Table 4.1 continued

Name	Country	Location	PV capacity [MW]	Installed or operated in
DeSoto Next Generation Solar Energy Center	USA	Arcadia, FL	30 (25AC)	2010
Gunthawada, Banaskantha district PV power plant	India	Gunthawada	30	2011
Parque Solar SPEX	Spain	Don Alvaro, Merida (Extremadura)	30	2008
FVE Sevetin	Czech Republic	Sevetink	29,9	2010
Solarpark Giebelstadt	Germany	Giebelstadt	28,3	2011
Solarpark Lonnewitz	Germany	Lonnewitz	28	2011
Solarpark Heideblick	Germany	Heldeblick	27,5	2011
Solarpark Eiche	Germany	Eiche	26,4	2010
Parque Fotovoltaico Fuente Álamo I, II, III	Spain	Fuente Álamo (Murcia)	26	2008
Solar electric power plant, Blythe	USA	Blythe, CA	26 (21AC)	2009
Solarpark Lauingen I, II, III	Germany	Lauingen	25,7	2010
Solarpark Georgsdorf	Germany	Bentheim	24,7	2011
Solarpark Finow	Germany	Finow	24,2	2010
Rideau Lakes PV power plant	Canada	Lombardy (Ontario)	24	2010
Solarpark Mixdorf	Germany	Oder-Spree	24	2011
Sinan power plant	Korea	Sinan	24	2008
Parc Solaire Saint-Symphorien	France	Saint-Symphorien	24	2011
Arnprior PV power plant	Canada	Arnprior (Ontario)	23,4	2009
Planta fotovoltaica de Lucainena de las Torres	Spain	Lucainena de las Torres (Almeria)	23,2	2008
Parque Fotovoltaico Abertura Solar	Spain	Abertura (Caceres)	23,1	2008
Parque Solar Nijar	Spain	Nijar (Andalucia)	23	2008
Parque Solar Hoya de Los Vincentes, Jumilla	Spain	Hoya de Los Vincentes, Jumilla (Murcia)	23	2008
Huerta Solar Almaraz	Spain	Almaraz (Caceres)	22,1	2008
Solarpark Pocking I,II	Germany	Pocking	22	2009
Solarpark Mengkofen	Germany	Mengkofen	21,7	2010
Parque Solar Parque solar El Coronil I, II	Spain	El Coronil (Andalucia)	21,5	2008
Solarpark Calavéron	Spain	Calavéron	21,2	2008
Parque Solar Villarrobledo	Spain	Villarrobledo (Castile La Mancha)	21	2008
Solarpark Ammerland	Germany	Wiefelstede	20,8	2011
Solarpark Rothenburg	Germany	Rothenburg (Sachsen)	20	2009
Seoul power plant	Korea	Seoul	20	2009
Parque Solar Fuentmayor	Spain	Fuentmayor (La Rioja)	20	2008
Parque Fotovoltaico SOLTEN I, II, Granadillas Industrial Estate	Spain	Granadilla de Abona (Canary Islands)	20	2008
Planta solar fotovoltaico Calasparra	Spain	Calasparra (Murcia)	20	2008
Solarpark Beneixama	Spain	Beneixama (Alicante)	20	2007
Parque Solar El Bonillo	Spain	El Bonillo (Albacete)	20	2008
Sheyang PV power plant	China	Sheyang (Jiangsu)	20	2010
Xuzhou City PV power plant	China	Xuzhou City (Jiangsu)	20	2010
Dafeng PV Plant	China	Dafeng (Jiangsu)	20	2010
Wulan PV Plant	China	Wulan (Qinghai)	20	2010
Golmud PV Plant	China	Golmud (Qinghai)	20	2010

(Data source: Resources Total System Co., Ltd., Mizuho Information & Research Institute, Inc., http://www.pvresources.com, etc.)

Table 4.2 Specific information on Perovo Solar Power Station

Plant name	Perovo Solar Power Station
Location	
Country	Ukraine
State, province, city, etc.	Kliuchi, Autonomous Republic of Crimea
Latitude	33° 55′ 57″ E
Longitude	47° 57′ 17″ N
PV power	100 MWp
Projected energy yield	1 314 kWh/kWp/year
Organization / Institution	Activ Solar GmbH
Developer	Activ Solar GmbH
Mounted type	Ground mounted, single row
Array type	Fixed
Installed or operated year	2011
Period of construction	May 2011 – December 2011
Total area of PV plant	200 Ha
No. of subfields	5 subfields of 20 MWp each
No. of PV modules	440 000
Type of PV cells/modules	25% monocrystalline Si, 75% multicrystalline Si
Manufacturer of PV modules	Asian made
No. of inverters	120
Manufacturer of inverters	German made
Nominal grid voltage	10 kV output solar plant voltage, Grid voltage 110 kV
Tilt angle of fixed structures	25 degrees
Structure materials	Piles: steel, Substructures: aluminum
Type of foundation	Foundation of substructures: steel piles, Foundation of inverter stations: concrete (casted on site or pre-fabricated)
Expected lifetime	25 years

Copyright: Activ Solar GmbH

Table 4.3 Specific information on Sarnia Solar Farm

Plant name	Sarnia Solar Farm
Location	
Country	Canada
State, province, city, etc.	Ontario
Latitude	42° 56′ 48″ N
Longitude	82° 20′ 02″ W
PV power	80 MWac (97 MWDC)
Projected energy yield	~120 000 MWh/year
Organization / Institution	Enbridge Pipelines
Developer	First Solar Inc.
Mounted type	Ground mounted
Array type	Fixed
Installed or operated year	2009 – 2010
Period of construction	Dec. 2009 – Oct. 2010
Total area of PV plant	~1 100 acres (~445 ha)
No. of subfields	8
No. of PV modules	~1 300 000
Type of PV cells/modules	CdTe
Manufacturer of PV modules	First Solar Inc.
No. of inverters	160
Manufacturer of inverters	Xantrex and SMA
Nominal grid voltage	27,6 kV
Tilt angle of fixed structures	25 degrees
Structure materials	Steel
Type of foundation	Steel posts
Expected lifetime	25+ years

Copyright: First Solar Inc. (http://www.firstsolar.com)

Table 4.4 Specific information on Ohotnikovo Solar Power Station

Plant name	Ohotnikovo Solar Power Station
Location	
Country	Ukraine
State, province, city, etc.	Karyerne, Saki Region, Autonomous Republic of Crimea
Latitude	33° 35′ 34″ E
Longitude	45° 14′ 20″ N
PV power	80 MWp
Projected energy yield	1 258kWh/kWp/year
Organization / Institution	Activ Solar GmbH
Developer	Activ Solar GmbH
Mounted type	Ground Mounted, double row
Array type (Fixed / tracking)	Fixed
Installed or operated year	2011
Period of construction	March 2011 – October 2011
Total area of PV plant	160 ha
No. of subfields	4 subfields of 20 MWp each
No. of PV modules	347 800
Type of PV cells	Multicrystalline Si
Manufacturer of PV modules	Asian made
No. of inverters	140
Manufacturer of inverters	German made
Nominal grid voltage	10 kV output solar plant voltage,
Grid voltage –110 kV	
Tilt angle of fixed structures	25 degrees
Structure materials	Piles-Steel, Substructures-aluminum
Type of foundation	Foundation of Substructures-steel piles
	Foundation of Inverter stations-concrete (casted on site or pre-fabricated)
Expected lifetime	25 years

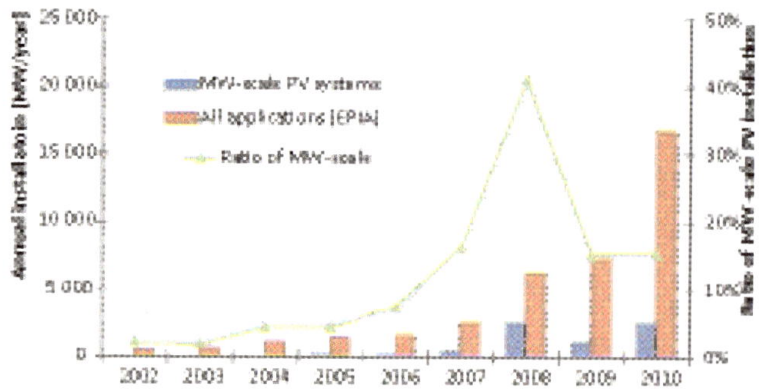

Figure 4.2 Installed capacity of MW-scale PV systems in each year, compared with total installed capacity by all applications in the world

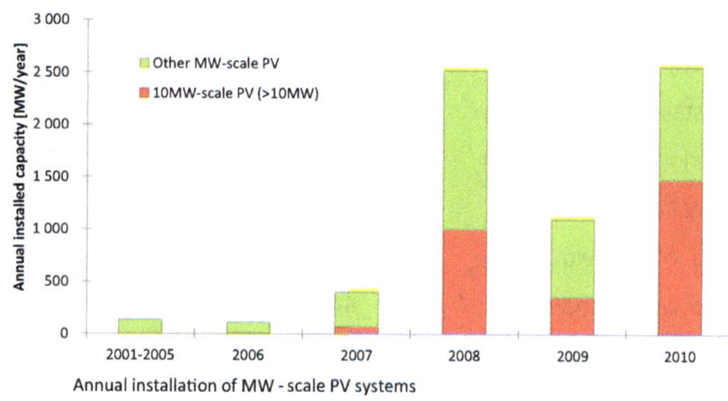

Figure 4.3 Trends in annual installed capacity of MW-scale PV systems by two categories, 10 MW-scale PV (>10 MW) and other MW-scale PV (<10 MW)

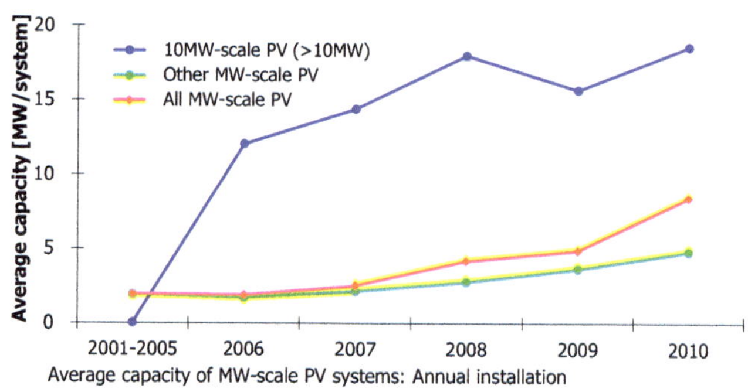

Figure 4.4 Trends in average capacity of MW-scale PV systems in each year

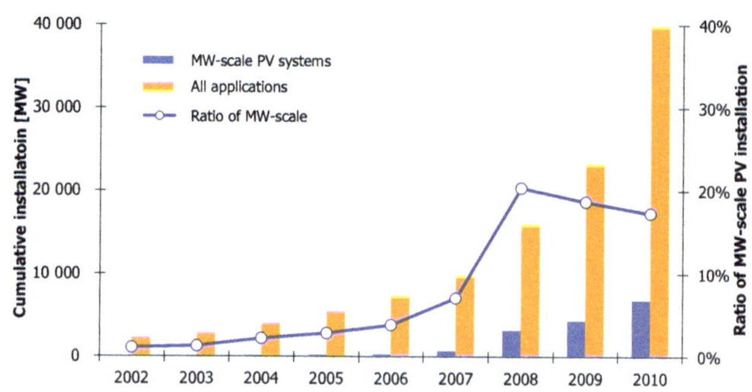

Figure 4.5 Cumulative installed capacity of MW-scale PV systems, compared with total installed capacity by all applications in the world

Further, summing up MW-scale PV systems installed by 2011, the total capacity of MW-scale PV systems installed has reached more than 8 GW, and their capacity occupies a significant share of the cumulative installed capacity of all PV applications in the world.

It is obvious that MW-scale PV systems are already a major PV application. As the capacity of the MW-scale PV systems expands year by year, the maximum capacity is achieving 100 MW and some PV plants which have a capacity of hundreds of MW are being prepared. Eventually, a GW-scale PV plant consisting of several 100 MW-scale PV systems is expected to be realized within a few decades. For the PV industry, worldwide MW-scale systems are increasingly becoming a basic load of its production capacity.

MW-scale and VLS-PV systems should be one of the promising options for large-scale deployment of PV systems. In the near future, VLS-PV systems will become an option for many large remote and desert regions in the world where conventional power plants have never been built.

4.1.2 Current status of concentrator photovoltaic (CPV)

(1) CPV TECHNOLOGY OVERVIEW

Concentrator photovoltaic (CPV) concentrates sunlight onto small but high efficiency cells. Unlike conventional flat-plate photovoltaic, crystalline silicon solar panels for example, CPV systems require concentrator optics and usually trackers. Although these optics and trackers are additional components, depending on design and climate conditions, the high conversion efficiency of CPV may effectively compensate them. Both trackers and optics are important technology parts of CPV as well as social aspects. This argument is discussed in a later section.

1) TRACKERS

Trackers align the optics to the sun so that they focus the sunlight onto cells. In most cases, 2-axes trackers containing two actuators and gears revolving in different directions and on different axes are used. Since the number of degrees of freedom of the pointing vector in 3-dimensional space is 2, any 2-axes trackers anywhere in the world theoretically can point to the sun except when both axes themselves do not lie in the direction of the sun on specific dates and times. If the concentrator optics have a sufficiently wide acceptance angle, 1-axis tracking may be used. However, the maximum tracking error of the ideally designed and adjusted 1-axis tracker is 23,45 degrees; any CPV system with a concentration of more than around 10 should expect a chance of losing the sun out of the acceptance area of concentrator optics.

Trackers can be classified by the selection of axis (see Table 4.5) and support structure (see Table 4.6). The most common tracker is the azimuth-declination type in the selection of rotation axis and a single pedestal structure in support structure category. There are several reasons for this. The most common reason is ease of installation and assembly. Most of the main structure can be made in the steel works. The module array supported by the rack can be

Table 4.5 Classification of trackers by means of rotation axis

Direction of 1st axis	Rotation of 1st axis	Direction of 2nd axis at sun noon	Rotation of 2nd axis at sun noon	Comment
Earth axis	Time angle	Declination	East-West	1st axis constant velocity. Correction of the 2nd axis may be once a day. Suitable to small trackers.
Perpendicular axis	Azimuth	East-West	Slope	Most common. Stable structure. Easy to install. Low position of center of gravity. Maximum azimuth speed is high in latitude zone.
Skewed axis	Azimuth in skewed plane	East-West	Slope	Reduction of maximum azimuth speed
East-West	Slope	Approx. earth axis	Approx. time angle	Advantage to big trackers.
Ball bearing				Simple structure. Limitation to tracking area.

Table 4.6 Classification of trackers by structure

	Foundation	Support	Components	Field work	Deflection of frame
Pedestal	1	Center	A few	Small	Big
Gate	2	Double	Moderate	Moderate	Moderate
Ring	Center and Rail	Double	Many	Big	Small

transported easily. The main gear system can be mounted onto the horizontal plane. The lower rim of the panel is always parallel to the ground level and this reduces the height of the support. In spite of the above-mentioned advantages, it is important to be cautious when implementing in the low latitude areas where many deserts are located. The angle between the first axis that rotates horizontally and the pointing vector to the sun becomes small with the increase of the sun-height. This means the first axis needs to rotate more and faster. In theory, the tracking speed of the azimuth motion will be infinity when the sun reaches the summit of the sphere. In view of the tracker design, the motor power needs to be enhanced and sometimes it is necessary to compromise to accept large tracking error at the highest position of the sun. One possible solution is to decline the 1st axis to avoid the parallelism of the 1st axis and the pointing vector of the sun.

2) OPTICS

Basically there are three types of primary concentrator optics that concentrates sunlight onto the absorber (see Table 4.7). The most common optics for CPV is a lens system. It is difficult to fabricate big lenses. Mirror concentrators are mainly used in CSP (Concentrated Solar Power) applications. The choice of optics mainly stems from preference of the size of optics. For CPV, smaller optics are preferred mainly because the heat flux is proportional to the size of the optics and CPV needs to avoid overheating of the cell. For CSP, the working temperature at the absorber is much higher and bigger optics are favored to reduce the number of components per watt. Therefore, to solve temperature problems some medium concentration CPV or active cooling systems use a mirror system to reduce the number of components per watt. The main drawbacks of the mirror system are much higher sensitivity to soiling [7] and deformation than the lens system. The mirror array system typically used in Heliostat CSP and CPV systems can partially compensate this drawback by allowing adjustment of individual mirrors, obviously leading to more operation and maintenance costs.

In many cases, the most common optics for CPV is a Fresnel lens. Two types of material may be used: PMMA and SOG (silicon on glass). PMMA is a highly transparent plastic typically used in automobile lamps, LCDs, windows and aquariums. The most common ways of processing the lens are injection molding and thermal pressing. The drawback is abrasiveness by sand. This will be discussed in a later section. In SOG lenses the prism is defined by transparent silicone resin on a glass substrate. Because the lens surface is made from glass with a harder surface, the degradation of the surface by sand abrasiveness is less than that of PMMA lenses. The drawback for desert applications is higher thermal degradation because silicone is a soft material and thermal mismatching from the glass substrate deforms the prism and thus defocuses the concentration [8]. Another disadvantage is that silicone is a sticky material and the lens has more chance of inside contamination that cannot be cleaned unless the lens is replaced.

CPV uses secondary optical elements especially when spectrum-sensitive multi-junction cells are used [9]. The main function of the secondary optics is not to concentrate the sunlight but additional functions that primary optics cannot perform. The most typical and important one is homogenizing the intensity and spectrum [10]. Other functions include beam-shaping, light confinement, shifting the focal point for the protection of off-axis damage, protection of cell apertures and expanding alignment tolerance [10]. In some designs, the secondary optics further concentrates the sunlight but not much. Different from the primary optics, the secondary optics are exposed to highly concentrated sunlight. Since the function and the operating conditions are different, the materials, requirements and design approaches are completely different (see Table 4.8). The main material is water-durable glass.

3) CONCENTRATOR CELLS

Basically, any cell can be a concentrator cell. However, a CPV system requires optics and trackers and in addition, for low efficiency cells it is difficult to justify the additional cost. Another consideration is resistance loss. The resistance loss inside the solar cell is proportional to the square of the output current, so it is preferable to use high voltage cells. This is the reason multi-junction solar cells are mainly used.

As a matter of course, some manufacturers use silicon cells, favoring the relatively lower cell price. The recent advancement of a back-contact silicon cell that can reduce the internal cell resistance and thus can increase the concentration ratio seems an encouraging idea [11].

Table 4.7 Types of concentrator primary optics

	Lens	Mirror	Multiple mirrors for central receivers
High concentration	Difficult by chromatic aberration	Easy	
Accuracy requirement	Low	High	Moderate
Sensitivity to deflection and distortion	Low	High	Moderate
Optical efficiency	Low	High	High
Sensitivity to soiling	Low	High	High

Table 4.8 Difference between primary and secondary optics

	Primary optics	Secondary optics
Function	Concentrate sunlight to absorber with minimum loss.	Additional functions including color mixing, homogenizing, expansion of tolerance
Material	Low cost but enables high accuracy processing	High quality, endurance but high accuracy is not required
Design	Geometrical calculation	Trial and error with ray-tracing simulation
Consideration	Ease of assembly and sealing and structure	Stress on cells

The main type of concentrator is the multi-junction cell. The most common one is a III-V 3-junction cell of lattice matching type, specifically InGaP/Ga(In)As/Ge structure. The efficiency in commercial products reaches up to 40% [12]. This lattice matching cell has high-quality epitaxial grown crystalline quality. However the band-gap energy combination, namely the balance of output current among three junctions, is not ideally matched to the standard direct solar spectrum. Specifically, the output current is constrained to the top junction and the current from the bottom junction is excessive [12]. This means there is more chance of an increase of output voltage by increasing the band-gap of the bottom cell and increasing current by decreasing the band-gap energy of the top cell. The motivation for the lattice-mismatching cell is to widen the freedom of combination of energy gap, and thus to attain better matching of the solar direct spectrum [12]. Some manufacturers have started production of spectrum mismatching cells and most of the R&D work is devoted to lattice mismatching cells. Sooner or later, the various problems of lattice mismatching cells including reliability and yield will be solved and lattice mismatching cells will be majority of CPV products.

(2) CPV TECHNOLOGIES SPECIFICALLY DEVELOPED FO DESERT AREAS

Needless to say, the climate and environment of desert areas are different from those of other areas. The technology developed out of a desert area cannot always be applied in a desert and keep good performance and reliability. Thanks to recent research and development as well as experiences in field tests in dry areas, several problems and challenges arising in applications in desert have been revealed. This section discusses specific technologies important to implementation in desert areas.

1) ARCHITECTURE FOR LOCAL ASSEMBLY

Assembly of CPV products does not require special facilities except for some limited components such as concentrator cells. Increasing the thickness of the module brought by free space for light deflection by concentration decreases the number of modules in the transportation bed and thus increases the cost of transportation. Trackers are also difficult to transport. In view of the overall system cost, local assembly rather than importing the complete modules and systems saves costs and is the preferred approach.

A CPV system has more components than a flat-plate PV system. This is definitely a drawback simply in view of the energy costs. However, turning to the development of the local economy, this complexity is desirable to encourage local industries. Possibly, solar cells and some of the electro-mechanical components used in the tracker need to be imported, but the rest of the components do not require special facilities and can be produced by local industries after appropriate technology transfer is done.

Recently, some CPV manufacturers have given serious consideration to a business strategy of local assembly and have started discussions on technology transfer in desert countries [7]. As CPV technology does not rely on specfic production facilities, it seems that manufacturing industries such as the commodity industry and automobile industry can go into local CPV assembling with less risk of local investment. Indeed, despite assembly outside of the manufacturer, the installed CPV system have an excellent performance record and are cost competitive.

2) ANTI-SOILING OF OPTICS

One of the well-known problems not only for CPV but also for CSP and PV implemented in desert areas is soiling of optics [7]. Modules are easily covered by dust and sand and there is much less chance of wash-out by rainfall. The influence of soiling on lens-based CPV is less than for mirror-based concentrators but more than flat-plate PV. For flat-plate PV, the main cause of performance loss by soiling is absorption, but that of CPV includes scattering as well. The soiling reduces both in-line and total transmittance. Reduction of in-line transmittance is caused by surface scattering. PV captures a wide range of incident light and reduction of in-line transmittance is not a problem. CPV captures only direct sunlight within typically less than a degree of acceptance angle. The reduction of in-line transmittance directly hampers the performance of CPV. The fact that the reduction of in-line transmittance is dominant in the blue region in the visible spectrum implies that the influence will be enhanced for CPV systems using lattice-matched triple junction cells where the top cell constrains current output. Obviously, a CPV system that has a smaller acceptance angle receives more light and its

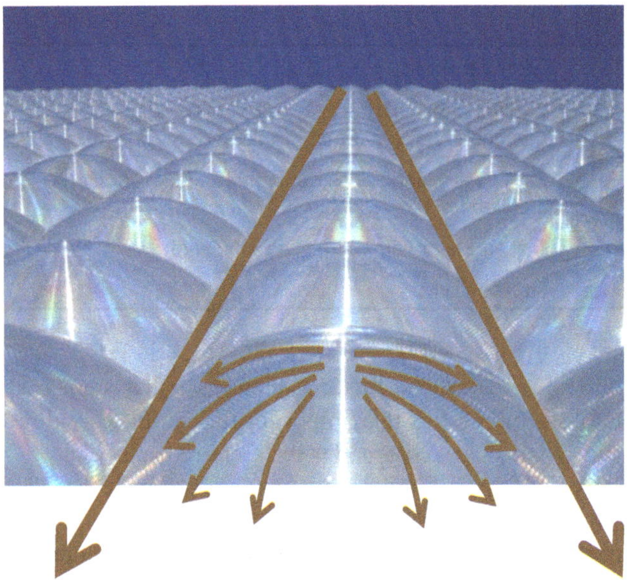

Figure 4.6 Shaped Fresnel lens showing water flow that helps save the amount of water in the cleaning of optics

performance is affected more by soiling than that of a CPV system with a wider acceptance angle [7].

An experiment with various outdoor exposed mirror-based and lens-based CPV systems showed that the mirror-based system received much more influence of soiling and the magnitude of degradation increases with the increase of concentration ratio [7].

One of the effective techniques that are applied to commercial products is helping rain cleaning by defining water flow out of the circumference region of optics (see Figure 4.6). The trench structure around the dome-shaped Fresnel lens collects washed dust and prevents recontamination onto the adjacent lenses [14]. The CPV system was implemented in a coastal area and high performance was achieved for more than 2,5 years without cleaning (see Figure 4.7). It is true that this method relies on the rainfall and the efficiency gradually decreases according to the number of days after rainfall, but the efficiency recovers every time after a small amount of the rain. This fact suggests that the amount of water used for cleaning of optics will be significantly reduced.

Another promising technology that recently developed is anti-soiling coating to the PMMA lens [15]. The reason for dust accumulation rather than dropping by gravity is adhesive organic materials on the surface. The coating of light catalysis material removes adhesive organics and thus prevents dust from staying on the lens surface. The basic structure is shown in Figure 4.8. The buffer layer helps good adhesion of the light catalysis layer to the PMMA base while preventing light catalysis attack to the lens itself. The outdoor exposure test showed promising results for keeping the optics surface clean (see Figure 4.9).

3) SAND ABRASIVENESS

CPV systems are much more sensitive to sand abrasiveness than flat-panel PV systems because they can only use a direct beam of sunlight. It is important to know how the materials of optics are damaged by sandstorms and other abrasiveness by sand. It is also important to estimate the degradation rate by understanding physical degradation mechanisms and to anticipate lifetime and replacement durations.

Figure 4.10 shows the surface of a PMMA plate after being attacked by flux of sand [16]. The surface was excavated by the blown sand. Like the soiling effect, the in-line transmittance, especially in short wavelength region is significantly reduced by the rugged surface.

By intensive analysis after varying abrasive conditions including energy and momentum of sand flux, the main factor that affects degradation was identified as the total momentum of sands (see Figure 4.11) [16]. This implies once the sand density and average wind velocity is acknowledged, it may be possible to anticipate the lifetime and replacement rate of the lens parquet.

4.1.3 Current status of concentrated solar power (CSP)

Another technology for large-scale solar energy utilization is concentrated solar power (CSP), in other words, solar thermal electric power, and its installation has been increasing recently.

As a technology for large-scale solar energy utilization, CSP has a longer history than PV technology. The largest and oldest CSP plant in the world is the SEGS plant in the Mojave desert, California, USA, which has a capacity of 354 MW and has been operational since the 1980s. After

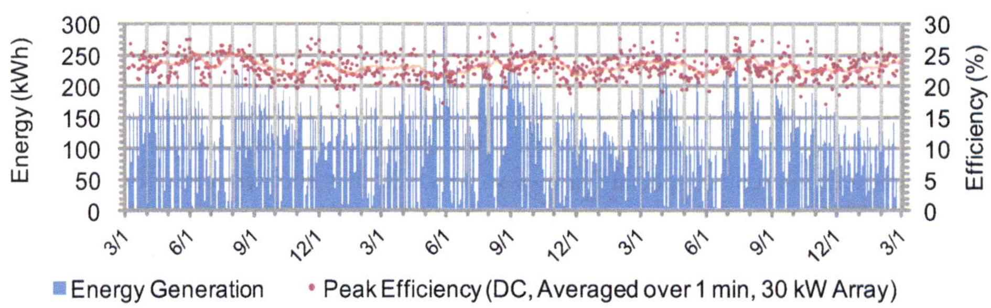

Figure 4.7 Energy and efficiency trend of CPV system equipped with shaped Fresnel lens that can utilize rainfall for cleaning (The efficiency remains high without relying on artificial cleaning using detergents and high-pressure water.)

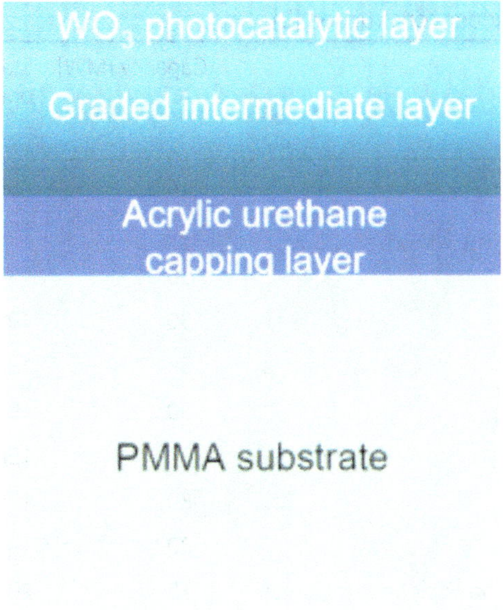

Figure 4.8 Structure of anti-soiling coating

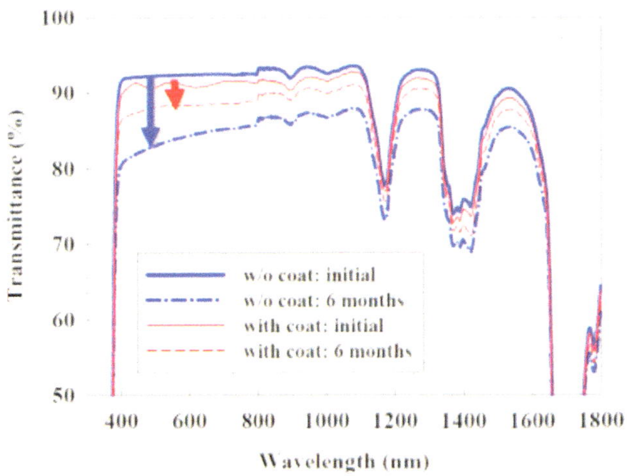

Figure 4.9 In-line transmittance after 6 months of exposure of samples with and without anti-soiling coating

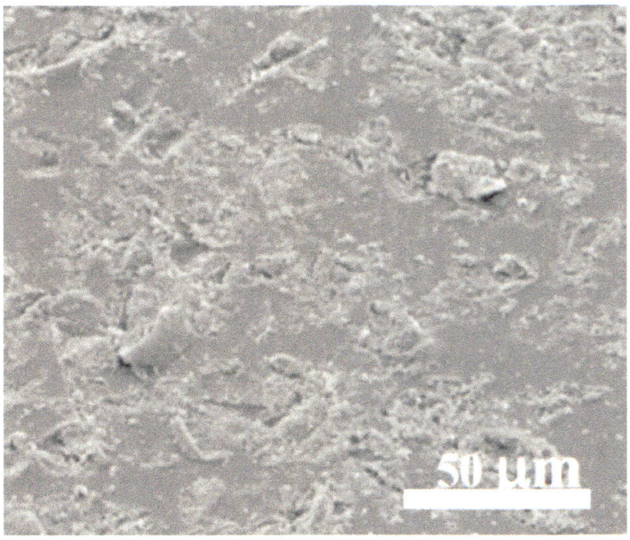

Figure 4.10 SEM image (45° view) of sample surface after sandblasting with 3,3 g of sand at wind velocity of 16 m/s [16]

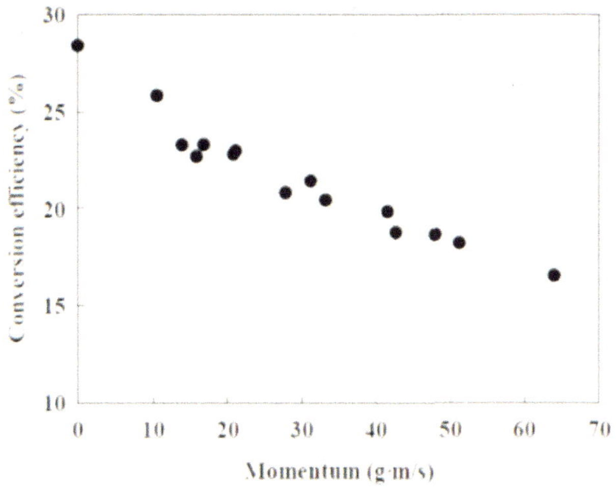

Figure 4.11 Momentum dependence of calculated conversion efficiency of CPV module [16]

the SEGS plant was built, new installations of CSP plants were not realized in a commercial way; however, since the latter half of the 2010s CSP technology has been focused on again.

CSP technology can be classified into four types: parabolic trough, power tower, Fresnel and dish Stirling. The parabolic trough type is already proven and most current CSP plants are of this type. Other types are still in the demonstration or R&D phase, however the power tower is likely to be proven next. Some CSP plants under construction are of the power tower type.

Most CSP plants in operation and under construction are located in the USA and Spain. Other potential markets for CSP are the MENA region, China, India, etc. In Morocco, Algeria and Egypt, CSP technology is used as ISSC, a hybrid with a fossil fuel power station. Such an application would be useful for developing countries, although the impact on reducing CO_2 emissions would be small compared with an independent solar power plant.

Until the end of 2011, the total capacity of operational plants was approximately 1,7 GW, and CSP plants under construction was about 2,5 GW. In addition, the total capacity of CSP plants under development was approximately 7 GW [18].

Table 4.9 shows examples of CSP plants in the USA and Spain. The capacities of CSP plants under construction in the USA are up to a few hundred MW. On the other hand, those in Spain are set at 50 MW because of an upper limit of the support scheme in Spain.

Table 4.10 shows information on expected annual generation and required area for some CSP plants.

For parabolic trough, expected annual generation would be 2 621 MWh/MW/year, and in case of with storage it would be 2 970 MWh/MW/year while without storage it would be 2227 MWh/MW/year. The required mirror area would be 8 358 m^2/MW (0,8 ha/MW) and land area would be 30 012 m^2/MW (3 ha/MW), which would be about 3,6 times the mirror area. In the case of CSP plants with storage, the mirror area would be 9 452 m^2/MW and

Table 4.9 Examples of CSP plants in the USA and Spain: operational and under construction [2]

Country (including plan)	Type of technology	Status	Site/Name	Capacity [MW]	Starting operation
USA	Parabolic trough	Operational	SEGS I	14	1984
			SEGS II	30	1985
			SEGS III	30	1985
			SEGS IV	30	1989
			SEGS V	30	1989
			SEGS VI	30	1989
			SEGS VII	30	1989
			SEGS VIII (hybrid)	80	1989
			SEGS IX	80	1990
			Martin Next Generation Solar Energy Center	75	2010
			Nevada Solar One	64	2007
			Holaniku, Keyhole Point	2	2009
			Colorado Integrated Solar Project (hybrid)	2	2010
			Saguaro Power Plant	1	2006
		Under construction	Solana	250	2013
			Mojave Solar Project	250	2013
			Genesis Solar Energy Project	250	2013
	Power tower	Operational	Chevron Solar-to-Steam Demonstration	13	2011
			Sierra SunTower	5	2009
		Under construction	Ivanpah Electric Generation Station	370	2013
			Crescent Dunes Solar Energy Project	110	2013
	Fresnel	Operational	Kimberlina Solar Thermal Power Plant	5	2008
	Dish Stirling	Operational	Maricopa Solar Project	1,5	2008
Spain	Parabolic trough	Operational	Andasol 1	50	2008
			Andasol 2	50	2009
			Andasol 3	50	2011
			Extresol 1	50	2009
			Extresol 2	50	2010
			HelioEnergy 1	50	2011
			Ibersol Puertollano	50	2009
			La Dehesa	50	2010
			La Risca	50	2009
			La Florida	50	2010
			Lebrija 1	50	2011
			Majadas	50	2010
			Manchasol 1	50	2010
			Manchasol 2	50	2011
			Palma del Rio 1	50	2011
			Palma del Rio 2	50	2010
			Solnova 1	50	2010
			Solnova 3	50	2010

Country (including plan)	Type of technology	Status	Site/Name	Capacity [MW]	Starting operation
			Solnova 4	50	2010
			Velle 1	50	2011
			Velle 2	50	2011
		Under construction	Arenales	50	2013
			Aste 1a	50	2012
			Aete 2b	50	2012
			Astexol	50	2012
			Casablanca	50	2013
			Enestar	50	2013
			Extremasol 1	50	2013
			Extresol 3	50	2012
			Guzman	50	2012
			Helios 1	50	2012
			Helios 2	50	2012
			Heliosenergy 2	50	2012
			La Aftricana	50	2012
			Moron	50	2012
			Orellana	50	2012
			Olivenza 1	50	2012
			Solaben 2	50	2012
			Solaben 3	50	2012
			Solacor 1	50	2012
			Solacor 2	50	2012
			Termesol 1	50	2013
			Termosol 2	50	2013
			Thermo Borges (hybrid)	22,5	2012
	Power tower	Operational	Gemasolar	20	2011
			PS10	10	2006
			PS20	20	2009
	Fresnel	Operational	Puerto Errado 1	1,4	2009
		Under construction	Puerto Errado 2	30	2012
	Dish Stirling	Operational	Casa de los Pinos	1	2011

Table 4.10 Expected annual generation and required area for some CSP plants [19]

Type of technology	Site/Name	Capacity [MW]	Storage	Expected annual generation		Mirror area		Land area	
				[MWh/year]	[MWh/MW/year]	[m²]	[m²/MW]	[m²]	[m²/MW]
Parabolic trough	SEGS (USA)	354	no	–	2 750	2 314 000	6 540	8 080 000	22 838
	Nevada Solar One (USA)	64	yes	134 000	2 094	357 000	5 578	1 620 000	25 313
	Holaniku at Keahole Point (USA)	2	no	–	–	–	–	12 141	6 071
	Saguaro Power Plant (USA)	1	no	2 000	2 000	10 000	10 000	40 000	40 000
	Andasol 1 (Spain)	50	yes	158 000	3 160	510 120	10 202	2 000 000	40 000
	Andasol 2 (Spain)	50	yes	158 000	3 160	510 120	10 202	2 000 000	40 000
	Extresol 1 (Spain)	50	yes	158 000	3 160	510 120	10 202	2 000 000	40 000
	Extresol 2 (Spain)	50	yes	158 000	3 160	510 120	10 202	2 000 000	40 000
	Ibersol Puertollano (Spain)	50	no	103 000	2 060	287 760	5 755	1 500 000	30 000
	La Dehesa (Spain)	50	yes	175 000	3 500	552 750	11 055	2 000 000	40 000
	La Florida (Spain)	50	yes	175 000	3 500	552 750	11 055	2 000 000	40 000
	La Risca (Spain)	50	no	105 200	2 104	–	–	1 350 000	27 000
	Majadas (Spain)	50	no	104 500	2 090	–	–	1 350 000	27 000
	Manchasol 1 (Spain)	50	yes	158 000	3 160	510 120	10 202	2 000 000	40 000
	Palma del Rio 2 (Spain)	50	no	–	–	–	–	1 350 000	27 000
	Solnova 1 (Spain)	50	no	113 520	2 270	300 000	6 000	1 150 000	23 000
	Solnova 3 (Spain)	50	no	113 520	2 270	300 000	6 000	1 150 000	23 000
	Solnova 4 (Spain)	50	no	113 520	2 270	300 000	6 000	1 150 000	23 000
	Archimede (Italy)	5	yes	9 200	1 840	31 860	6 372	80 000	16 000
	Average				2 621		8 358		30 012
			yes		2 970		9 452		35 701
			no		2 227		6 716		24 891
Power tower	PS 10 (Spain)	11	yes	23 400	2 127	75 216	6 838	550 000	50 000
	PS 20 (Spain)	20	yes	48 000	2 400	150 000	7 500	800 000	40 000
	Solar Tower Julich (Germany)	1,5	yes	–	–	18 000	12 000	180 000	120 000
	Average				2 264		8 779		70 000

land area would be 35 701 m^2/MW. On the other hand, the mirror area without storage would be 6 716 m^2/MW and land area would be 24 891 m^2/MW. Thus, CSP plants with storage require a large land area.

For power tower with storage, the mirror area would be 8 779 m^2/MW (0,9 ha/MW) and land area would be 70 000 m^2/MW (7 ha/MW). Compared with parabolic trough, the land area required for power tower is much larger.

4.2 A COMPREHENSIVE COMPARISON BETWEEN SOLAR-POWERED VLS RENEWABLE ENERGY TECHNOLOGIES

4.2.1 Introduction

Within the world of solar energy research, a kind of competition has arisen between adherents to the photovoltaic (PV) approach and those to the solar-thermal, better known as concentrated solar power (CSP) method of generating electricity. This is probably due to the different scientific backgrounds of their respective practitioners: the former being grounded in solid state physics and the latter in mechanical engineering. Such a state of affairs is unfortunate because it can easily lead to confusion on the part of decision makers and among the public. The purpose of this section is accordingly to provide a dispassionate review of the two approaches, emphasizing both their strengths, and their weaknesses *at the present state of their respective arts*. The review will also include what is to a certain extent a combination of the two: concentrator photovoltaics (CPV). We shall also touch upon one of the relevant properties of wind technology. Our emphasis will be on the degree to which these developing technologies may be regarded as complementary to one another rather than in competition. To the extent that we shall point out weaknesses, our attitude is that effective engineering solutions will, in all probability, ultimately be found for all of them.

4.2.2 CSP

The CSP approach to generating electricity concentrates sunlight onto a heat transfer fluid (HTF) in order to produce high temperatures that will ultimately (i.e. either directly or via a heat exchanger) drive a conventional electricity generator. In this approach, high temperatures are critical in order to achieve high thermodynamic efficiencies. Historically, attempts to operate CSP systems at the kind of temperatures that solar technology is capable of achieving resulted in serious materials failure. However, with the development of modern materials that are capable of withstanding much higher temperatures than in the past, many of the early CSP ideas that failed in the 1970s are being re-tried.

(1) DISH CONCENTRATORS

From the optical point of view a dish-shaped concentrator has the highest possible efficiency. This is because the optical axis of the dish is always directed towards the sun, and the receiver is mounted on that line. This results in minimal cosine losses compared with other types of concentrator. There are, of course some cosine losses from the outer parts of the dish but these are relatively small compared with those associated with the other technologies that will be discussed below. Dish concentrators also have the highest possible concentration ratio – in theory, some 45 000 times, but in practice, material imperfections limit this to approximately 10 000 times. With today's materials, 1 000 times gives maximal efficiency, but if materials and an HTF could be developed that would operate at 1 800 K, then a concentration of 10 000 times could yield even higher efficiency.

A general problem associated with the employment of concentrator dishes is that each dish is mechanically limited as regards the size of the power unit that can be placed at its focus. Such systems are accordingly best suited to relatively small stand-alone systems of up to about 100 kW rating. Early attempts to employ a field of many concentrator dishes for powering a large central generator tended to suffer from excessive heat losses from the HTF flow lines.

For operation as a single stand-alone system, a number of generator types have been tried with dish-concentrators [20]. These have been of the steam (Rankine cycle) variety, the air-heating (Brayton cycle) type, Stirling engines, and hybrid combinations of solar with conventional fuels. Probably the chief attraction of dish-concentrator CSP systems at the present time is their modularity and relative compactness. This renders them a potentially low-cost investment for many situations in which systems of up to approximately 100 kW are required. For this reason a number of companies are already reaching the pre-commercial stage with demonstration systems. For example, Figure 4.12 shows a novel air-heating CSP receiver powered by a 500 m^2 aperture solar dish operating at a concentration factor of 2 000 times.

A recent larger-scale commercial dish-based CSP solar field is the 1,4 MW Maricopa Solar Project at Peoria, Arizona, USA, which became operational in 2010. It consists of 60 separate dishes, each illuminating a 25 kW Stirling engine. The entire system occupies 15 acres (6,1 ha) of land.

(2) TROUGH CONCENTRATORS

Less efficient than dish concentrators, but more practical for large-scale power production, are parabolic trough concentrators. In sun-belt locations, the horizontal axes of these collectors point north-south, and the mirrored collector troughs perform daily rotations from east to west

Figure 4.12 The 420 KW HelioBoosterTM, powered by HelioFocus's 500 m^2, 2 000 times solar dish at Park Rotem in the Negev desert [courtesy Heliofocus Corp.]

Figure 4.13 An 80 times parabolic trough CSP collector under test at Sede Boqer in the Negev desert. Collector length = 95 m, width = 5,75 m [courtesy Siemens Solar Corp.]

about these axes. Figure 4.13 shows such a collector being tested at Ben-Gurion University's Sede Boqer test site.

One immediately apparent advantage that such troughs have over dishes is that a large fraction of the HTF piping constitutes the collector itself. Therefore, relatively short lengths of inter-connecting piping need to be insulated. For this reason, heat losses are easier to minimize for large solar fields than is the case for the long lines of insulated piping that would be needed if a field of solar dishes were to provide heat to a central power unit.

But that is not the whole story as regards heat loss prevention. The steel receiver tube at the focus of the collector trough has a spectrally-selective coating to reduce radiative losses and it is housed within an evacuated glass tube to reduce conductive/convective losses. This enables an average annual solar-to-heat efficiency of typically 50% to be achieved at the HTF operating temperature. However, because of their intrinsic two-dimensional geometry, the theoretical concentration of mirrored troughs is limited to 213 times, and in practice, to 50–100 times. This tends to restrict to a certain extent the temperatures at which these collectors can operate, 350°C being a typical operating temperature. Furthermore, it is only from mid-spring through mid-autumn that such collectors can receive the incoming radiation at normal incidence – and only once in the morning and once again in the afternoon. For the winter half of the year, even though the collectors track the sun's apparent motion, this tracking being one-dimensional, there is always a non-zero angle of incidence whose cosine reduces the effect of concentration.

The trough kind of CSP system first came into prominence during the early 1980s when the Luz Corporation constructed a series of multi-MW plants in California: first, a 14 MW plant at Daggett, followed by another of 30 MW rating at the same site. Five further 30 MW plants were constructed at Kramer Junction, and two 80 MW plants at Harper Lake [21]. In spite of the subsequent demise of Luz, their CSP plants have continued to operate ever since. They are the only solar power installations from that era that are still operational and they accordingly provide a benchmark against which all other solar plants are judged.

With the current revival of interest in CSP, several new companies have arisen, "in the ashes of Luz", so to speak,and many new parabolic trough systems have been built. A comprehensive list of the trough systems completed to date, which are fully operational, was included in Table 4.10 (IEA SolarPACES [22] is also available). At the time of writing it is still too early to assess the extent to which the efficiency of the modern parabolic trough plants improves upon that of the original Luz plants.

(3) CENTRAL RECEIVER CONCENTRATORS

The third CSP technology type, which aims to overcome some basic weaknesses of both dish-type and trough-type technologies, is that of the central receiver with its associated surrounding field of heliostat mirrors. In this kind of geometry, the receiver is mounted atop a tower, and illuminated by the light from many ground-located, sun-tracking mirrors – the so-called *heliostats*. Figure 4.14 shows such a receiver and some of the associated heliostats.

Although the optical efficiency of central receiver systems is limited, owing to cosine losses and large spaces between the heliostats, there is effectively no limit to the desired concentration that may be achieved and hence, the temperature that may be reached at the receiver. This is because any number of heliostats may be employed. In this manner it is in principle possible to obtain higher thermodynamic efficiencies for power generation than is possible from parabolic trough systems. Furthermore, because the energy transfer from the heliostat field to the central receiver is carried out via optical means, there are no heat losses of the kind that limit the size of parabolic dish fields.

During the 1970s, central receiver power plants in the MWe scale were built and tested in several countries.

Figure 4.14 A small central receiver CSP plant at kibbutz Samar in the Negev desert [courtesy Aora Corp.]

Figure 4.15 Part of the 30 MW Puerto Errado 1 linear Fresnel CSP plant in Spain [courtesy Novatec Solar Corp.]

Construction details, costs and performance figures were all conveniently summarized in an important series of papers that appeared in a single issue of *Transactions of the ASME Journal of Solar Energy Engineering*: in the USA [23]; in Italy [24]; in Spain [25]; in France [26]; and in Japan [27]. The performance of these early central receiver plants was limited to relatively low temperatures by the materials that were available at that time. However, with modern high-temperature materials, there is renewed interest in central receiver technology. A number of new demonstration systems have been constructed and larger-scale systems are presently under consideration. A comprehensive list of the central receiver type systems completed to date, that are fully operational, was included in Table 4.10.

(4) OTHER CSP VARIANTS

A number of variations on the three main geometries outlined above were examined in the 1980s with a view to reducing system costs. These included: stationary dishes with moving receivers; asymmetric dishes with off-axis receivers; parabolic troughs with polar sun-tracking axes, and several others. One technology, which has lately received renewed interest, involves the employment of an upward-facing sun-tracking array of horizontal mirror strips (a so-called *Fresnel array*) that illuminates an elevated, stationary receiver tube. Figure 4.15 shows such a system, which operates according to this principle.

4.2.3 PV

One of the great attractions of photovoltaics is its great range of modularity as a useful power source. Wrist watches and small home appliances are good examples at the mW scale. City parking meters and street lighting are larger, more visible examples at the level of watts to hundreds of watts. Rooftop systems can provide some kW, etc. In this review we shall focus attention on the largest grid-connected systems, presently approaching 100 MW in rating, as these may be compared as regards their relative advantages and disadvantages with the CSP systems discussed earlier. Figure 4.16 shows part of the 97 $MWp_{(DC)}$ Sarnia plant in Canada.

The first point to comment upon is the "p" subscript, which stands for "peak". This rating is the power that a PV panel could deliver at the maximum power point of its current-voltage operating curve if the panel were exposed at normal incidence to irradiance at 1 000 W m^{-2}, with a spectral designation known as AM1,5G and if the cell temperature were 25°C. These are the so-called "standard test conditions" (STC) even though a panel would rarely, if ever, operate under such conditions when in actual use! The STC rating of one panel, multiplied by the number of panels in the system becomes the STC rating of the system. By contrast, CSP systems have no need to employ such a "p" subscript. However, their output also varies with solar intensity, angle of incidence, receiver temperature and inter-connection losses. Therefore, an 80 MW CSP system cannot be compared directly with an 80 MWp PV system. In both cases, it is necessary to inquire how much annual

Fig 4.16 Part of the 97 MWp(DC) Sarnia PV plant in Canada [photo credit: First Solar Inc.]

energy they give. Furthermore, if the comparison is to be meaningful, it must be made at the same place and over the same period of time. For this reason, the performance figures that are quoted in this chapter are estimates only. With very few exceptions have actual performance figures been documented.

(1) STATIC VS. SUN-TRACKING

At any given location the output of a solar power producing system of any specified rating depends upon the tracking strategy employed for its operation. Figs. 4.17a, 4.17b and 4.17c show examples of fixed tilt, 1-axis sun-tracking and 2-axes sun-tracking systems, respectively.

The relative advantage of the different tracking strategies will, in turn, depend on geographic latitude. Table 4.11 shows the simulated total irradiance available to a number of stationary and sun-tracking PV systems for a variety of geographic locations in the USA ranging from approximately 20°N to 60°N at 10° steps in latitude. The isotropic approximation is used for calculating the diffuse and reflected light contributions, together with the generic PV incidence angle modifier measured by King *et al.* [28]. A ground albedo of 0.3 is assumed at all sites. The irradiance data upon which these simulations are based are taken from the US National Solar Radiation Data Base [29].

When examining trends in Table 4.11 it is convenient to take "fixed tilt = latitude" as a base-line case because this is the simplest and most widely-employed field layout. It corresponds to the fixed angle that would theoretically receive the most irradiance in the course of a year in a perfectly clear climate, but it is also approximately the case for sunny and moderately sunny climates. The table indicates that: (a) the closer one is to the Equator, the less advantage this configuration has over horizontal upward-facing modules; (b) at all latitudes, single-axis tracking about a NS horizontal axis provides some improvement compared with static systems – up to about 17% at latitudes around 30°N, but diminishing towards higher latitudes; (c) the advantage of single-axis tracking about an EW horizontal axis increases with increasing geographic latitude but remains less than the advantage of a NS tracking axis; (d) dual axis tracking provides the greatest improvement, becoming successively larger the higher the geographic latitude. All of these features are intuitively obvious, but Table 4.11 quantifies them.

Figure 4.17a A fixed-tilt PV system (part of the Masdar PV plant in Abu Dhabi) [courtesy Masdar Solar Corp.]

Figure 4.17b Part of an array of 1-axis SunPower T0 Trackers at the 250-megawatt California Valley Solar Ranch in San Luis Obispo County, California (at noon, as in the photo, the PV panel planes are parallel to the ground) [courtesy SunPower Corp.]

Fig 4.17c Part of a 2-axes sun-tracking PV system at Alcolea del Tajo (Toledo), Spain

(2) SOLAR CELL MATERIALS

Most large PV systems employ modules of mono-crystalline or polycrystalline silicon cells. These modules have proven their stability over many years of use, degrading at a typical rate of 1% per year [30]. Other module types are of interest because of the potentially lower manufacturing costs that are involved. For example, so-called "thin-film" materials

Table 4.11 Simulated annual plane-of-array irradiance seen by a number of stationary and sun-tracking PV modules at a variety of geographic locations in the USA

Site	Horizontal upward-facing module	Fixed S-facing module with tilt angle = latitude	Module tracking about a NS horizontal axis	Module tracking about an EW horizontal axis	Dual axis sun-tracking module
Hilo, HI (19,7°N)	1 588 (96)	1 654 (100)	1 866 (113)	1 713 (104)	1 965 (119)
Austin, TX (30,3°N)	1 610 (94)	1 708 (100)	2 000 (117)	1 794 (105)	2 148 (126)
Springfield, IL (39,9°N)	1 451 (89)	1 639 (100)	1 876 (114)	1 740 (106)	2 098 (128)
Minot, ND (48,3°N)	1 356 (86)	1 579 (100)	1 780 (113)	1 687 (107)	2 048 (130)
Seward, AK (60,1°N)	819 (84)	971 (100)	1 074 (111)	1 058 (109)	1 271 (131)

Units are kWh m^{-2}. Numbers in parenthesis are percentages relative to a fixed, south-facing module with tilt angle equal to the latitude of the site.

[mainly amorphous silicon (a-Si), cadmium telluride (CdTe), copper-indium-diselenide (CIS) or copper-indium-gallium-diselenide (CIGS)] have lower efficiency than their crystalline silicon counterparts but are generating increased interest. Their long-term stability properties are as yet uncertain. Still newer solar cell technologies [mainly dye-sensitized ("Graetzel") solar cells and organic (OPV) solar cells] are emerging, but it is not clear whether the latter will eventually aim at the large-scale power market rather than for use in consumer electronics, domestic appliances, clothing, etc.

Having now discussed the salient features of PV plants, the reader may care to refer to the comprehensive list of multi-megawatt PV plants that have been completed to date in Table 4.1.

4.2.4 CPV

We have drawn attention to the intriguing potential of high-concentration photovoltaics [31]. Here, the basic idea is to concentrate sunlight many hundreds of times in order to increase the power output and thereby reduce the amount of costly PV material needed for power generation. In addition, it turns out that PV cells, if appropriately designed, can reach efficiencies under concentrated sunlight that greatly exceed those of flat-panel crystalline silicon. As a result, a solar concentration by a few hundred times can reduce the PV material requirement by three orders of magnitude. At the time of writing STC efficiencies of CPV cells are comfortably in excess of 40% and they are expected to become even higher in a relatively short amount of time. By contrast, the best silicon cells have barely reached 30% efficiency, even under concentrated sunlight. The downside of CPV technology is that, like CSP, it is only suitable for "sunny" climates. Moreover, the requirement of accurate sun-tracking will have to be factored into the long-term reliability and economics of such systems.

From the technology standpoint, there are two approaches to CPV. One employs so-called *passive cooling*. Here the cells are mounted on air-cooled heat sinks. In order for passive cooling to be effective, the cells must be small (typically 1 cm or less in linear dimension) and each cell must have its own optical concentrator (typically a Fresnel lens or a small mirrored dish). Figs. 4.18a and 4.18b show two passively-cooled CPV systems, one employing Fresnel lenses and the other, modules of mini-dishes.

If larger area CPV cells or modules are used, active cooling is necessary. Such systems typically employ a single, large concentrating mirrored dish to illuminate a module of CPV cells. An example is shown in Figure 4.19.

Of course, with active cooling, what would otherwise be waste heat is actually captured and can be used in addition to the electrical output of the CPV system. This considerably raises the effective efficiency of the CPV system.

4.2.5 Discussion

There is much debate among entrepreneurs of solar power systems and the public about the relative advantages and disadvantages of the various technologies: CSP or PV, Stirling or Rankin cycles; and indeed, even about the components of each technology: troughs or central-receivers; c-Si or thin films, etc. What is not usually emphasized is that any specific solar power plant would be the result of say, 10 characteristic design features coupled with 100 features aimed at reducing costs and arriving at a reliable system. In the long run it is reliability that counts. As a result, very few solar technologies have yet been around long enough to pass that grueling test – grueling as witnessed by the huge number of promising companies, with their associated technologies, that filled the pages of solar conference proceedings from the 1980s but who are, alas, no longer with us.

With this in mind we need to realize that at the present state of the art, there are only two benchmarks for reliability.

Figure 4.18a A passively-cooled "Fresnel-lens" CPV system at Ein Atrun, Egypt [courtesy Soitec Corp.]

Figure 4.18b A passively-cooled "mini-dish" CPV system at Victor Valley College, Victorville, California [courtesy SolFocus, Inc.]

Fig 4.19 An actively-cooled CPV system at kibbutz Kvutzat Yavneh [courtesy ZenithSolar Corp.]

The first are the CSP parabolic trough *systems* that were built by Luz corporation back in the 1980s and which have continued to operate until this day with only relatively minor improvements – a lasting memorial to the successful efforts of a company that did not last long enough to bask in the glory of its achievement. Secondly, although no large PV *systems* from that era are still operative, many of the crystalline silicon PV modules from the 1980s are still in existence and show sufficiently high efficiencies under test to enable today's manufacturers to offer a 30-year performance warranty on modern modules [note *modules*, not systems].

So we have two benchmarks for reliability: one CSP and one PV. Back in the 1980s there was also some pioneering work in CPV [32], but the technology was not taken seriously enough at that time to enable any of the original test systems to have left any benchmarks for today's emerging systems. With these facts in mind, we must weigh very carefully the claimed advantages and disadvantages of PV, CPV and CSP. If all of them eventually work out, we shall have three kinds of very large scale solar technologies with which to be able to replace fossil fuel. It is therefore worthwhile to review the advantages of each from the standpoint of how they can enable solar energy, in all of its manifest varieties, to become a major part of mankind's future.

(1) PARTICULAR VIRTUES OF VLS-CSP

The developed world has a highly-developed electric power infrastructure that is currently powered by heat from either fossil fuel or nuclear sources. CSP, because it too generates heat, fits more smoothly into that infrastructure than does any other renewable energy technology.

To this end, another virtue of VLS-CSP is that the heat it generates is easier to store for short periods (i.e. in order to overcome the effect of passing clouds, or for peak load shifting) than electricity. In fact, owing to the relatively large thermal mass of the HTF, CSP automatically contains a certain amount of built-in storage.

Yet a third virtue of VLS-CSP is that it can power conventional turbines. There are accordingly no problems having to do with the removal of harmonics of the sort that can arise in PV systems when a DC-to-AC inverter is less than perfect.

Last, but by no means least, the fact that the original Luz parabolic trough CSP systems are still operative provides so-called *bankability* to this kind of technology probably more readily than to any other.

Disadvantages of CSP at the present time include the need for cooling water (for efficient operation); an as-yet-incomplete knowledge about the efficacy of thermal storage – without which some form of fossil backup is necessary; and, of course, the need for a "sunny" climate.

(2) PARTICULAR VIRTUES OF VLS-PV

First and foremost, such plants can operate in cloudy climates, which are prevalent in most of the developed world.

For the developing part of the world, VLS-PV is most likely to be the technology of choice because of (a) the increasing local familiarity with PV technology owing to the ever more widespread employment of stand-alone-systems; and, related to this, (b) its relative simplicity when compared with CSP systems.

Inter-connected VLS-PV systems, when strung out across different time zones in the developing world, will go a long way to overcoming the storage problem that is inherent to PV technology.

In many regions, such systems will have an advantage over similarly located VLS-CSP systems by virtue of another feature of their simplicity: only the addition of power lines (without the need for additional supply concerns such as cooling water and perhaps backup fuel) being needed to create an electricity grid.

The downside of PV is its intrinsic lack of storage that, in contrast to inter-connected VLS-PV plants, will have to be developed if such plants are to perform as major power sources in isolation.

(3) PARTICULAR VIRTUES OF VLS-CPV

CPV has the advantage over PV that it requires typically three orders of magnitude less PV material. This has economic implications on both system cost and on the size needed for a fabrication plant (also three orders of magnitude smaller per MWp); on the continued availability of possibly rare materials; and on the energy payback time of the PV cells (approximately zero!).

Another advantage is that the actively-cooled variety of CPV plants can provide thermal energy for a variety of associated processes (desalination, air-conditioning, etc.). The prospect of employing the thermal output for additional electricity production is poor at the output temperatures presently allowed by CPV receivers. However, with improved materials, it is possible that VLS-CPV plants might in the future provide both PV and CSP power.

The downside of CPV is that, as in the case of PV, appropriate storage systems will have to be developed and, as in the case of CSP, "sunny" climates are necessary.

(4) SPACE CONSIDERATIONS

One of the main contributors to the cost of solar energy is the large size required for the plants. This constraint is brought about by the relative diluteness of solar energy compared with fossil-embedded energy. For example, in desert regions it takes approximately 1 year for 1 m^2 of ground to receive from the sun the energy equivalent of one barrel of oil. It is, accordingly, instructive to discuss the amount of land area required for the various solar technologies.

VLS-PV

The Springerville, Arizona, USA PV plant includes (*inter alia*) 26 subsystems of 135 kWp rating employing crystalline silicon modules. Each subsystem powers a 150 kW inverter. The entire crystalline silicon part of the system consequently has a power rating of 3,51 MWp. From the average reported performance during the system's first three years of operation [33] one can infer that the land requirement for this system is 346 kWp/ha, and that the land productivity was 0,6 GWh(AC)/year/ha. By up-scaling these numbers, we arrive at a land requirement of 28,9 km^2 for a 1 GWp VLS-PV plant, or 16,7 km^2 of land per TWh(AC)/year. It is noteworthy that the Springerville PV system is the only multi-MW solar power plant for which detailed performance data have been published. It therefore serves as a valuable benchmark for the estimated performances of other systems.

It is accordingly instructive to compare this *actual* system performance with the *estimated* performance figures for the plants in Table 4.1. The latter fall in the range 0,321 GWh/y/ha (Lieberose) to 0,625 GWh/y/ha (Ohotnikovo).

Now, under standard test conditions, the laboratory efficiency of the PV modules employed at Springerville is approximately 13%. Nowadays, the highest-efficiency commercial modules are close to 20% efficiency. Thus, a Springerville-like VLS-PV plant with 20% efficient modules need only occupy 65% of the space described above.

VLS-CSP

The 150 MW parabolic trough CSP system at Kramer Junction, California, USA [21], consists of five solar electricity generating stations (SEGS), each of power rating 30 MW. It occupies 2,53 km^2 of land and produces 2,35 GWh (AC)/year/MW (this is the solar part of its output). Thus the annual production is 1,39 GWh (AC) per ha of land, or 7,2 km^2 of land per TWh (AC). This may be compared with the *estimated* range of 0,687–0,987 GWh/y/ha for the newer systems in Table 4.10. On the other hand, the annual DNI at Kramer Junction is considerably higher than at any of the other sites in Table 4.10. Therefore the present estimates are probably reasonable assessments of what may be expected from these newer plants.

Now, the efficiency of the old Luz plants at Kramer Junction was estimated to be approximately 12%. It is likely that the more recent parabolic trough plants could have an annual average efficiency of 15%. Thus land requirements should be about 20% less than at Kramer Junction (for systems located in similarly sun-rich areas).

Tower-type CSP plants will probably require more land per TWh than the parabolic trough variety because of the larger spacing needed to prevent their optical units from shading one another. This assessment is borne out by the Table 4.10 land productivity estimates for such plants, which fall in the range 0,425–0,600 GWh/y/ha.

For dish-type CSP plants there are, as yet, no published data for actual plants. However, for equal conversion efficiencies the land requirements and productivity should be comparable to the CPV estimates discussed below.

VLS-CPV

Because no large CPV plants have yet been built, this paragraph represents an estimate of the land requirements. We have discussed a VLS-CPV plant previously [34]. There we estimated that at Sede Boqer in the Negev desert, a VLS-CPV plant would yield 1,67 GWh (AC) per ha of land, or would need to occupy 6 km^2 of land per 1 TWh (AC) output. That estimate was based, *inter alia*, on CPV cells having 32% efficiency under standard laboratory test conditions, which reduced to 25% AC system efficiency after including estimates for all system losses. For 25% efficient dish-powered CSP systems, the land requirements and productivity should be comparable to these CPV estimates.

However, current CPV cells have efficiencies well in excess of 40% [35]. This would raise the system efficiency to more than 31%. Consequently the previous land requirement estimates would be reduced by at least 20%.

VLS-WIND

Although wind power is not strictly within the mandate of our studies, it *is* a derivative of solar energy and it *is* a renewable energy resource. Moreover, perhaps more important is the fact that wind technology achieved engineering maturity and economy of scale far sooner than did solar technology. As a result, it is already widely used, and individual plants are already nearing the GW scale. It is accordingly of interest to compare the land requirements of large wind farms with those of VLS solar plants. On the one hand, individual turbines may have ratings close to 1 MW, and require no more land than the footprint of their support tower. However, for larger systems, enough space has to be allowed to prevent the individual turbines from shielding one another from the wind. Table 4.12 accordingly lists the relevant features of the 10 largest wind farms in the USA [36].

Table 4.12 Land requirements for the 10 largest wind farms in the USA

Plant name	Name plate rating [MW]	Land requirement [ha/MW]	Yield @ CF = 1/3 [GWh/y/km^2]
Shepherd's Flat	910	14,3	20,4
Roscoe	782	36,3	8,1
Horse Hollow	733	26,0	11,2
Whispering Willow	500	74,5	**3,9**
Maiden	494	10,7	**27,4**
Lone Star	400	20,3	14,4
Biglow Canyon I	400	25,3	11,5
Fowler Ridge	400	55,7	5,2
Twin Groves/Arrowsmith	396	22,5	13,0
Crystal Lake	350	48,1	6,1
Average		33,4	12,1

Table 4.12 shows that, under an assumed capacity factor of 1/3 (because no actual performance details have been documented), the annual yield varies from 3,9 GWh/km^2 of land (at Whispering Willow) to 27,4 GWh/km^2 (at Maiden).

We have previously emphasized that in order to compare the output of two different solar technologies the comparison must be made at the same place and during the same time period. One should therefore not rush to draw relative conclusions from previous descriptions. Nevertheless, from simulations performed by us at a number of suitable sites in the Negev desert, the relative ordering in terms of increasing land requirements is: VLS-CPV, VLS-CSP, VLS-PV and VLS-WIND. With expected improvements in technology, VLS-PV and VLS-CSP should become comparable in their land requirements, VLS-CPV should be better, and wind appears to be considerably worse, except at particularly suitable sites. Nevertheless, a redeeming feature of wind farms is the fact that they allow dual usage of land. For example, in arable terrain, such as occurs widely in northern Europe, widely spaced wind turbines are to be found on farmland shared with crops and animal grazing. In sunny deserts, however, such dual usage may not be feasible, and economic factors will probably dictate in favor of VLS-PV rather than VLS-WIND. Another important feature of wind power is that it is, in principle, available 24 hours per day. Therefore, in tandem with VLS-PV the two technologies overcome, to a certain extent, the storage problem [37].

4.2.6 Conclusions

In this section we have reviewed the salient features of the four principal renewable power sources for the future very large-scale electricity production: CSP, PV, CPV and wind. For each of them we have emphasized attractive features that the others lack and we have pointed out weaknesses that will almost certainly be solved by engineering ingenuity. Ultimately, therefore, we may envisage a worldwide grid of perhaps ultrahigh voltage DC power lines (or possibly ambient-temperature superconducting cables) fed by the various VLS technologies, each located in climatic regions appropriate to its most favorable economic performance. In summary, at the present stage of their respective developments, all VLS renewable energy technologies look as if they will be complementary rather than competitive with one another.

REFERENCES

1 Kurokawa, K. (2003) *Energy from the Desert: Feasibility of Very Large Scale Photovoltaic Power Generation (VLS-PV) Systems*, James & James, London.
2 Information from Resources Total System Co., Ltd.
3 http://www.pvresources.com
4 Information from Mizuho Information & Research Institute, Inc.
5 IEA PVPS (2001) *Trends in Photovoltaic Applications, Survey report of selected IEA countries between 1992 and 2010*, Report IEA-PVPS T1-20.
6 EPIA (2011) *Global Market Outlook for Photovoltaics until 2010.*
7 M. Vivar, R. Herrero, F. Martínez-Moreno, R. Moretón, I. Antón, and G. Sala (2008) *Proc. 23rd European Photovoltaic Solar Energy Conference*, 142.
8 Hornunga, T., A. Bachmaiera, P. Nitza, and A. Gombert, (2010) *Proc. CPV-6*, 81.
9 Araki, K. and M. Yamaguchi (2001) *Proc. 17th European Photovoltaic Solar Energy Conference*, 2187.
10 Araki, K., R. Leutz, M. Kondo, A. Akisawa, T. Kashiwagi and M. Yamaguchi (2 002) *Proc. 29th IEEE PVSC*, 2002, 1572.
11 Fellmeth, T., S. Fritz, M. Menkoe, F. Clement, D. Biro and R. Preu, *Proc. CPV-6*, 47.
12 Guter, W., M. Meusel, W. Köstler, R. Kern, G. Siefer, R. Kellenbenz and F. Dimroth (2010) *Proc. CPV-6*, 4.
13 Araki, K. (2007) *Proc. 4th ICSC*, 73.
14 Araki, K., T. Yano, and Y. Kuroda (2010) *Energy Express*, 26 April, Vol. 18, No. S1, OPTICS EXPRESS A53.
15 Nishioka, K., K. Nabemoto, Y. Sakurada, Y. Ota, K. Takami, H. Nagai and K. Araki (2011) "Anti-soiling layer coated on PMMA Fresnel lens for concentrator photovoltaic modules", *37th IEEE PVSC.*
16 Nishioka, K., K. Ikematsu, Y. Ota, and K. Araki (2010) "Influence of Sandblasting in Desert Area on Transmittance of Fresnel Lens for Concentrator Photovoltaic Modules", *CPV-7.*
17 http://www.nrel.gov/csp/solarpaces/by_project.cfm
18 Photon International, December 2011.
19 Survey by Mizuho Information & Research Institute, Inc.
20 Poullikkas, A., G. Kourtis and I. Hadjipaschalis (2010) "Parametric analysis for the installation of solar dish technologies in Mediterranean regions". *Renewable and Sustainable Energy Reviews* 14, 9: 2772–83.
21 Frier, S.D. and R.G. Cable (1999). "An overview and operation optimization of the Kramer Junction Solar Electric Generating Systems" *Proc. ISES World Congress*, Jerusalem, July 4–9, 1999, 1: 241–46.
22 IEA SolarPACES (2011) http://www.solarpaces.org/News/Projects/projects.htm
23 Bartel, J.J. and P.E. Skvarna (1984) "10-MWe Solar Thermal Central Receiver Pilot Plant". *Transactions ASME J. Solar Energy Engineering* 106: 50–58.
24 Borgese, D., G. Dinelli, J.J. Faure, J. Gretz and G. Schober, (1984) "Eurelios, The 1-MW(el) Heliocentric Power Plant of the European Community Program". *Transactions ASME J. Solar Energy Engineering* 106: 66–77.
25 Munoz Torralbo, A., C. Hernandez Gonzalvez, C. Ortiz Roses, J. Avellaner Lacal and F. Sanchez (1984) "A Spanish 'Power Tower' Solar System: Project CESA-1". *Transactions ASME J. Solar Energy Engineering* 106: 78–82.
26 Drouot, L.P. and M.J. Hillairet (1984) "The Themis Program and the 2500-kW Themis Solar Power Station at Targasonne". *Transactions ASME J. Solar Energy Engineering* 106: 83–89.
27 Hirono, T. and T. Horigome (1984) "A 1-MWe Central Receiver Type Solar Thermal Electric Power Pilot Plant". *Transactions ASME J. Solar Energy Engineering* 106: 90–97 .
28 King, D.L., W.E. Boyson and J.A. Kratochvill (2004) "Photovoltaic Array Performance Model" (*Sandia Report* SAND2004-3535) 14.
29 US National Solar Radiation Data Base: (1991–2005) Update; Typical Meteorological Year 3. <http://rredc.nrel.gov/solar/old_data/nsrdb/1991-2005/tmy3/by_state_and_city.html>
30 Faiman, D., D. Bokobza, A. Dafni, S. Kabalo, L. Levy, V. Melnichak and A. Razon (2009) "Rapid Outdoor Quantification of Degradation For c-Si Photovoltaic Modules" 24th European Photovoltaic Solar Energy Conference, 21–25 September, Hamburg, Germany. 3290–3.
31 Kurokawa, K. (ed.) (2003) *Energy From the Desert: Feasibility of Very Large Scale Photovoltaic Power Generation (VLS-PV) Systems.* (James & James, London, UK).
32 Salim, A.A., F.S. Huraib, B. Khoshaum, N.N. Eugenio and N.R. Rao (1985) "Four-year performance summary of 350 kW concentrating photovoltaic power sustem". *Proc. 18th IEEE PVSEC.*
33 Moore, L.M. and H.N. Post (2008). "Five years of operating experience at a large, utility-scale photovoltaic generating plant" *Prog. Photovolt: Res. Appl.* 16: 249–59.
34 Kurokawa, K., K. Komoto, P. van der Vleuten and D. Faiman (eds) (2007) *Energy From the Desert: Practical Proposals for Very Large Scale Photovoltaic Systems.* (Earthscan, London, UK).
35 Wojtczuk, S., P. Chiu, X. Zhang, D. Derkacs, C. Harris, D. Pulver and M. Timmons (2010) "InGaP/GaAs/InGaAs 41% concentrator cells using bi-facial epigrowth. *Proc. IEEE PVSEC, Hawaii, 2010* 1259–64.
36 Denholm, P., M. Hand, M. Jackson and S. Ong (2009.)"Land-use requirements of modern wind power plants in the United States. *Technical Report* NREL/TP-6A2-45834.
38 Solomon, A.A., D. Faiman and G. Meron (2010) "Grid matching of large-scale wind energy conversion systems, alone and in tandem with large-scale photovoltaic systems: An Israeli case study". *Energy Policy* 38: 7070–81.

CHAPTER FIVE

Environmental aspects of VLS-PV

5.1 INTRODUCTION

According to reports by the Intergovernmental Panel on Climate Change (IPCC), global warming brings a variety of adverse effects including record high temperatures, flooding due to increased rainfall, expansion of arid areas and a higher risk of drought, and stronger typhoons. Accordingly, it is necessary to mitigate emissions of greenhouse gases (GHG; CO_2, CH_4, N_2O and others), which cause global warming. However, as GHG are invisible, the amounts in which they are released are generally unclear.

Life-cycle assessment (LCA) – the main topic of this chapter – is necessary in calculating emissions during the entire life-cycle of a product or a system. Although it is not ideally suited for evaluation on a macro scale (investigation from a global viewpoint, for example), it is highly appropriate for micro-scale analysis (e.g., consideration of products and generation systems). The results of LCA can clarify major emissions, thereby enabling consideration of measures for their reduction.

This chapter discusses LCA in relation to photovoltaic (PV) systems. First, an overview is given and the scheme of LCA is described, and evaluation indices, LCA limitations, inventory analysis, impact assessment and interpretation are outlined. Then, guidelines for LCA in regard to PV systems are discussed with a focus on important matters for related evaluation. Next, the collection of LCA data is outlined, and finally, calculations from example papers and the VLS-PV system are introduced in relation to LCA for PV modules, PV systems and balance of system (BOS) technologies.

In addition, Ecological Footprint (EF) is discussed. EF is a measure of the biological productive area that humanity uses to produce resources and to absorb waste. EF analysis compares the Ecological Footprint with Biocapacity (BC), that is the Earth's available biological capacity to produce useful biological materials and to absorb waste.

5.2 ENVIRONMENTAL ISSUES

Countries are promoting PV installation and production to mitigate environmental issues, especially global warming. It is caused by greenhouse gases (GHG), that is CO_2, CH_4, N_2O, HFCs, PFCs, SF_6 and so on. Because it is not easy to quantify emissions and predict amounts, many researchers are working on this. The IPCC has already introduced informative data summarizing research results. Figure 5.1 shows concentrations of greenhouse gases from year 0 to 2005. After the industrial revolution, CO_2, CH_4, and N_2O increased drastically. Figure 5.2 shows global temperatures. The temperature is also increasing along with the emissions. The GHG is molecular structure resonates light from the sun and the Earth. Figure 5.3 shows the radiative forcing of climate between 1750 and 2005. Usually, aerosol is not good for human health, but in view of global warming, it has good effect for cooling the Earth. However, the global warming effect of CO_2 and CH_4 is larger than it. Therefore, the total radiative forcing is positive, about 1,5 W/m^2.

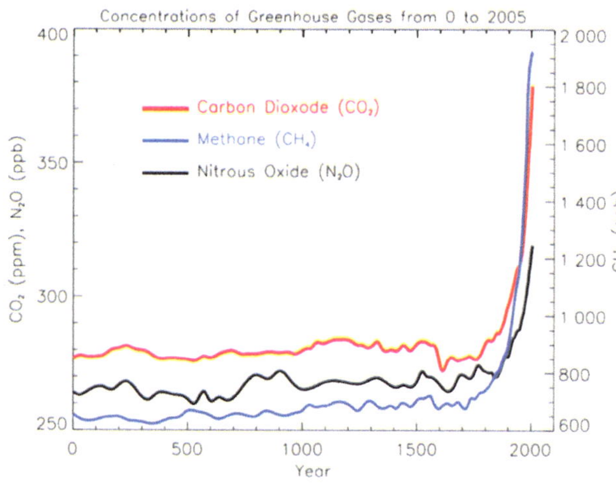

Figure 5.1 Concentrations of greenhouse gases, CO_2, CH_4, and N_2O [1]

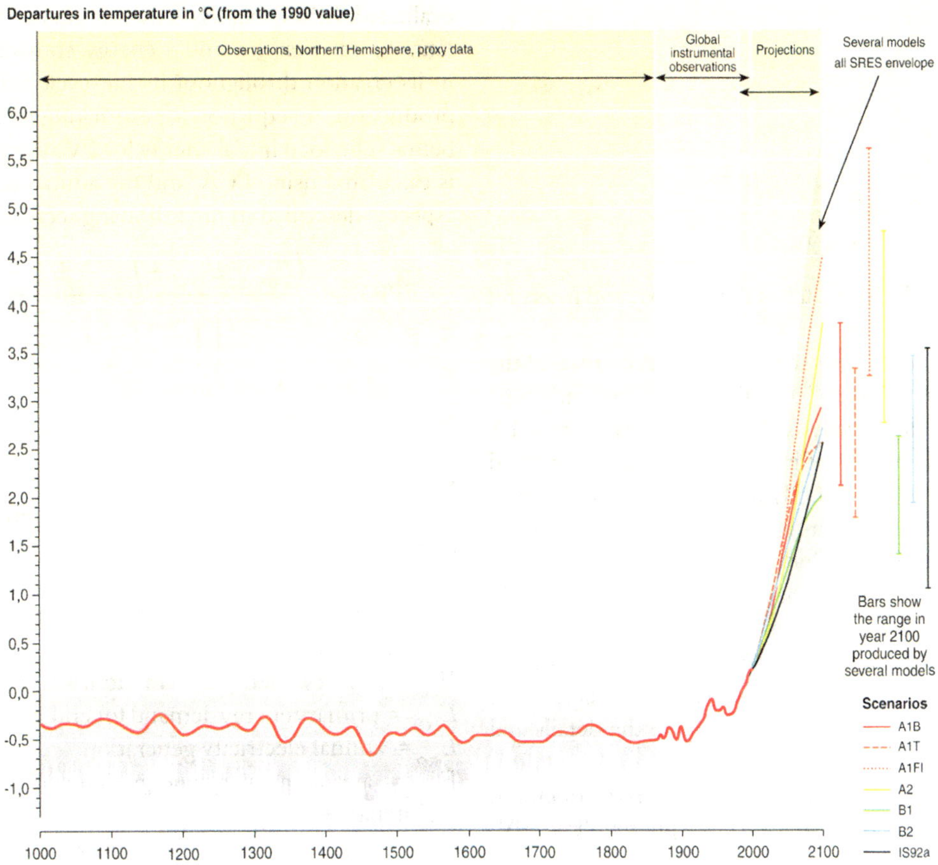

Figure 5.2 Global temperature from 1000 to 2100 [2]

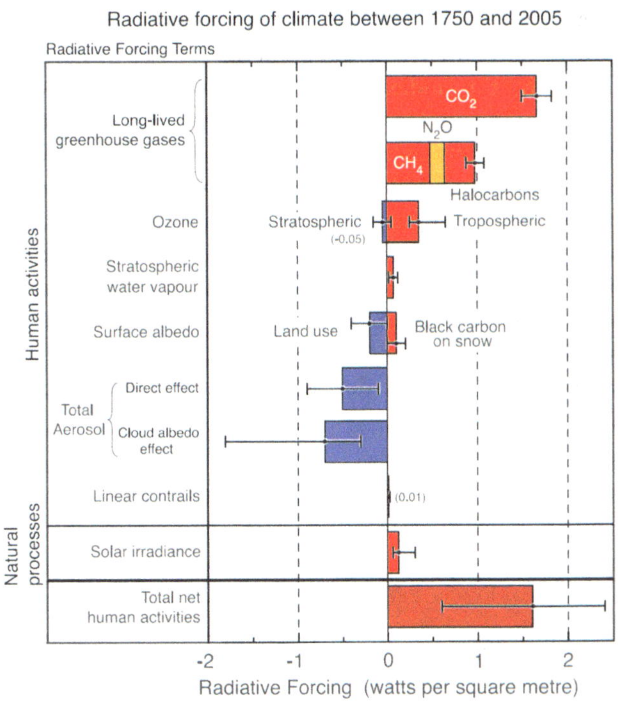

Figure 5.3 Radiative forcing of climate between 1750 and 2005 [1]

The climate change effect has been evaluated by an economics expert. Lord Nicholas Stern, former vice-president of World Bank, produced the Stern Review on the Economics of Climate Change [3] which was prepared for the Prime Minister and Chancellor in the UK. The report is about the economic impact caused by climate change and the cost to stabilize GHG. In the case of BAU (Business as Usual), it is expected that temperature changes may exceed 2–3 °C by the end of this century, and 5–6 °C in the next century. This will cost 5%/year of GDP at least, and 20%/year may be needed. However, if GHG concentration is stabilized in the range 450–550 ppm, the cost will be decreased. The cost is expected to be about 1%/year of GDP by 2050.

5.3 LCA FOR VLS-PV

Life-cycle assessment is an approach to environmental management system implementation involving the quantitative evaluation of a product's overall environmental impact. Energy requirements and CO_2 emissions throughout the whole life-cycle of the product (including its manufacture, transport, use, disposal, etc.) are estimated in order to enable such evaluation, and the results can be used for related environmental assessment. However, since life-cycle is related to a broad range of variables and is complicated, it is difficult to comprehend the exact significance of the results. Accordingly, it is very important to set a purpose for the evaluation. An LCA operator should

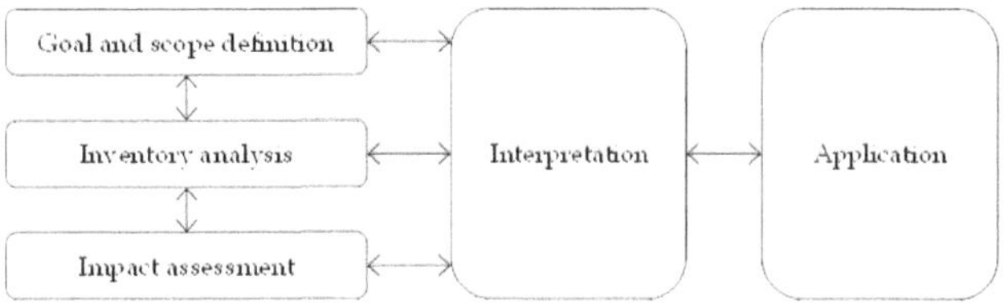

Figure 5.4 Scheme of LCA

implement research that matches the purpose and interpret the outcomes appropriately.

The research and analysis scheme for LCA consists of the four stages shown in Figure 5.4 as follows: 1. goal and scope definition; 2. inventory analysis; 3. impact assessment; and 4. interpretation. The results of inventory analysis are referred to as life-cycle inventory (LCI) data. LCA is applicable to any product or service, but its results are affected by objects, assumptions, data availability and accuracy. Hence, it is impossible to generalize the method in a very clear way. As a result, LCA operators and users must properly understand the limitations of LCA and the assumptions that can be drawn from its results. The essentials of LCA are standardized in ISO 14040 and ISO 14044, which stipulate the details and basic points of the approach.

In any LCA study, the purpose depends on the operator. However, when the operator evaluates a photovoltaic (PV) system, the main research point or characteristic relates to energy generation. This is a significant difference between PV systems and other products. When a building developer discusses new energy supply systems (e.g., in relation to buildings with low carbon emissions and high energy efficiency), LCA can highlight the potential of PV systems and useful materials. This is expected to provide two advantages, the first of which is PV system optimization. When a developer studies the installation of a PV system, the environment of the installation site must be considered. To ensure optimization, a variety of variables (e.g., cost and CO_2 emissions) are discussed. If LCA is used, the system can be optimized from an environmental viewpoint.

The second advantage is comparability. When comparing energy generation technologies (e.g., when researching the possible installation of a PV system as a supply of alternative energy as opposed to other generation systems, or when installing energy supply systems based on multiple generation technologies), the evaluation methods and rules applied must be uniform. In such cases, LCA can provide quantitative results, thereby enabling comparison of each technology on an equal footing.

5.3.1 Scheme of VLS-PV's LCA

(1) EVALUATION INDICES

In an LCA study, evaluation indices are decided based on the purpose at hand. As PV systems generate electricity, the new index of energy payback time (EPT or EPBT) can be evaluated. EPBT expresses the number of years the system takes to recover the initial energy consumption involved in its creation throughout its life-cycle via its own energy production. An equation for estimating EPBT [4] is shown below. The total initial energy for PV systems in Eq. 5.1 [4] is calculated using LCA, and the annual power generation aspect is described in the following sections.

$$\mathrm{EPBT} = \frac{\left(E_{mat} + E_{manuf} + E_{trans} + E_{inst} + E_{EOL}\right)}{\left(\left(\dfrac{E_{agen}}{n_G}\right) - E_{O\&M}\right)}$$

(Eq. 5.1)

where

E_{mat} = primary energy demand to produce materials comprising PV system

E_{manuf}= primary energy demand to manufacture PV system

E_{trans}= primary energy demand to transport materials used during the life-cycle

E_{inst} = primary energy demand to install the system

E_{EOL} = primary energy demand for end-of-life management

E_{agen} = annual electricity generation

$E_{O\&M}$= annual primary energy demand for operation and maintenance

n_G = grid efficiency, the average primary energy to electricity conversion efficiency at the demand side.

The CO_2 emission rate is a useful index for determining how effective a PV system is in terms of global warming. Generally, this index is used for comparison between generation technologies. As a PV system does not operate in the same way as a tree, there is no payback of CO_2 emissions as such. However, some comparisons between PV systems and other fossil fuel generation technologies have used CO_2 payback time as a metric. In these studies, PV systems were viewed as an alternative to fossil fuels and as offering a corresponding reduction in CO_2 emissions, which allowed calculation of the CO_2 payback time. However, this chapter does not deal with the concept of CO_2 payback time.

$$g_{co2} = \frac{G_{input}}{E_{gen} \cdot L_{PV}} = \frac{G_{man} + G_{trans} + G_{inst} + G_{use} + G_{decomm}}{E_{gen} \cdot L_{PV}}$$

(Eq. 5.2)

where

g_{co2} = life-cycle CO_2 emission intensity [g-C/kWh]

G_{input} = CO_2 emissions during system life-cycle [g-C]

G_{man}= CO_2 emissions due to manufacture of each system component [g-C]

G_{trans}= CO_2 emissions for transportation of the system components [g-C]

G_{inst}= CO_2 emissions for installation of the system components [g-C]

G_{use} = CO_2 emissions during system operation and maintenance [g-C]

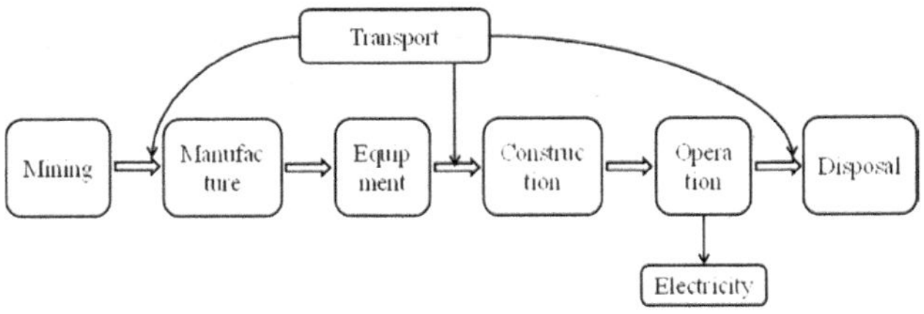

Figure 5.5 Boundaries of LCA for a PV system

G_{decomm} = CO_2 emissions during system decommissioning
E_{gen} = annual AC power generation [kWh]
L_{PV} = lifetime of the PV system [years].

(2) BOUNDARIES OF LCA

As using different boundaries obviously creates different results, defining and making boundaries known is important. Figure 5.5 shows typical boundaries for LCA of a PV system from the mining of its raw materials to its final disposal. The next consideration is the boundary for each stage. Boundaries involve products and services related to the item's life-cycle. As the details vary in each case, it is important to fit the definition to the purpose of the product. For example, factors including the type of PV module used, efficiency, array, foundation, installation method and operation method should be identified to build a suitable system. Indirect factors should also be considered as much as possible.

(3) INVENTORY ANALYSIS

Inventory analysis is performed to evaluate the amounts of environment-influencing materials consumed or produced during the object's life-cycle. It involves pinpointing the processes involved in the life-cycle and evaluating them quantitatively, then identifying all related environment-influencing materials. The object's data are subsequently evaluated as a whole. However, as it is difficult to collect all information on related processes, the results may have simplified or missing data. Accordingly, it is important to understand the applicable boundaries, the quality of data and the assumptions involved in calculation when performing an LCA study.

(4) IMPACT ASSESSMENT

Impact assessment consists of three processes; classification, characterization and weighting. In classification, environment-influencing materials are categorized in terms of related influence events. For example, CO_2 will be categorized as producing global warming, sulfur oxide (SO_x) will be categorized as producing acid rain, affecting public health and so on. Impact potential is calculated based on inventory analysis. In research on energy payback time, the amount of energy consumed is calculated and classified. In research on CO_2 emission rates, emissions are calculated and classified into a suitable category.

In characterization, amounts of output materials are calculated with characterization factors to produce impact category indicators. In particular, input energy is calculated in terms of electricity or calorific value. Greenhouse gas emissions are calculated in terms of CO_2 equivalents (CO_{2eq}) using global warming potential (GWP) figures as defined by the IPCC. For example, in the case of a power conditioning system (PCS), the weight of each material would be determined as relevant data, and the energy requirements/CO_2 emissions of the production process would be ascertained. Then, input and output data would be calculated using inventory analysis, and the results indicating the energy requirement and CO_{2eq} values would be calculated to provide the impact category indicator.

Weighting is not stipulated in international standardization because it is considered difficult to form a single indicator for the different areas of global warming potential and ozone depletion potential. However, a simple comparison method is still needed. The two possible methods for this are damage evaluation and environmental category weighting by estimation. Whichever is used, the weighting must be transparent.

(5) INTERPRETATION

The results of LCA may depend on research boundaries and approaches to inventory analysis. Accordingly, in related interpretation, the effects of operation methods should be discussed. Usually, the data used in LCA include estimates and referred information. For this reason, if the data affect the results significantly, sensitivity analysis should be included.

5.3.2 LCA guidelines for PV systems

A set of LCA guidelines for PV systems titled "Methodology Guidelines on Life-cycle Assessment of Photovoltaic Electricity" [4] was published by the IEA-PVPS Task 12. This is an informative and useful resource for LCA operators of PV systems that helps with evaluation difficulties. This section describes important considerations covered in the guidelines for evaluating PV systems.

(1) LIFETIME

Lifetime is difficult to quantify because most PV systems introduced are still in operation or were produced in the early stages of the technology's development. However, many researchers have studied the life expectancy of PV systems. The guidelines follow the results of papers outlining such research, and set the lifetimes shown in Table 5.1.

Table 5.1 List of lifetimes (data from IEA PVPS Task 12)

PV modules	30 years for mature module technologies
Inverters	15 years for small plants or residential PV systems; 30 years with 10% part replacement every 10 years for large plants
Structure	30 years for rooftop- and facade-mounted units, and between 30 to 60 years for ground-mounted installations on metal supports. Sensitivity analysis should be performed
Cabling	30 years

(2) IRRADIATION DATA

Irradiation data depend on the location and tilt angle of PV modules. Accordingly, the two main recommendations given are analysis of industry averages / best-case systems and analysis of average systems installed on the grid network.

(3) PERFORMANCE RATIO

The performance ratio (PR) depends on the type of installation. In general, the value rises with lower temperatures and monitoring of PV systems for early detection of defects. IEA PVPS Task 12's recommendation [4] is 75% for rooftop-mounted and 80% for ground-mounted latitude-optimal installations. Alternatively, actual performance data can be used where available.

(4) DEGRADATION

Most PV modules degrade year by year to an extent that is still an active topic of research, especially for thin-film PV systems. However, 0,5% per year seems to be a typical number for crystalline silicon PV modules. Accordingly, the guidelines set the degradation rate for flat-plate PV modules. Mature module technologies are considered to maintain 80% of their initial efficiency at the end of the 30-year lifetime under the assumption of linear degradation during this time.

5.3.3 Collection of LCA data

LCA data are usually categorized into foreground and background types. Foreground data relates to the materials from which products are made, such as arrays, foundations and cables. Background data relate to materials that are indirectly involved, such as array steel, foundation cement and cable copper. Foreground data are usually provided by producers, while database values are used for background data due to the difficulty of collecting such information. Such databases summarize the input and output data for various materials. For example, LCA data for galvanized steel in an LCA database would show that the unit is 1 kg; the input data are the weight of coal, limestone, iron ore, natural gas, crude oil and so on used in production; and the output data are the weight of related emissions of CO_2, nitrogen oxide (NO_x), SO_x, biochemical oxygen demand (BOD) and so on.

These data can be obtained from an LCA database or by using LCA software. Ecoinvent (Switzerland) and the Life-cycle Assessment Society of Japan (JLCA) have well-known LCA databases. The Ecoinvent resource is an inventory database with more than 4 000 entries developed from research for the companies' environment reports, summaries of references and questionnaire surveys. The JLCA database includes inventory data, impact category indicators and reference data, which are based on a five-year project implemented by the New Energy and Industrial Technology Development Organization (NEDO). Although inventory data are limited to about 280 entries, these are typical data obtained in collaboration with industry associations, thus making them highly reliable. There are also approximately 300 reference data entries made by industry associations themselves.

Calculation for small systems or products can be performed manually, but this is difficult for large systems. Accordingly, LCA software is produced to support such operations. As this type of software generally already includes LCA data, the operator does not need to input individual values. SimaPro developed by PRé Consultants, GaBi Software by PE International and MiLCA by the Japan Environmental Management Association for Industry are examples of such programs.

Irradiation data are also required for LCA calculations in regard to PV systems. If it is possible to use actual long-term generation data for such systems, there is no need for irradiation data. However, environmental reporting is needed before a PV system is installed. If irradiation data are available, PV system generation can be estimated and pre-LCA can be evaluated. Meteonorm developed by Meteotest (Switzerland) is a well-known irradiation database. It also provides a function to calculate irradiation in relation to tilted planes, thereby eliminating the need to use complex metrological models. A further resource is the System Advisor Model (SAM) energy analysis software developed by the National Renewable Energy Laboratory (NREL, USA), which also includes a function for calculating PV system generation. METPV and MONSOLA developed by NEDO are other irradiation databases with data related exclusively to Japan. Table 5.2 gives a summary of these databases.

5.3.4 LCA calculations from example papers

This section introduces four interesting papers on PV system LCA and their results. The studies in question addressed PV modules, rooftop systems, balance of system (BOS) technology and large PV systems.

Table 5.2 List of databases

LCA databases	Ecoinvent, JLCA
LCA software	SimaPro, GaBi, MiLCA
Irradiation databases	Meteonorm, System Advisor Model, METPV

(1) LCA STUDY ON PV MODULES

This paper [5] (see Table 5.3) describes PV module LCA with a focus on emissions, including not only greenhouse gases (GHG) but also NO_x, SO_x, cadmium (Cd) and heavy metals. The results for GHG are summarized, and heavy metals form the main topic of the paper.

The use of cadmium telluride (CdTe) PV modules is growing rapidly because of their high efficiency and low price. Cadmium can have adverse health effects, and there is now a tide of concern regarding the safety of CdTe PV modules. However, this paper indicates that emissions from such modules are much lower than those of oil power plants on a like-for-like basis. LCA is a good method for highlighting this type of finding.

Data on GHG, NO_x and SO_x are summarized in the paper assuming three cases: Case 1: current electricity mixture for silicon (Si) production from the CrystalClear project and the Ecoinvent database; Case 2: combination of the Co-ordination of Transmission of Electricity (UCTE) grid mixture and the Ecoinvent database; and Case 3: the US grid mixture and the Franklin database. In Case 1, GHG emissions of Si modules for the year 2004 are 30–45 g CO_{2eq}/kWh, and the EPBT is 1,7–2,7 years. These figures are for rooftop installation. The GHG emissions and EPBT of a CdTe frame without PV modules are 24 g CO_{2eq}/kWh and 1.1 years for ground-mounted installations. CdTe has about half the GHG emissions of crystalline Si. A summary is shown in Table 5.4.

Life-cycle atmospheric Cd emissions for PV systems from electricity and fuel consumption are also evaluated for ribbon-Si, mc-Si, mono-Si, CdTe, hard coal, lignite, natural gas, oil, nuclear, hydro, and UCTE average, and the results are given as 0,8, 0,9, 0,9, 0,3, 3,1, 6,2, 0,2, 43,3, 0,5, 0,03 and 4,1 g/GWh (10^9Wh), respectively. Compared with the emissions from oil at 43,3 g/GWh, PV system emissions are much lower.

Table 5.3 Summary of the paper [5]

Paper title	Emissions from photovoltaic life cycles
Author(s)	Fthenakis, V.M., Kim, H.C. and Alsema, E.
Journal	*Environmental Science & Technology* 2008; 42 (6): 2168–2174
Irradiation	1 700 kWh/m²/year, 1 800 kWh/m²/year
PV type	ribbon-Si, mc-Si, sc-Si, CdTe
System configuration	0,75–0,8 performance ratio, rooftop- and ground-mounted
Lifetime	30 years
Results	20–55 g CO_{2eq}/kWh, 40–190 mg NO_x/kWh, 60–380 mg SO_x/kWh (reading from figure)
Year	2006

Table 5.4 GHG emissions and EPBT

PV type	Assumption	GHG emissions	EPBT
Si modules	Rooftop-mounted, 0,75 PR, 1 700 kWh/m²/yr	30–45 g CO_{2eq}/kWh	1,7–2,7 years
CdTe	Ground-mounted, 0,8 PR, 1 800 kWh/m²/yr 30-year lifetime	24 g CO_{2eq}/ kWh	1,1 years

Table 5.5 Atmospheric Cd emissions

PV type and fuel type	Atmospheric Cd emissions
Ribbon-Si	0,8 g/GWh
mc-Si	0,9 g/GWh
Mono-Si	0,9 g/GWh
CdTe	0,3 g/GWh
Hard coal	3,1 g/GWh
Lignite	6,2 g/GWh
Natural gas	0,2 g/GWh
Oil	43,3 g/GWh
Nuclear	0,5 g/GWh
Hydro	0,03 g/GWh
UCTE average	4,1 g/GWh

Atmospheric emissions of arsenic (As), cadmium (Cd), chromium (Cr), lead (Pb), mercury (Hg) and nickel (Ni) are also evaluated, with the results of CdTE shown in Table 5.5. The CdTe PV module shows the highest level of performance, and replacing the regular grid mix with it affords significant potential to reduce these atmospheric heavy-metal emissions.

(2) LCA STUDY ON BOS IN A 3,5 MW PV SYSTEM (USA)

This paper [6] (see Table 5.6) was published in 2006. At the time, there were not many large PV systems such as those operating at over a megawatt. Accordingly, this study provided worthwhile LCA results. Even now, it is difficult to find such a detailed LCA study focusing on BOS. The investigation did not include PV modules.

The 3,5 MW Tucson Electric Power (TEP) Springerville PV plant is located in eastern Arizona, USA. The high elevation of this site and its low-temperature environment enables higher efficiency for its PV modules, which are the mc-Si type. PV support structures are anchored to the ground with 30 cm nails, thereby eliminating the need for concrete foundations. The structures' design wind speed

Table 5.6 Summary of the paper [6]

Paper title	Energy payback and life-cycle CO_2 emissions of the BOS in an optimized 3.5 MW PV installation
Author(s)	Mason, J. E., Fthenakis, V. M., Hansen, T. and Kim, H.C.
Journal	*Progress in Photovoltaics*, 2006. 14 (2): 179–190
Location/ country	Springerville, AZ/USA
Irradiation	1 725 kWh/kW (corresponding to PR of 82%; actual performance data used for LCA), approx. 2 100 kWh/m²/yr (average)
PV capacity/ PV type	3,5 MW/mc-Si
System configuration	Ground-mounted fixed flat-plate system
Lifetime	PV metal support structure: 60 years; inverters and transformers: 30 years (parts: 10 years)
Results	BOS: 542 MJ/m², 29 kg CO_{2eq}/m², 0,21 years of EPBT
Year	2006

is 160 km/h. The annual average AC electricity output in 2004 was 1 730 kWh/kW, corresponding to a performance ratio of 82%. The modules are large (2.4 m²) and each weighs 46,6 kg (including 5,44 kg of Al frame), and have a rated efficiency of 12,2%.

The life expectancy of the PV metal support structures is assumed to be 60 years. Inverters and transformers are considered to have a life of 30 years, but parts amounting to 10% of the total mass must be replaced every 10 years.

The total primary energy in the BOS life-cycle is 542 MJ/m². Using the average US energy conversion efficiency of 33% produces an EPBT of 0,21. Under the average irradiation of the US (1 800 kWh/m²/year), the EPBT becomes 0,37 years.

(3) LCA STUDY ON THE 2 MW HOKUTO MEGA-SOLAR PLANT (JAPAN)

This paper [7] (see Table 5.7 and Figure 5.6) describes a comparative study on LCA for 20 different types of PV systems. Usually, comparative PV studies use different types of PV modules, but each module has only one or two pieces. However, in this PV project, each PV module is about 10 kW, making it necessary to evaluate the array size rather than the module size. On the other hand, the LCA data used were not for the PV modules themselves; they were from the NEDO PV project, which researched LCA for six types of PV modules including mono-crystalline silicon (sc-Si), amorphous silicon (a-Si)/sc-Si, multi-crystalline silicon (mc-Si), a-Si, micro-crystalline silicon (μc-Si)/a-Si and copper indium selenium (CIS). The data are listed in Table 5.8.

Table 5.7 Summary of the paper [7]

Paper title	A comparative study on life cycle analysis of 20 different PV modules installed at the Hokuto mega-solar plant
Author(s)	Ito, M., Kudo, M., Nagura, M., and Kurokawa, K.
Journal	*Progress in Photovoltaics: Research and Applications*, Volume 19, Issue 3
Location/ country	Hokuto City, Japan
Irradiation	1 725 kWh/m²/year at a 30-degree tilt angle
PV capacity/PV type	600 kW/mc-Si, sc-Si, a-Si/sc-Si, thin-film Si, CIS, μc-Si/a-Si
System configuration	Ground-mounted fixed flat-plate system
Lifetime	30 years; inverters: 15 years
Year	2011

Figure 5.6 The 2 MW Hokuto mega-solar plant

The installed PV modules are shown in Table 5.9. Six crystalline Si, one a-Si/sc-Si, seven mc-si, one a-Si, two μc-Si/a-Si, and two CIS PV modules were installed and evaluated. The table shows that mc-Si PV modules have average or higher efficiency, while sc-Si PV modules are lower than average. This should be noted and understood, as pointed out in the paper.

The results showed an energy requirement ranging from 19 to 48 GJ/kW and an energy payback time of between 1,4 and 3,8 years. CO_2 emissions were between 1,3 and 2,7 t CO_2/kW, and CO_2 emission rates ranged from 31 to 67 g CO_2/kWh. The mc-Si and CIS types showed good results. In particular, the CIS module generated more electricity than expected with catalogue efficiency. The sc-Si PV module did not produce good results because, considering the energy requirement, installed sc-Si PV modules do not have high efficiency.

Table 5.8 LCA data from NEDO, Japan, on PV modules

	Module efficiency in reference	Energy requirement	CO_2 emissions
PV module			
sc-Si	14,3%	3 986 MJ/m^2	193,5 kg CO_2/m^2
a-Si/sc-Si	16,6%	3 679 MJ/m^2	178,0 kg CO_2/m^2
mc-Si	13,9%	2 737 MJ/m^2	135,2 kg CO_2/m^2
a-Si (in 2000)	–	1 202 MJ/m^2	54,3 kg CO_2/m^2
a-Si/µc-Si	8,6%	1 210 MJ/m^2	67,8 kg CO_2/m^2
CIS	10,1%	1 105 MJ/m^2	67,5 kg CO_2/m^2
10 kW inverter		0,57 GJ/kW	43 kg CO_2/kW
Cable, conduit		1 068 GJ/600 kW	62,0 t CO_2/600 kW
Array (galvanized steel)		22,5 GJ/t	1,91 t CO_2/t

Table 5.9 List of installed PV modules

Type	Nominal power [W]	Module efficiency [%]	Capacity [kW]
sc-Si	84	13,2	30
a-Si/sc-Si	186	15,9	30
sc-Si	160	12,6	10
sc-Si	160	12,6	10
sc-Si	150	11,8	10
sc-Si	200	12,0	30
sc-Si	173	12,0	30
mc-Si	167	12,6	30
mc-Si	179	14,0	100
mc-Si	167	13,2	30
mc-Si	180	12,3	10
mc-Si	190	13,0	10
mc-Si	240	12,4	30
mc-Si	170	13,5	10
a-Si	60	6,1	30
µc-Si/a-Si	110	8,8	10
µc-Si/a-Si	130	8,3	10
CIS	70	8,8	30
CIS	125	11,2	3

5.4 LIFE-CYCLE ANALYSIS OF VARIOUS KINDS OF VLS-PV SYSTEMS FOR THE DESERT

The Life-Cycle Analysis (LCA) methodology is an appropriate measure to evaluate the potential of VLS-PV systems in detail, because the purpose of this methodology is to evaluate its input and output from cradle to grave.

This topic has been discussed in the books, *Energy from the Desert: Feasibility of Very Large Scale Photovoltaic Power Generation (VLS-PV) Systems* [8], *Energy from the Desert: Practical Proposals for Very Large Scale Photovoltaic Systems* [9], and *Energy from the Desert: Very Large Scale Photovoltaic Systems: Socio-economic, Financial, Technical and Environmental Aspects* [10]. The first book discussed a feasibility study of the LCA of the VLS-PV system in Gobi desert. The second book discussed LCA and economic analyses of six world deserts. The third book assumed a suitable size, 1 GW, for VLS-PV considering economic efficiency. An array structure with present technology and waste management was discussed. In this chapter, assumptions of LCA follow LCA guidelines [4] written by IEA PVPS Task12 team. A study with a CdTe PV module is under development. It will be available soon.

5.4.1 Assumptions

(1) LCA SCHEME

The LCA of VLS-PV systems are calculated by following six steps. The first is designing the array and foundation with geographic data and equipment, the second is estimation of in-plain irradiation because it is not usually measured in desert areas, the next is calculation of output in desert conditions, calculation of transmission loss for 100 km lines considering PV output curve [11], then the main part, LCA, and sensitivity analysis.

1 Design of array and foundation
2 Estimation of in-plain irradiation
3 Calculation of output
4 Transmission loss
5 LCA
6 Sensitivity analysis

(2) ENERGY PAY-BACK TIME AND CO_2 EMISSIONS RATE

In this study, Energy Payback Time (EPBT), and the CO_2 emission rate of the VLS-PV systems are evaluated with the LCA method. These indices are defined by the following equations. Details are found in Section 5.3.1.

$$\text{EPT (Year)} = \frac{\text{Total primary energy requirement of the PV system throughout its life-cycle [J]}}{\text{Annual power generation [kWh/year]}} \quad \text{(Eq. 5.3)}$$

$$CO_2 \text{ Emission rate (g } CO_2/\text{kWh)} = \frac{\text{Total } CO_2 \text{ emission of its life-cycle [g } CO_2]}{\text{Annual power generation [kWh/year]} \times \text{Lifetime [year]}} \quad \text{(Eq. 5.4)}$$

(3) CASES ASSUMED

Six types of PV modules were evaluated in this study. They are single crystalline silicon (sc-Si), multi crystalline silicon (mc-Si), amorphous silicon/single crystalline silicon hetero junction (a-Si/sc-Si), amorphous silicon/micro crystalline thin-film silicon (thin-film Si), and Copper Indium di-Selenide (CIS). The data were based on the NEDO database [12, 13] averaged with results from Japanese PV manufacturers. The data are shown in Table 5.10.

(4) GEOGRAPHICAL DATA

The site of the assumed installation is in Huhhot in the Gobi Desert in China's Inner Mongolia. The assumed irradiation is 1 702 kWh·m^{-2}yr.$^{-1}$, in-plain irradiation at 30 degree tilt angle is 2 017 kWh·m^{-2}yr.$^{-1}$. The annual average ambient temperature is 5,8 °C.

(5) VLS-PV DESIGN AND CONFIGURATION

The 100 MW system consists of 200 sets of 500 kW array units. In the mc-Si case, the 100 MW system requires 576 000 modules, 10 000 tonnes of steel, and 140 000 tonnes of concrete. The length of cable reaches 1 400 km. The system size is 1 km in length and 2 km in width. Buffer plants are set beside the system as windbreakers and for the environment. If it is 1 km in width, then the total 1 GW PV system will be 15 km from east to west and 2 km from north to south.

Table 5.11 shows the balance of systems (BOS) for one GW VLS-PV system. System life-time is assumed to be 30 years, and only inverters are assumed for 30 years with 10% replacement every 10 years.

(6) ARRAY STRUCTURES

The array structures were assumed to be conventional structures with 30 degree tilted steel array and concrete foundations.

(7) TRANSPORTATION

Most of equipment is manufactured in Japan, and transported by cargo ship. But the foundation, array, and transmission tower are manufactured in China. The equipment will be transported an estimated 600 km. Marine transport is estimated at 1 000 km. If LCA data of the PV module manufactured in China is obtained, this assumption will be changed.

(8) WASTE MANAGEMENT

This study assumed disposal. After their life-time, all equipment is transported to a wrecking yard and then to reclamation land. The estimated disposal transport is 300 km to the wrecking yard and 300 km to reclamation land.

(9) DATA PREPARATION FOR THIS CASE STUDY

For Life-cycle Analysis of VLS-PV systems, energy, and CO_2 emissions data in China and Japan were collected. Tables 5.12, 5.13 and 5.14 show energy consumption and CO_2 emissions of each material and product.

The inventory data of PV modules were referred to in a NEDO report in 2009 [12].

5.4.2 Results and sensitivity analysis

(1) ENERGY REQUIREMENT AND CO_2 EMISSIONS

Figure 5.7 and Figure 5.8 show the energy consumption and CO_2 emissions of the 1 GW VLS-PV systems. There are no big differences, but CIS has the best energy consumption, and also mc-Si is the best in view of CO_2 emissions. The efficiency of PV modules creates significant differences in energy and CO_2 emissions of array support and foundations. Because thin-film PV systems have lower efficiency than crystalline PV module, there are big differences in steel and concrete use. Figure 5.9 and Figure 5.10 show the energy

Table 5.10 PV modules for the LCA study

Cell type	mc-Si	sc-Si	a-Si/sc-Si	Thin-film Si	CIS
Nominal power (W)	186	165	195	37,5	80
Efficiency of module (%)	13,9	14,3	16,6	8,6	10,1
Flame	Aluminum	Aluminum	Aluminum	Aluminum	Aluminum

Table 5.11 Equipment for one GW VLS-PV system

PV module	mc-Si	sc-Si	a-Si/sc-Si	Thin-film Si	CIS
Capacity (MW)	1 071	1 024	1 021	1 036	1 034
Annual power generation [GWh]	1 687	1 596	1 592	1 615	1 604
Land requirement (km^2/GW)	21,3	20,7	18,0	34,5	29,2
Array support structure (10^3 tonne)	93	88	83	199	152
Foundation (10^3 m^3)	513	546	463	798	683
Cable (30-degree)					
600 V CV 2 mm^2 (km)	10 692	10 469	11 941	129 397	37 812
600 V CV 5,5–8 mm^2 double core (km)	1 717	1 838	1 734	4 095	2 183
600 V CV 60–150 mm^2 (km)	1 060	941	1 146	2 100	1 555
6,6 kV CV-T 22 mm^2 (km)	262	233	283	343	302
6,6 kV CV 200 mm^2 (km)	306	320	229	356	344
110 kV CV 150 mm^2 (km)	255	245	236	319	292
Trough (30-degree) (m^3)	346 307	345 816	316 393	669 445	529 150
Common apparatus					
Inverter with transformer (set)	4 040 (including replacement and maintenance)				
6,6 kV circuit breaker (set)	2 080				
110 kV/6,6kV transformer (set)	50				
110 kV disconnecting SW (set)	180				
110 kV GIS (set)	100				
SVC (set)	20				
Common power board (set)	10				
Power transmission corresponding to a 100 MW PV system					
Transmission line					
110 kV TACSR 410 mm^2 (km)	1 202				
AC 70 mm^2 (km)	100				
Tower (steel) (tonne)	22 044				
Foundation (tonne)	50 932				

Table 5.12 Energy consumption and CO_2 emission data of an average power plant (in kWh of electricity, in MJ of primary energy)

Item		Unit	Value
Japan			
Primary energy consumption for power generation		MJ/kWh	10,38
Average CO_2 emission rate of utility		g-C/kWh	114
China			
Primary energy consumption for power generation		MJ/kWh	12,01
Calorific value of Chinese standard coal		MJ/SCE-t	29 302
Common			
Coal	Calorific value	MJ/t	25 643
	CO_2 emission rate	g-C/MJ	24,7
Oil	Calorific value	MJ/t	45 000
	CO_2 emission rate	g-C/MJ	19,2
Diesel oil	Calorific value	MJ/l	38,5
	CO_2 emission rate	g-C/MJ	18,7
Heavy oil	Calorific value	MJ/l	40,6
	CO_2 emission rate	g-C/MJ	19,4

Table 5.13 Energy consumption and CO_2 emission data of PV modules

	Product	Energy contents (MJ)	CO_2 contents (kg-CO_2)
NEDO	mc-Si PV module [m²]	2 737	135,2
NEDO	sc-Si PV module [m²]	3 986	193,5
NEDO	a-Si/sc-Si PV module [m²]	3 679	178,0
NEDO	Thin-film Si [m²]	1 211	67,8
NEDO	CIS PV module [m²]	1 105	67,5

Table 5.14 Energy consumption and CO_2 emission data of products and disposal used in this study

	Product	Energy contents (GJ)	CO_2 contents (t-C)
Japan	Silicon steel (tonne)	13,4	0,321
	Aluminium (tonne)	227	2,13
	Copper (tonne)	46,5	0,771
	HDPE (tonne)	15,8	0,264
	PVC (tonne)	29,4	0,373
	Epoxy resin (tonne)	40,5	0,754
	FRP (tonne)	81,6	2,74
	Ceramics (tonne)	0,8	0,02
China	Steel (tonne)	26,6	0,657
	Zinc-plated steel (tonne)	53,1	1,31
	Cement (tonne)	5,3	0,270
Disposal	Wrecking yard (tonne)	0,027	0,006 1
	Reclamation (tonne)	0,060	0,001 3

pay-back time and the CO_2 emissions rate. They are higher than the recently published papers by Fthenakis et al. [14] or Mason et al. [6]. The reason for this is the efficiency of the manufacturing in developed and developing countries. The results of energy pay-back time are 2,1 to 2,8 years. This means that they can recover energy consumption in 3 years. The CO_2 emissions rates are 52 to 71 g CO_2/kWh.

Figs. 5.7 to 5.10 assume 2 017 kWh·m^{-2}·yr.$^{-1}$ irradiation in Hohhot in China. BOS and 100 km transmission lines on Japanese construction standards are produced in China, and PV modules and inverters are produced in Japan. PV module data are from NEDO. Module efficiencies are 13,9% (mc-Si), 14,3% (sc-Si), 16,6% (a-Si/sc-Si), Si 8,6% (thin-film Si), 10,1% (CIS).

(2) DISCUSSION

Estimations of CO_2 emissions rate of generating technologies were summarized by IPCC [15] as shown in Figure 5.11. They summarized lifecycle greenhouse gas emissions of conventional fossil fuel technologies and

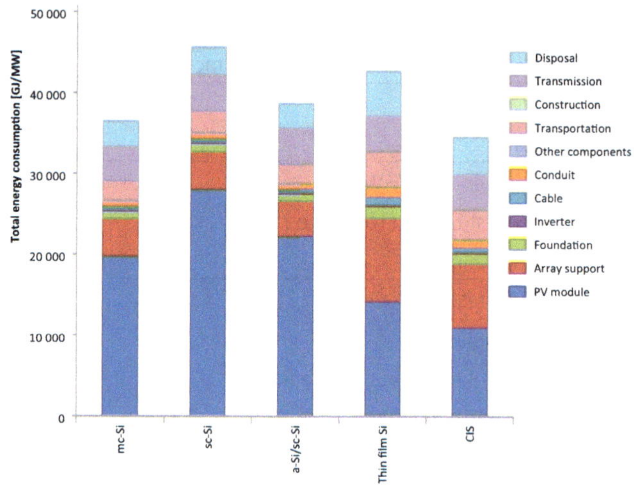

Figure 5.7 Energy consumption of the VLS-PV systems [GJ/MW]

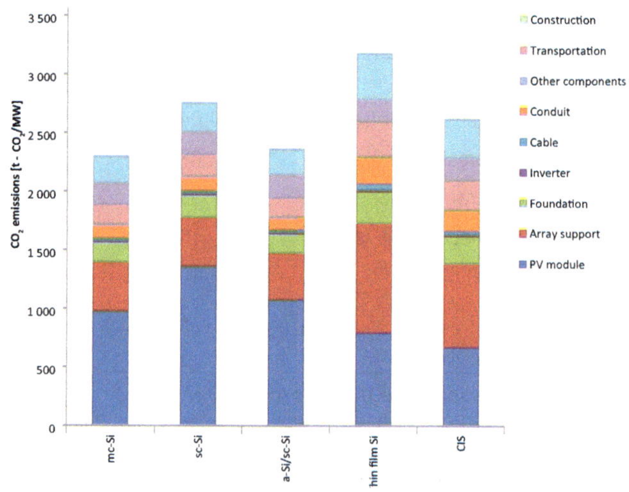

Figure 5.8 CO_2 emissions of the VLS-PV systems [t-CO_2/MW]

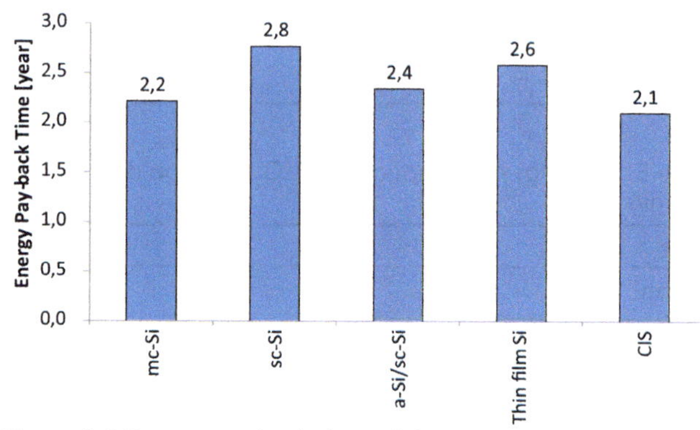

Figure 5.9 Energy pay-back time of the VLS-PV systems [year]

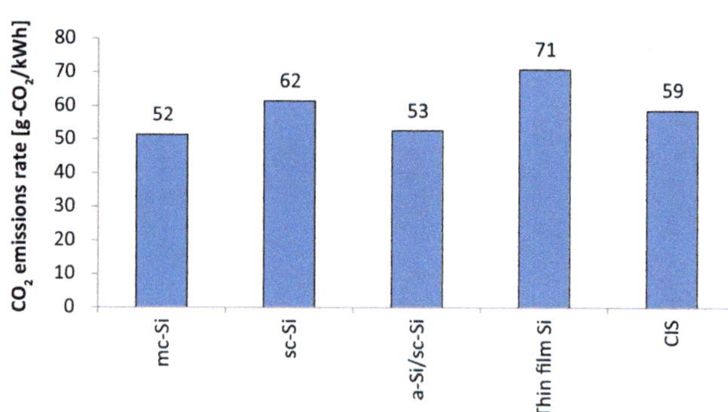

Figure 5.10 CO_2 emissions rate of the VLS-PV systems [g CO_2/kWh]

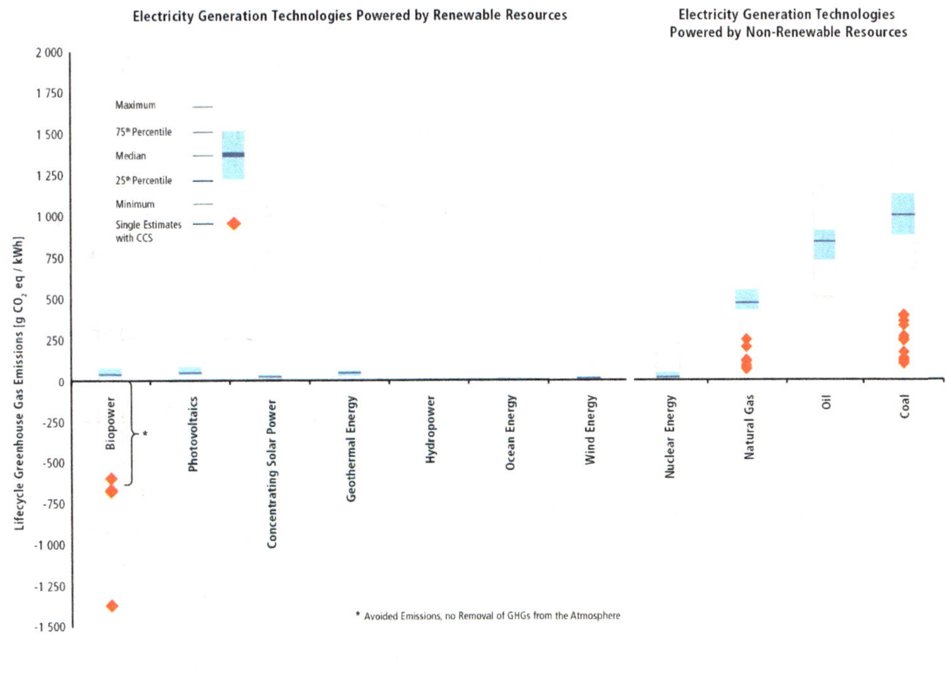

Count of Estimates	222(+4)	124	42	8	28	10	126	125	83(+7)	24	169(+12)
Count of References	52(+0)	26	13	6	11	5	49	32	36(+4)	10	50(+10)

Fig 5.11 Estimations of GHG emissions (g CO_{2eq}/kWh) from a large number of sources summarized by IPCC [15]

many types of renewable energies. It is easy to understand that there are many types of renewable technologies, and their values are much smaller than fossil fuel power plants.

Figure 5.12 shows the CO_2 emissions rate of estimated data, assuming 60 years lifetime of steel and concrete, and Japan's steel CO_2 emissions rate for comparison and sensitivity analysis. It may be possible to use arrays and foundations for more than 30 years. If they are assumed to have 60 years lifetime, the CO_2 emission rate will reduce 30% in thin-films and 20% in crystalline approximately. The steel CO_2 data from China is not recent data; it should be reduced at present. If the steel CO_2 data is reduced to the same as Japan, the CO_2 emissions rate will reduce 10–20%, especially for thin-films. The references in the figure are Fthenakis et al. [16], de Wild-Scholten [17], and CRIEPI [18]. Because they are assuming a rooftop PV system, the numbers are smaller than those in this chapter. From Figure 5.8, more than half, especially in thin-films, is BOS, and the numbers are much smaller in their papers.

Figure 5.13 and Figure 5.14 show EPBT and CO_2 emissions rate from a conference paper at EU-PVSEC in 2009 written by Fthenakis et al. [20]. The major part of a CdTe PV system is the PV part (blue), and other parts, BOS (red) are smaller than the PV part. Comparing the VLS-PV result shown in Figure 5.7 and Figure 5.8, the share of BOS is different. The reasons for this are related

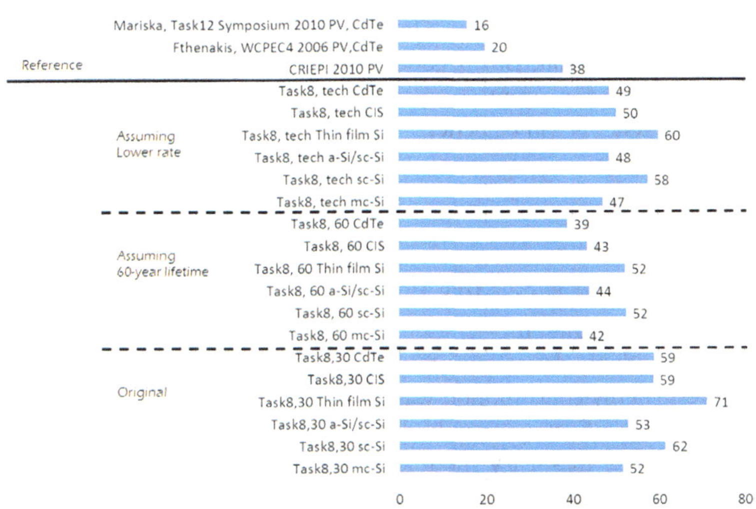

Figure 5.12 Sensitivity analysis assuming 60 years life-time or using Japan's steel CO_2 emissions rate, and from journals

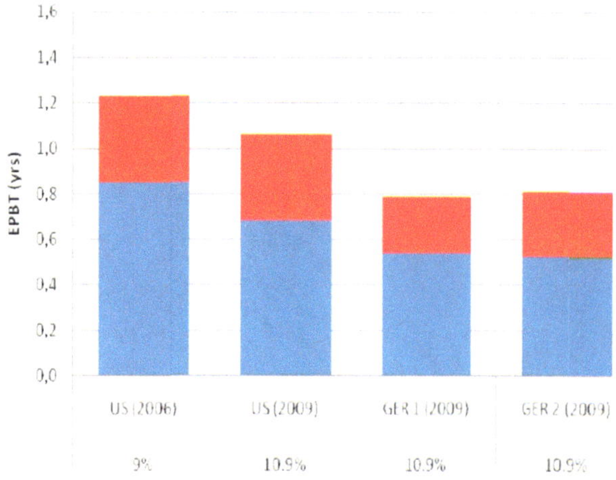

Figure 5.13 EPBT results of CdTe PV ground-mounted system from Fthenakis *et al.* 2009 [20]

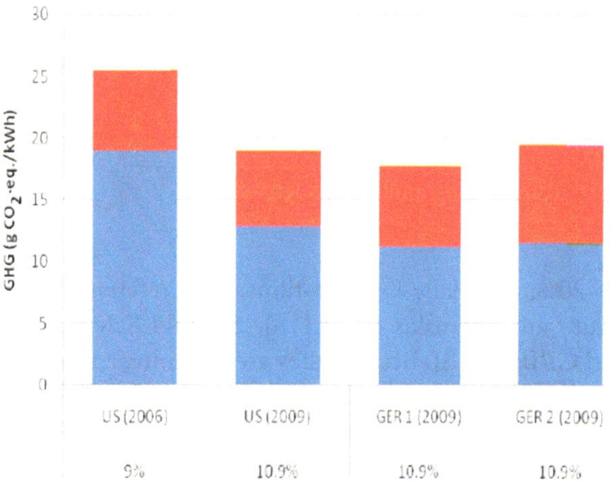

Figure 5.14 CO_2 emission results of CdTe PV ground-mounted system from EU-PVSEC conference paper in 2009 [19].

to desert conditions. BOS assumptions of the VLS-PV include 100 km transmission lines, 1 000 km marine and 600 km land transportation, disposal with 600 km land transportation and array support for weak sand condition. These large BOS parts are not included in the paper's result using actual installation system data by Fthenakis et al. The large BOS may be too great an assumption, But is good because if a system evaluated with hard assumptions gets good results, then this means real systems will be better.

5.4.3 Conclusions

In this section, the environmental potential of the VLS-PV system, which uses six types of PV modules, mc-Si, sc-Si, a-Si/sc-Si, thin-film Si and CIS with the latest inventory data were discussed. The lowest energy consumption was with a CIS module, with 2,1 years of energy pay-back time assuming desert conditions. The mc-Si PV module provided the lowest carbon emission, and a-Si/sc-Si module was at the same level. By the comparison of life-time, 60 years steel and concrete lifetime will reduce the CO_2 emissions rate by 20–30%, and the CO_2 emissions rate will decrease 10–20% by applying low CO_2 steel manufacturing technology.

5.5 ECOLOGICAL FOOTPRINT OF THE PV SYSTEM WITH AGRICULTURE

5.5.1 Introduction

(1) BACKGROUND AND OBJECTIVE

The world economy is rapidly developing because of the increase in the world population and in energy consumption – and they are expected to increase continually [20]. If we do not act, we will exceed the limit of our ecosystem. Actually, we can see many problems such as global warming and acid rain all over the world. And since the mid-1970s, we have been in a state of overshoot which means the Ecological Footprint (EF) has exceeded the Earth's Biocapacity (BC). For example the EF exceeded the BC by about 50% in 2007 [21] (see Figure 5.15).

(2) THE ECOLOGICAL FOOTPRINT (EF) AS AN INDICATOR OF SUSTAINABILITY

In this section, we utilize EF analysis for PV systems. The EF is a measure of how much biological productive land area humanity uses to produce resources and to absorb waste. The BC is the Earth's available biological capacity to produce useful biological materials and to absorb waste, shown in Figure 5.16.

If the EF exceeds the BC (i.e. ecological balance (BC – EF) is negative), this ecosystem is not sustainable. If we are able to reduce the EF or increase the BC, we can reduce this negative ecological balance.

In this section, methods of calculating the EF of PV systems and fossil fuel energy and the BC of the greening

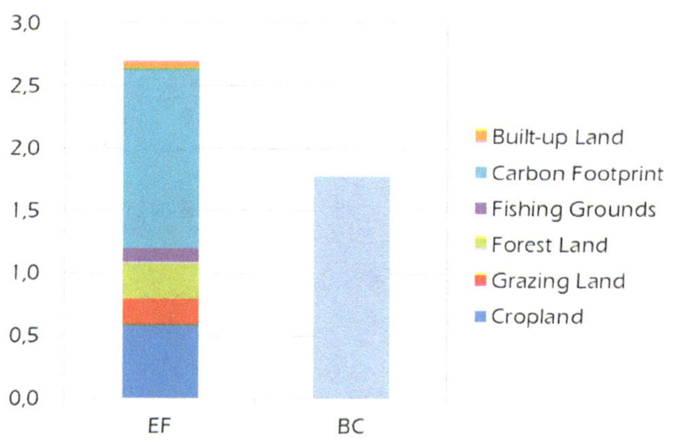

Figure 5.15 World Ecological Footprint and Biocapacity per capita (2007) [21]

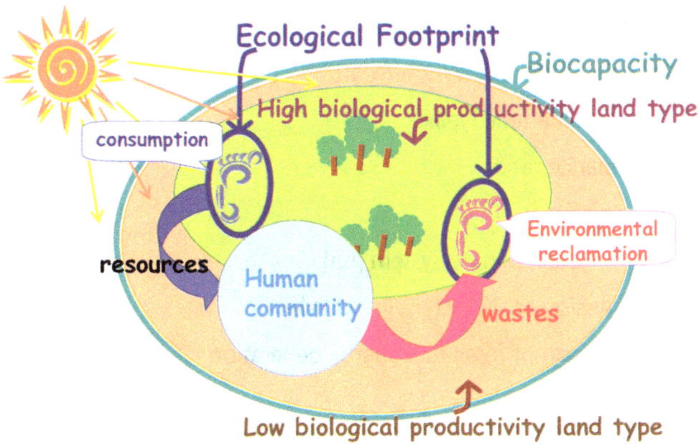

Figure 5.16 Image of the Ecological Footprint and Biocapacity

to reduce the ecological balance are developed and we show the results of a case study.

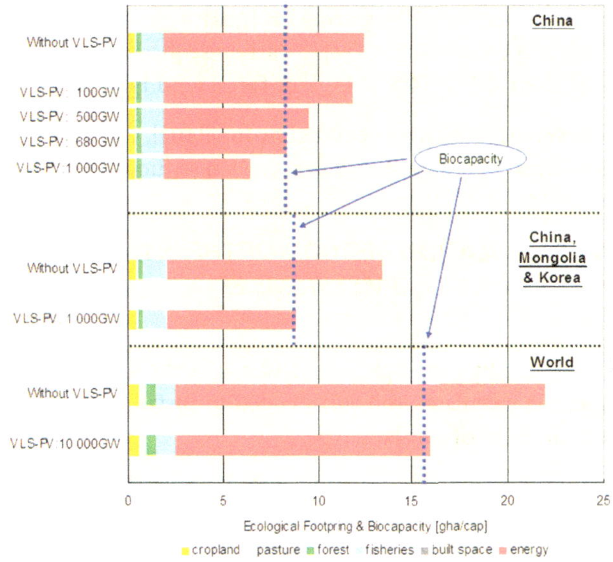

Figure 5.17 Results of estimating possible ecological impacts by VLS-PV project on the Gobi desert

(3) EQUIVALENCE FACTOR

The EF and BC are expressed in units of global hectares [gha]. The global hectare is normalized to the area weighted average productivity of a biological productive area, because different land types have different biological ability. The weighting factor is called the Equivalence Factor (EQF). We used Net Primary Productivity (NPP) as a basis for EQF [22]. NPP is a rate process which tracks the net flux of carbon from the atmosphere into green plants and oceans (Table 5.15).

The EQF of forest land is 15 times larger than that of desert land. This means that forest land has more biological productivity than deserts. Human area has the third lowest EQF of all land types.

Table 5.15 The Net Primary Productivity (NPP) and the Equivalence Factor (EQF)

Ecosystem type	Surface area (10^6 km^2)	NPP (Pg)	EQF (gha/ha)
Forest	31	48,7	3,57
Grassland	37	52,1	3,20
Deserts	30	3,1	0,23
Arctic-alpine	25	2,1	0,19
Cropland	16	15	2,13
Human area	2	0,4	0,45
Others*	6	10,7	4,05
Lakes and streams	2	0,8	0,91
Marine	361	91,6	0,58
Total	510	224,5	1

* Chaparral, bogs etc..

(4) ESTIMATION OF POSSIBLE IMPACTS OF VLS-PV DEVELOPMENT

Before the calculation of the main topic of this chapter, the last calculation in the previous edition of this book [10] should be introduced. It assumed that a PV system produced alternative energy to fossil fuel electricity, and did not assume greening. There were three cases: China; China, Mongolia and Korea; World; shown in Figure 5.17. In case of China, it would be balanced by installing 680 GW VLS-PV, 1 000 GW in the case of the three countries, and 10 000 GW in the case of world.

5.5.2 Method

In order to evaluate the environmental potential of PV systems, we calculated Photovoltaic (PV) system EF, fossil fuel EF and BC by greening. PV systems use an area for installation and their manufacture emits carbon emissions. Therefore, we calculated installation EF and carbon emission EF for PV EF.

(1) CALCULATION METHOD OF THE MANUFACTURING PV SYSTEMS EF

PV systems emit carbon dioxide when they are manufactured. Therefore, Manufacturing PV systems EF are calculated by estimating the forest land area which assimilates carbon dioxide by manufacturing PV systems and subtracting the percentage of this carbon dioxide sequestered by oceans. Carbon absorbed by oceans is 26% [23].

Manufacturing PV systems EF [gha] = carbon emission [t-C/yr] ÷ carbon sequestration [ha/yr/t-C] × 74% × EQF of forest land [gha/ha]

For every ton of carbon emitted by fossil fuel, we apportion 1.05 hectare per ton of carbon as carbon sequestration [24].

(2) CALCULATION METHOD OF THE INSTALLATION PV SYSTEMS EF

The PV EF by installation is calculated by the installed land type. We calculated the area needed to install PV systems and the EQF of this land type.

Installation of PV system EF [gha] = installation area [ha] × EQF of installed land type [gha/ha]

This equation shows that if you want to reduce the PV EF for installation, you should select a less biologically productive land type such as arctic-alpine, deserts and human areas.

(3) CALCULATION METHOD OF USE OF THE FOSSIL FUELS EF

When burning fossil fuels, carbon is produced. Therefore, the calculation of this effect is the same as manufacturing PV systems.

Fossil fuel EF [gha] = carbon emission [t-C/yr] ÷ carbon sequestration [ha/yr/t-C] × 74% × EQF of forest land [gha/ha]

(4) CALCULATION METHOD OF THE BC BY GREENING

The biocapacity by greening is calculated by the area and the EQF for each land type such as forest land and grassland.

BC by greening [gha] = area [ha] × EQF [gha/ha]

5.5.3 Case study 1: PV and greening

(1) ASSUMPTION

Here, we assume that 100 GW PV systems are installed in the city (the irradiation is 1 295 kWh/m^2/year). This system's lifetime is 20 years, the performance ratio is 80% and the efficiency of this module is 14,2%. CO_2 emission of this system is 12,14 [g-C/kWh] by Life-Cycle Analysis (LCA) methodology [25, 26].

A crystalline silicon PV module is used for the calculation. The electricity produced by PV systems will replace the same amount of electricity produced by fossil fuel. In this paper, we calculated the EF of annual electricity by 100 GW PV systems and the EF of equivalent electricity by fossil fuel energy. Table 5.16 gives the data for this case study.

Table 5.16 The data for case study 1

Annual irradiation [kWh/m^2/year]	1 295
PV system life time [year]	20
Performance ratio	80 %
Module efficiency	14,2 %
Area of installation PV system [ha]	70 000
CO_2 emission of PV system [g-C/kWh]	12,14
CO_2 emission of fossil fuel energy generation [g-C/kWh]	192,0
Annual electricity by this PV system [kWh]	$1{,}04 \times 10^{11}$

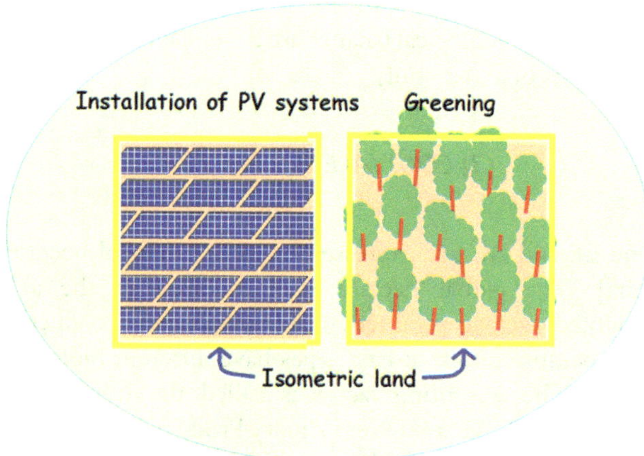

Figure 5.18 Image of case study 1 assumption

We assumed the module efficiency of PV systems is 14,2%. Therefore, the area of installed PV systems is 70 000 [ha] at least. We calculated that the EF of installed PV systems and that of greening in the same amount of area (70 000 [ha]) (Figure 5.18). We assumed that the land type of installation PV systems is human area. We calculated the BC of greening (forest land and grass land).

(2) RESULTS

Table 5.17 gives the results of this case study. The EF of PV systems was 3 520 000 [gha] (manufacturing PV systems EF was 3 490 000 [gha] and installation PV systems EF was 31 500 [gha] in human area).

100 GW PV systems can generate $1{,}04 \times 10^{11}$ [kWh/yr] electricity. We calculated the EF of equivalent electricity by fossil fuel energy to be 55 180 000 [gha].

The BC by greening was 250 000 [gha] (forest land) and 224 000 [gha] (grassland).

Replacing electricity generated by fossil fuel with the electricity from PV systems reduces the EF. Greening increases the BC because of the increase of forest land and grassland (Table 5.18). As a result, both the greening

Table 5.17 Results of case study 1 assuming 70 000 ha for PV installation

		EF [gha]	BC [gha]
PV	Manufacturing	3 490 000	
	Installation	31 500	
Fossil fuel		55 180 000	
Greening	Forest land		250 000
	Grassland		224 000

Table 5.18 The decrease of EF and the increase of BC

	PV system EF – fossil fuel EF	Greening	
		Forest	Grass
EF [gha]	–51 620 000		
BC [gha]		+250 000	+224 000

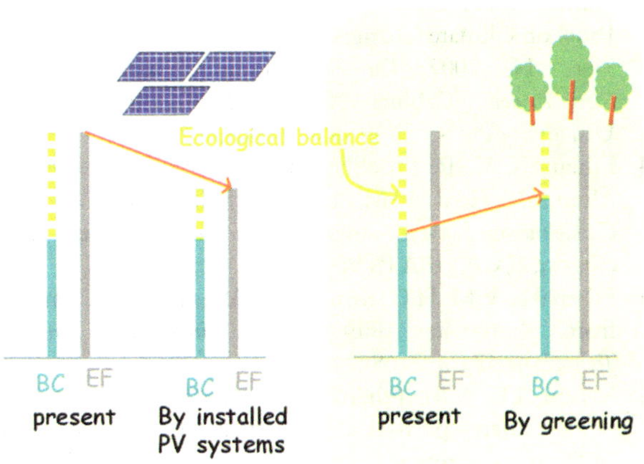

Figure 5.19 Image of the ecological balance by installation of PV systems and greening

and installation of PV systems can reduce the negative ecological balance (= BC – EF) (Figure 5.19).

This result indicates two effects. First, both effects can reduce the negative ecological balance. Second, the installation of a PV system can reduce by as much as 200 times the negative ecological balance of greening. That is to say, a PV system is more effective in reducing the negative ecological balance than forest land which has the best natural biocapacity of all.

5.5.4 Case study 2: PV and sustainable agriculture

Agriculture is a growing technology, but the BC does not increase because it requires energy from diesel oil for tractors and fertilizer manufacturing. Also agriculture in desert regions needs irrigation. This case study focuses on sustainable agriculture by using PV systems. The PV energy is assumed to be used for fertilizer manufacturing and irrigation, and/or tractors. An image of sustainable agriculture is showing in Figure 5.20.

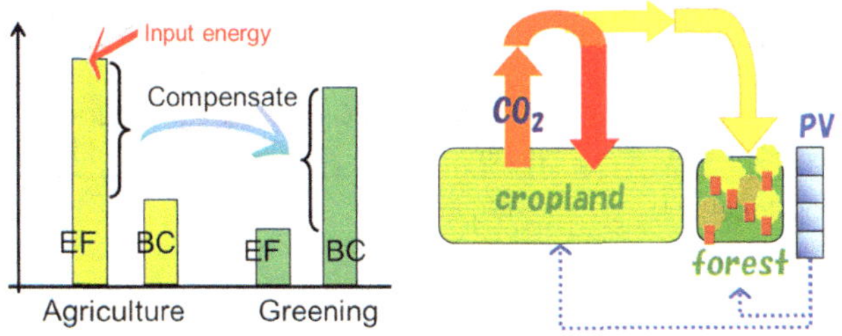

Figure 5.20 Image of sustainable agriculture: greening compensates CO_2 emissions from cropland and agricultural development

Table 5.19 Energy input for greening willows

	Energy input [MJ/ha]
Diesel used in farm equipment	45 218
Nursery operations	8 847
Farm machinery manufacture	2 949
Agri. chemical transport	1 966
Fertilizer manufacture	36 371
Herbicide manufacture	2 949
Total	98 300

(1) ASSUMPTION

The EF of greening is calculated based on an LCA report [27] about greening of willows in New York. The energy input is shown in Table 5.19. Because data from New York may not match that from deserts, this should be updated. The ecological balance of greening is still negative if it uses fossil fuel only. Therefore, PV energy is assumed for energy consumption of greening in this case study. Agriculture data is from a USDA report [28] about corn in the northern part of the USA. Because rainfall in the USA and deserts is different, irrigation data was also added from research on water saving irrigation in the west of China by Prof. Nishimura who is a member of the Japanese Task8 team.

(2) RESULTS

Table 5.20 shows the amount of greening and the PV area in cases of fossil fuel only, fossil fuel and PV, and PV only, for one hectare of agriculture. In case of fossil fuel only, the total required greening area was 62 ha. But if PV electricity is used for irrigation and fertilizer, the total area is reduced to 6,9 ha which is the area for PV system and greening. And if a tractor is powered by PV, the total area decreases to 2,3 ha. Therefore, introducing a PV system greatly reduces the required area for sustainable agriculture.

Table 5.20 Results of sustainable agriculture in cases of fossil fuel and PV

Energy source for agricultural development	Agricultural development BC-DF [gha]	PV capacity for agricultural development [kW]	Forest area for sustainment [ha]	PV capacity for greening [kW]	Total area [ha]
Fossil fuel only	−120,88	0	61,19	47,00	62,30
Fossil fuel and PV	−15,25	22,36	5,87	0,42	6,93
PV only	−3,84	25,69	1,27	0,78	2,33

5.5.5 Conclusions

The environmental potential of PV systems and that of greening are evaluated in this section. This section shows that the effectiveness of PV for Earth's ecosystem is better than greening. This was calculated by the estimation of Ecological Footprint analysis.

However, it is important that greening can increase the BC, but PV systems cannot increase it. And by using PV electricity, the required area reduces to 10%. Therefore, PV systems have a potential to realize sustainable agriculture.

5.6 SUMMARY

This chapter described life-cycle assessment (LCA) of PV systems. An overview of LCA was given in Section 5.4.3 outlining the method for quantitatively determining the environmental effects of products. Section 5.4.4 described LCA for PV systems, outlining evaluation indices, boundaries of LCA, inventory analysis, impact assessment and interpretation. Section 5.4.5 explained Ecological Footprint and PV system potentials with sustainable agriculture.

Section 5.3.4 presented LCA calculation by introducing related example papers. LCA of PV modules, PV systems and BOS was described. According to Fthenakis et al. (2008), greenhouse gas (GHG) emissions from Si modules were 30–45 g CO_{2eq}/kWh, and the EPBT of such modules was 1,7–2,7 years. The corresponding figures for CdTe frames without PV modules were 24 g CO_{2eq}/kWh and 1,1 years for ground-mounted installations. The results of calculation in Section 5.4 showed an energy requirement ranging from 34 000 to 46 000 GJ/MW and an energy payback time of between 2,1 and 2,8 years. CO_2 emissions were from 2 300 to 3 200 t CO_2/kW, and CO_2 emission rates ranged from 52 to 71 g CO_2/kWh.

Section 5.5 presented PV system values from the point of view of Ecological Footprint. By reducing the use of fossil fuel by PV systems, the ecological balance would possibly come to zero. And it was found that it would be difficult to realize sustainable agriculture without PV systems.

Since the lifetime of PV systems exceeds 20 years, a low EPBT means that a system can recover the energy required to pay for itself more quickly. These figures for CO_2 emission rates are also much lower than those for fossil fuel plants. It can therefore be concluded that PV systems have a significant potential to mitigate global warming.

REFERENCES

1 Solomon, S., D. Qin, M. Manning, Z. Chen, M. Marquis, K.B. Averyt, M. Tignor and H.L. Miller (eds), (2007) Contribution of Working Group I to the *Fourth Assessment Report of the Intergovernmental Panel on Climate Change*, Cambridge University Press, Cambridge, UK and New York.

2 Watson, R.T. *et al.* (2001) Summary for Policymakers, Climate Change 2001: Synthesis Report, Intergovernmental Panel on Climate Change,

3 Stern, N. (2007) *The Economics of Climate Change, The Stern Review*, Cabinet Office – HM Treasury, Cambridge University Press.

4 Fthenakis, V., R. Frischknecht, M. Raugei, H.C. Kim, E. Alsema, M. Held and M. de Wild-Scholten, 2011, *Methodology Guidelines on Life-cycle Assessment of Photovoltaic Electricity*, 2nd edition, "LCA", IEA PVPS Task 12.

5 Fthenakis, V. M. H.C. Kim, and E. Alsema (2008). Emissions from photovoltaic life-cycles. *Environmental Science & Technology*; 42 (6): 2168–74.

6 Mason, J.E., V.M. Fthenakis, T. Hansen, and H. C. Kim (2006). Energy payback and life-cycle CO_2 emissions of the BOS in an optimized 3.5 MW PV installation. *Progress in Photovoltaics*; 14 (2): 179–90.

7 Ito, M., M. Kudo, M. Nagura, and K. Kurokawa K. (2011). A comparative study on life-cycle analysis of 20 different PV modules installed at the Hokuto mega-solar plant, *Progress in Photovoltaics: Research and Applications*, 19(7).

8 K. Kurokawa (2003) *Energy from the Desert: Feasibility of Very Large Scale Power Generation (VLS-PV) Systems*, James and James.

9 Kurokawa, Kosuke, Keiichi Komoto, Peter van der Vleuten, and David Faiman (2006) *Energy from the Desert: Practical Proposals for Very Large Scale Photovoltaic Systems*, Earthscan.

10 Komoto, Keiichi, Masakazu Ito, Peter Van Der Vleuten, David Faiman, and Kosuke Kurokawa (2009) *Energy from the Desert: Very Large Scale Photovoltaic Systems: Socio-economic, Financial, Technical and Environmental Aspects*, Earthscan.

11 Ito, M., K. Kato, K. Komoto, T. Kichimi, and K. Kurokawa (2005) Analysis of transmission losses of Very Large-Scale Photovoltaic power generation systems (VLS-PV) in world desert, *31st IEEE-PVSC*, P2.6.1.10, Orlando, FL.

12 NEDO, (2009) Research and development of fabrication technologies for Life-Cycle Assessment of PV systems.

13 Komoto, K., H. Uchida, M. Ito, K. Kurokawa, and A. Inaba, A. (2008). Estimation of energy payback time and CO_2 emissions of various kind of PV systems. *Proceedings of 23rd EUPVSEC*, 3833–35.

14 Fthenakis, V.M., H.C. Kim, and E. Alsema (2008) Emissions from photovoltaic life-cycles. *Environmental Science & Technology* 42: 2168–74.

15 IPCC, 2011: Summary for Policymakers. In: *IPCC Special Report on Renewable Energy Sources and Climate Change Mitigation,* O. Edenhofer, R. Pichs-Madruga, Y. Sokona, K. Seyboth, P. Matschoss, S. Kadner, T. Zwickel, P. Eickemeier, G. Hansen, S. Schlömer, C. von Stechow (eds), Cambridge University Press. Figure SPM.XX.
16 Fthenakis, V. and E. Alsema (2006) Photovoltaics energy payback times, greenhouse gas emissions and external costs: 2004 – early 2005 status. *Progress in Photovoltaics*. 14(3): 275–80.
17 de Wild-Scholten, Mariska (2010) Life-cycle Assessment of Photovoltaics: from cradle to cradle, *1st International Conference on PV Module Recycling, Berlin*, 26 January.
18 Imamura, E. and K. Nagano (2010) *Evaluation of Life-cycle CO_2 Emissions of Power Generation Technologies – Update for State-of-the-art Plants*. Socio-economic Research Center, Central Research Institute of Electric Power Industry. Y09027.
19 Fthenakis, V.M., H.C. Kim, et al. (2009) Update of PV Energy Payback Times and Life-Cycle Greenhouse Gas Emissions. *24th European Photovoltaic Solar Energy Conference*, Hamburg, Germany, pp.4412–4416.
20 International Energy Agency, *World Energy Outlook 2011*, 9 November 2011.
21 Global Footprint Network, *Ecological Footprint Atlas 2010*, 13 October 2010.
22 Venetoulis, Jason and John Talberth, (2007) Refining the Ecological Footprint, *Environ Dev Sustain*, DOI 10. 1007/ s10668-006-9074-z.
23 IPCC. The Carbon Cycle and Atmospheric Carbon Dioxide, in *Climate Change 2001: The Scientific Basis*, 190.
24 International Energy Agency, (2003) CO_2 emissions from Fuel Combustion 1960–2001.
25 Yamada, Koichi and Hiroshi Komiyama, (2002) *Photovoltaic Engineering*, Nikkei BP.
26 Ito, Masakazu, Kazuhiko Kato, Keiichi Komoto, Tetsuo Kichimi, and Kosuke Kurokawa, (2007) A comparative study on cost and Life-Cycle Analysis for 100 MW Very Large-Scale PV (VLS-PV) Systems in Deserts Using m-Si, a-Si, CdTe and CIS modules, *Progress in Photovoltaics: Research and Applications*, 16(1): 17–30,
27 Heller, Martin C., Gregory A. Keoleian, and Timothy A. Volk, (2003) Life-cycle assessment of a willow bioenergy cropping system, *Biomass and Bioenergy* 25: 147–165.
28 Shapouri, Housein, James A. Duffield, and Michael Wang, (2002) *The Energy Balance of Corn Ethanol: An Update*, United States Department of Agriculture, July.
29 Yamashita, Naoko, Masakazu Ito, Keiichi Komoto and Kosuke Kurokawa, (2007) An Environmental Potential of PV systems in Japan by utilizing the Ecological Footprint, PVSEC-17, Fukuoka, Dec. 2007, 6O-C9-04.

PART II

Engineering and financial guidelines for VLS-PV systems

CHAPTER SIX

Technical engineering guidelines for VLS-PV

6.1 INTRODUCTION

Since the beginning of industrialisation electrical energy has been the driver of economic development towards today's high standard of living in developed countries as well as in other countries around the world.

Utilities were formed who built up infrastructures making electrical connections to towns and many other locations. For production of electrical power many different kinds of power plants were developed and applied using different energy sources. The first choice at the beginning was water power as a renewable energy source. With little delay coal was used, then oil too allowing larger power plants to be built. The circuit of users was extended constantly. More and more energy was used not only for homes but also for industry.

Exploration of crude oil and natural gas fields and their use led to new options extending the dissemination of power plants. In the mid-1960s nuclear power plants added to the variety of power generation plants.

All technical power generation plants, except hydro-electric plants, are dependent on limited fossil and uranium resources. The consequences of running power plants are drilling holes in the planet, the destruction of its surface and exploiting it in way that has never been done by mankind before – in a blink of an eye in relation to the period mankind has existed on Earth. What is taken out is simply burned on a large scale by power plants and nuclear waste remains. The products of burning processes are – beside effective energy e.g. for electrical appliances – pollution inducing climate change and waste. The increasing amount of nuclear waste is a dangerous and unsolved burden for mankind's future.

It is now commonly understood around the world that the business-as-usual scenario can no longer be accepted. All societies need to take part on a restructuring of the energy landscape. Utilities as one of main industry sectors concerned have to recognise sustainable solutions. Renewable energy, in the following solar energy, offers a key to a transition to sustainable power generation. Very Large Scale Photovoltaic Power Plants, described as VLS-PV, are predestined to play a main role within this transition scenario.

6.2 TRANSITION OF UTILITY POWER SUPPLY TO SUSTAINABLE SOLUTIONS

One of the most promising large scale renewable energy technologies for investors – like utilities or private equity – is photovoltaics used on a large and utility scale, like VLS-PV applications. VLS-PV is defined as a multiple of 10 MW of installed power. Countries having high solar irradiation can benefit more from the sun compared with countries in the northern hemisphere. VLS-PV power plants have already proved their reliability in hundreds of applications in many countries around the world. During daytime they partly substitute fossil fired power plants, leading to a reduction of pollution.

Since around 2006/07 the number of installations has increased remarkably year by year. Currently, more than 1 000 PV power plants with a rated power of more than one megawatt are in operation. It is worth mentioning the different feed-in tariff schemes and tax benefits etc. which have encouraged investors stepping into the business.

Falling prices of the main PV products accompanied by increasing prices of fossil fuels are helping VLS-PV projects to become more attractive for investors. In a lot of countries this price development will soon reach a so-called grid-parity. That means that the real price of PV electricity generated will achieve the market price of conventional produced electrical power. In the mid-term and no doubt long-term, PV electricity will more and more undercut conventional produced electricity costs. Considering only the electricity price of large

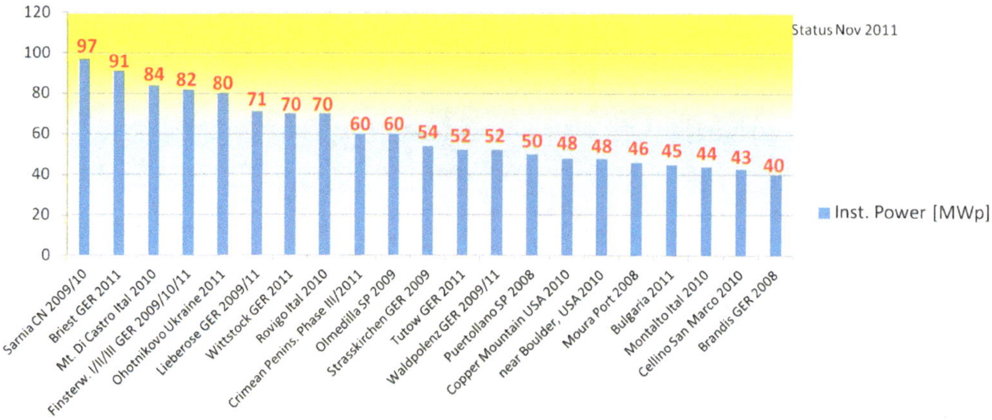

Figure 6.1 Selection of VLS-PV installations ≥40 MWp worldwide

grid-connected PV applications, the generation of electricity is already more economical compared with CSP (Concentrating Solar Power) plants. The price of electricity produced by VLS-PV is going to beat or in some countries has already beaten the price of conventionally produced electricity.

A selection of realized VLS-PV projects with a rated power of 40 MW and more is shown in Figure 6.1.

6.3 COST OF VLS-PV ELECTRICITY

It is a matter of fact that in many countries the price of electricity is not always calculated on a real costs basis. It is subsidised to make electrical power affordable for the broad mass of consumers, no matter whether for ordinary people or for industry. Either primary energy must be imported at high prices or the income of the public is not sufficient to bear the real cost. As a consequence the national budgets in this concern are not balanced.

Table 6.1 provides an overview of the cost situation of VLS-PV as medium values reflecting 2009 to 2011. It shows a continuous decrease in plant cost. The main cost factor "PV module" is expected to decline depending on the international market situation / supply demand. Setting values in relation to the expected yield, the cost of kWh / MWh generation can be calculated.

The table only refers to the plant erection cost. Other cost factors like cost of development of the site, later operational costs etc. are not considered here since they differ from site to site and country to country. Putting them together with the erection cost may serve as a first approach to the real production cost of PV kWh / PV MWh.

6.4 CONVINCING ADVANTAGES OF VLS-PV

National grids are mostly owned by governmental utilities. Integration of VLS-PV matches perfectly to local and national grids. The main benefits of VLS-PV are especially for utility use:

- Load discharging of existing conventional power plants during the day
    - Load shedding is accompanied by fuel saving and reduction of pollution discharging the local climate / environment
    - Substituting fuel in conventional power plants = reduction of operating cost
- Process of approval and construction of VLS-PV is much less complicated
    - Preparations of sites are marginal, construction is uncomplicated, only foundations and some cabling are put in the ground
- Flexibility = low requirements on the site
    - No water needed for operation, except in regions where frequent cleaning of the PV modules is required to maintain performance
- Modularity of VLS-PV
    - Power blocks of e.g. 1,5–3 MW can be operated in parallel. Limitation factors are mainly the land area and the capacity of the local MV or HV grid.
- Noise-free operation, except some fan noise for airflow for internal power electronics
- Relatively easy to operate by well-trained personnel
    - Option: unmanned and remote controlled operation or only some operators on site
- Low maintenance required since no rotating parts.

This can only be a mid-term view. In the long run a combination of several renewable energy techniques using different energy sources will complement one another and lead to a sustainable master plan. Large PV power plants will mainly be responsible for mid and peak load supply while CSP would be one of the base load contributors. This role may be more differentiated or will even change significantly when suitable electricity storage systems become available.

Figure 6.2 shows expectations on VLS-PV projects.

Table 6.1 Plant cost development

	Influenced by		2009 Euro/Wp	2010 Euro/Wp	2011 Euro/Wp
1	Area	Modules (c-Si)	1,7	1,5	1,2
2	Area	Support structures, foundations	0,23	0,23	0,23
3	Area	DC low voltage components, cabling	0,08	0,09	0,09
4	Area	Logistics	0,05	0,05	0,05
5	Area	Installation labour	0,25	0,25	0,25
6	Power	Inverters	0,23	0,21	0,18
7	Power	MV components	0,09	0,09	0,09
8	Power	SCADA	0,04	0,03	0,03
9	Power	Engineering labour (mechanical/electrical/other)	0,13	0,12	0,12
10		Taxes, permits, legal, insurance	0,25	0,25	0,25
		Costs	3,05	2,82	2,49
		Module share %	55,7	53,2	48,2
Remarks on:					
2	Uncertain	Market price of aluminum			
2	Not a good basis	Reason: the efficiency of a module is not really reflected in the price, comparison is much better by euro/m²			
3	Uncertain	Market price of copper			
7	Uncertain	Market development			

VLS-PV projects (Megawatt range) are utility scale projects:

Main expectations:

- Profitably of stakeholder's investment
  ROI as calculated (or higher)
- High availability, some decades of operation
- Easy to operate, low maintenance cost

Utility's aspects:

- Unproblematic feed-in partner
- Being a reliable part of the regional energy supply system

Figure 6.2 Expectations on VLS-PV projects

6.5 A PRACTICAL APPROACH

6.5.1 Project development and evaluation of integration

a. Investigation and assessment of the national energy / electricity master plan and management in regard of integration VLS-PV
b. Identification of grid sections VLS-PV which are optimal matching with existing and mid-term power demand profiles
c. Identification of potential areas for VLS-PV, as a comprehensive approach; may be significant – interests of some local authorities are already existing
d. Clarification of property ownerships
e. Preparation of PV technology profiles
f. Technology descriptions and cost profiles
g. Technology advantages and disadvantages in regard of the country
h. Feasibility study / Pre-study for feasibility assessment.

6.5.2 Possible follow-up procedure in direction of implementation

After finalising the sequence described above decisions for carrying out projects are to be made. There are several possibilities, e.g.

- BOO model (build, own, operate): The project is based on public-private partnership (PPP).

- BOOT model (build, own, operate, transfer): The public-sector partner within this model contracts with a private developer – typically a large company or consortium with specific expertise – to design and implement a VLS-PV project.
- Tendering the projects launched by a local or governmental authority: Since in public entities the expertise of large scale PV projects is normally not yet available elaboration of tenders is commonly done by experienced consultants who are familiar with state-of-the-art PV system technology.
- The responsible public authority / utility already has the special technical knowhow available. Then the complete project will be executed by them: project development, design and execution, commissioning and operation.

6.6 OVERALL DESIGN APPROACH

The following description provides an overall impression of the design phases of VLS-PV power plants. It describes relevant principles and makes the design procedure more transparent for a better understanding of main VLS-PV plant issues to be considered.

It is aimed at groups such as engineers as well as decision makers involved at e.g. the political, environmental or commercial level.

An example of a pragmatic approach and action plan is shown in Figure 6.3.

A concept proposal based on a feasibility study will provide decision makers information on principle technical feasibility of the VLS-PV accompanied by magnitudes / first indications of cost and other commercial information. The key for getting basic concept values is an evaluation and rough estimation of all main technical and non-technical issues and costs of relevance for assessing the economics of the investment. The deciding factor normally is economic viability. But other issues may be of importance too like mid-term and long-term social and environmental and other impacts of the investment.

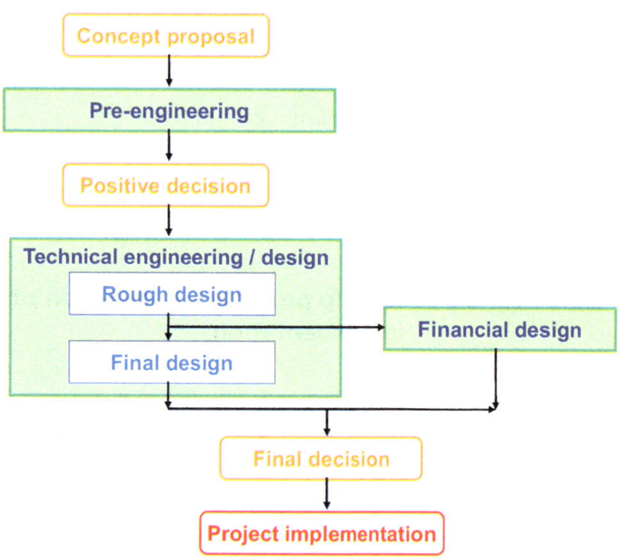

Figure 6.3 Engineering approach

A preliminary positive investment decision is among others the starting point for beginning the first technical approach / planning.

6.7 ROUGH DESIGN

This is the initial design step on the way to the final design. The technical rough design focuses on the expected cost structure of the project and provides financial experts with a cost estimation for proceeding in contacting financial possibilities and resources. The accuracy of calculation considering the current and mid-term market situation should normally be in the range of ±5%.

Since the module price still is the main cost factor, the fluctuating market price situation has to be taken into consideration.

Only the final technical design presents the real, respectively direct project cost for integrating in the overall project cost. It can be achieved either by design and execution of the realization of the entire VLS-PV by the owner or by getting quotations from EPCs (companies offering engineering, procurement and construction).

In the following, design and planning is described, independently whether the owner or an EPC would do it. The procedure is mostly the same.

6.8 FINAL DESIGN

The principle design and planning process is shown as Figure 6.4.

6.8.1 Quality assurance strategy and planning

At the beginning of the planning for all project phases general goals should be discussed, defined and established. A risk assessment should be included. Detailed checklists should be then available. For each project phase, quality features can be described together with the kind of control. Project phases are e.g.:

- Engineering design phase
- Construction phase, including quality checks of incoming material and execution of mechanical and electrical works
- Commissioning phase.

A person in charge of the execution must assure that the quality goals are kept.

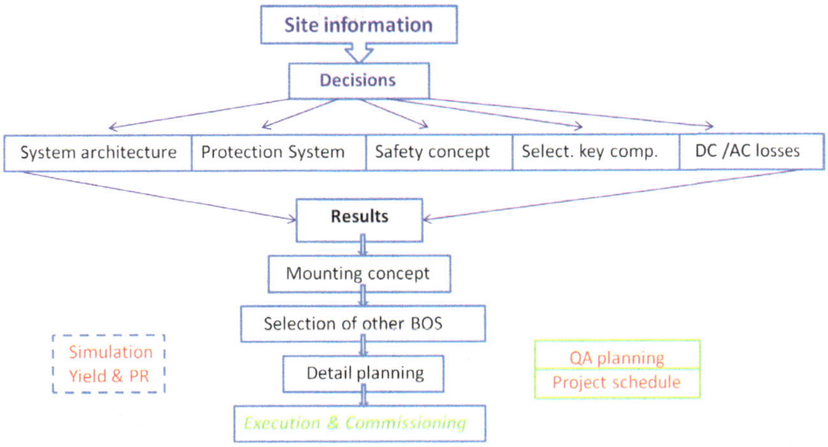

Figure 6.4 Final design approach

6.8.2 Site information

The selection of a VLS-PV suitable site must take several different views and approaches into account. A successful evaluation takes the special needs of PV into account for qualifying it for VLS-PV use. Table 6.2 shows a synopsis of the main headings to be looked at.

During the evaluation process some characteristics may lead to the decision to reject the site in mind.

(1) CLARIFICATION OF OWNERSHIP

The site to be looked at has to be analysed and verified from several viewpoints. First of all the situation of ownership has to be clarified. Could the area be leased or bought or are other options viable for a mid- or long-term operation of the VLS-PV? The operation period depending on, for example, a valid feed-in period is the basis for a project calculation. Sometimes it is of interest to use the property longer than initially was expected. And last but not least the question of the end of operation and dismantling of the complete installation should be considered. In this concern another and different scenario is described in Section 6.23.

Table 6.2 Site information

- Clarification of ownership
- Access
- Longitude and latitude
- Solar irradiation data
- Weather data (temperature, wind, etc.)
- Area [ha]
- Topology/land profile
- Soil information
- Archeological items in ground?
- Distance to MC/HV connection point
- External shading factors (if relevant)
- Distance to the sea (if relevant)

(2) ACCESS

Next the access to the property is an important issue regarding logistics. It has to take into account that a lot of material has to be transported there by long and heavy trucks; containers for power electronics and transformers require heavy cranes.

(3) LONGITUDE AND LATITUDE, SOLAR IRRADIATION, WEATHER DATA

The longitude and latitude have to be identified as they constitute part of the basic information needed for processing the simulation of the solar irradiation data.

For generating solar irradiation data, several data sources are available, e.g.:

- DWD (D) (German weather service)
- Meteonorm (CH) private institute Meteotest, having access to worldwide meteorological stations
- SODA (F) satellite based data bank, use of satellites Helioclim1, 2 and 3 (Ecole des Mines, Paris)
- PVGIS (EU)data for Europe and Northern Africa; for Europe mainly ground station measurement data used, for Northern Africa mainly data from geostationary weather satellites, converted by European Joint Research Centre ISPRA
- NCDC, NREL (USA) National Renewable Energy Laboratory
- GEBA (ETH Zurich, CH) Global Energy Balance Archive, etc.

For meteorological values like relative humidity and ambient temperature, the WMO (World Metrological Organisation) database is usually used.

Solar irradiation as well as other meteorological data are processed in PV sizing / simulation programs. The chosen irradiation data source among others influences the results of energy yield simulation later on since data differ from source to source more or less significantly. Thus commonly two different simulation programs are used for cross checking

(4) AREA AND TOPOGRAPHY

The area and topography of the site have an influence on the installed PV power. A site tilted to South allows distributing the support structures nearer to each other which increases the installed power for a given area. Different site profiles sometimes require a levelling of the ground to secure equal shading of the PV modules.

Some types of thin film modules compared with common crystalline modules allow more shading. Shading losses are less significant in this regard.

(5) SOIL INFORMATION, HISTORICAL MONUMENTS

Soil information about the physical and chemical composition of the site influences the choice of design of the foundations. An aggressive soil could require cathodic protection measures to avoid increased corrosion of the metallic foundation. Concrete bars as foundation put on the ground avoid the corrosion problem but increase cost and are a less good solution from the environment perspective.

If the soil report brings historical monuments to light then one should consider stopping the project. Here issues of national historical interest should dominate.

(6) DISTANCE TO THE MV / HV CONNECTION POINT

The distance to the nearest suitable MV or HV grid connection point also influences the decision whether to carrying on works for developing the site. The longer the distance, the higher the connection cost. Costs in this regard depend also on whether a free-line or a direct-buried cable connection is required or can be chosen. In some countries the latter provides security against theft.

(7) EXTERNAL SHADING

Another concern when looking at the site and its optimal ability for VLS-PV application is the external shading aspect. Initially a project is normally designed without considering external shading; sometimes horizontal shading, e.g. caused by a hill chain, is unavoidable. But if there is e.g. a young palm or tree plantation nearby, it has to be evaluated how the growing of the palms or trees could cause mid-term or long-term shading followed by energy losses for the VLS-PV.

If the mid-term planning of the local authority for developing the region is considering a conventional power plant, cement mill or other industrial settlements with higher buildings, external shading and possibly emissions may lead to power losses at the PV.

(8) DISTANCE TO THE SEA

A site not far from the sea often has a moderate climate and ambient temperature which is normally welcomed for PV application. In regions near to the sea the air has a high salt content. This aggressive content must be considered when a project in this region is being considered. Limitations in technology could discourage such a project. But it would still be of interest if certain approvals could be given: approvals from the PV module producer that their modules are long-term resistant against the salt content of the air; approvals by the inverter and electronics suppliers guaranteeing protection measures accordingly. What remains is the corrosion problem of the countless contacts in the PV field.

6.8.3 Technical feasibility of grid-integration

Beside collection and verification of physical and geographic information on the site, the technical feasibility of grid-integration of a VLS-PV must be clarified. A counterpart for the project holder for this evaluation is the local grid operator.

The defined grid requirements, codes and settings by the operator are to be considered by the inverters. They are – apart from the MV/HV switch gears – the power conditioning units preparing the grid-friendly AC power.

- As they are of a defined grid frequency, inverters automatically reduce the feed-in of active power according to the definable characteristic curve. This control contributes to stabilisation of the grid frequency.
- To stabilize the grid voltage, state-of-the-art inverters feed reactive power (leading or lagging) into the grid. By occurrence of a LVRT (Low Voltage Ride Through) the inverters stay connected to the grid during voltage drops and support the grid by a feed-in of reactive power.
- Another feasible feature – if required by the grid operator – is remote controlled power reduction in case of a grid overload (power level adjustment: PLA). The grid operator presets a nominal power value which the inverter will implement within 60 seconds (reaction time). The signals are received by a ripple control receiver in combination with a special control box transferring the command to the inverters.

The grid-study evaluates the capability of the existing grid for the VLS-PV feeder. Sometimes an extension of the grid may be the final solution for realization of a large PV power plant. Nevertheless it would be a valuable and logical

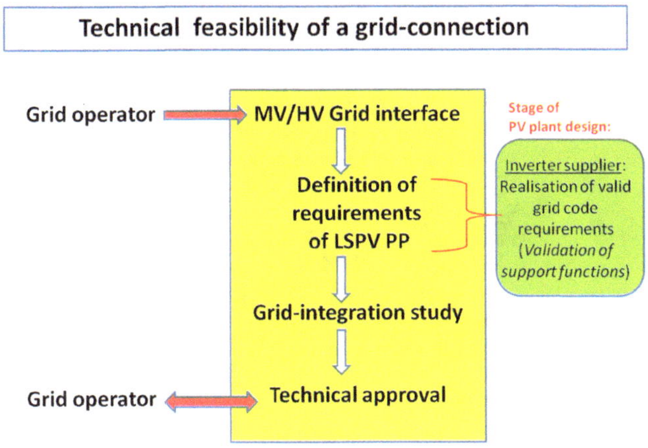

Figure 6.5 Example of a technical approval approach

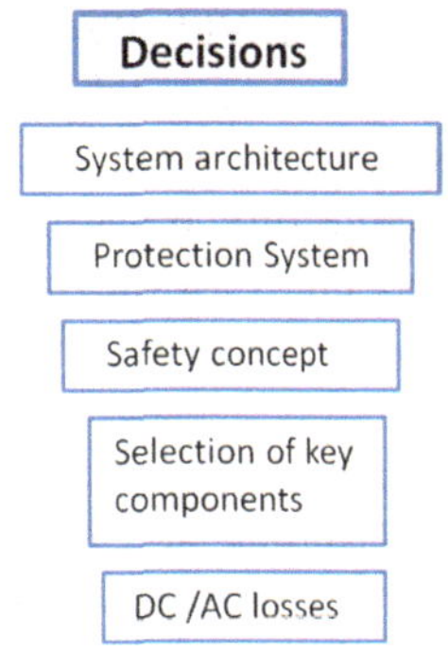

Figure 6.6 Decision steps

future investment. The study itself could be conducted by an experienced consultant or by experts of the grid operator (see Figure 6.5).

6.9 FINAL DESIGN AND PLANNING

Another precondition for starting final planning / design of the VLS-PV is the provision of initial files and reports of the site. Those files are as follows (see Section 6.8.2 (3)):

- Plan of the site with exact borders
- Topographic file showing the site profile
- Soil investigation report.

6.9.1 Engineering know-how

Final planning requires the determination of a lot of details and system decisions to be professionally combined into an overall professionally coherent concept.

Setting up the concept for a VLS-PV requires beside qualified knowledge also being familiar with international, national standards and local requirements of the country in question. Thus sometime a cooperation or consortium having a local partner company involved could be an optimal solution.

On the other hand, experienced independent consultants can support the planner by transferring their know-how and experience gained by operating a lot of installations. Their support is not representing any company's interest and could substantially back the interest of the owner.

Figure 6.6 provides an overview of conceptual decisions to be made. Execution of these actions will cause a lot of parallel work which needs to be coordinated. Matching of all "puzzle bricks" can secure a qualified and professional engineering plant design.

6.9.2 Expectations of investors

The planning company's engineers need to fulfil the expectations of the investor. These expectations are expressed as success factors below:

- High safety standard of personnel and adequate security for the plant and its operation
- Compliance with valid technical and environmental standards
- State-of-the art system design
- High efficiency, reliability and availability
- Final yield / PR meets or exceeds the estimated values
- Low service and maintenance cost
- ROI as estimated or higher.

Each issue mentioned above is a challenge for the planning engineers. They have to find a balance of technical and economical cost securing the high overall quality of the plant.

As in all electrical installations, the safety of personnel has the highest priority. In the history of electric power plants VLS-PV power plants are newcomers; in particular, their high DC current application is one of the new items engineers are facing and have to handle. International standards do not yet cover safety issues in all details for large PV power plants. Thus safety aspects require outstanding attention.

6.9.3 System architecture of a VLS-PV

(1) TRACKING SYSTEMS

The main idea of applying tracking systems is the reduction of investment cost. The real competition with fixed installation is the overall cost including service, maintenance, and repair costs.

Tracking systems are divided into:

- One-axis systems (east–west or north–south)
- Two-axie systems
- CPV (concentrator PV) systems.

The majority of realized VLS-PV are based on fixed systems. They are characterized by PV fixed installed module support structures. Such an installation is the most simple field architecture. It has the lowest service and maintenance demand and is for the operator less complex to operate since the PV generator is a static construction; it has no moving parts.

Tracking systems as shown in Figures 6.7 and 6.8 require more sophisticated planning and design. The land use factor is less than for fixed systems. The benefit expected is a higher energy yield together with a lower investment compared with fixed PV modules. Between the tracking systems, two-axis systems have a higher yield than one-axis ones. However, the advantage of higher yield goes along higher operational costs for service, maintenance and spare part demand of the support structures. Service requires skilled and trained personnel with mechanical and electrical knowledge. The inverter concepts partly differ. Some tracker suppliers recommend decentral inverters linked to each tracker, others prefer central inverters.

Figure 6.7 One-axis tracking system

Figure 6.8 Two-axis tracker

Tracking systems in general are more challenging PV modules than fixed systems. The thermal stress of the modules will increase, caused by the increase of collection of solar irradiation by tracking. This additional thermal affect may affect their technical lifetime. In any case the module supplier should give his approval for the particular application. That means the product warranty and power output guarantee should be guaranteed.

Another developed PV tracking system in the market of large and very large PV systems on the utility scale is concentrating PV (CPV) as shown in Figure 6.9. It is developed for regions with a very high content of direct solar irradiation. Several concentration systems exist. All have in common the permanent concentration of the solar irradiation precisely on a small and high efficient solar cell. The idea behind CPV is a reduction of plant cost. This is achieved by a dramatic reduction of the size of the solar cell and by use of low cost concentrator material.

Figure 6.9 CPV tracking system

A challenge is the accuracy of the tracking. Concentration factors vary according to the kind of tracking system. Variation could be in the range of 2 to 500 times or more. The higher the concentration factor the higher is the need for accuracy in staying always in a precise position to the sun. At a certain concentration factor, active cooling of the solar cells is required. Currently special inverters are being developed for high concentration appliances.

Thus, tracking systems in VLS-PV utility scale projects may promise higher energy yields. In contrast to fixed systems and from the mechanical point of view they are costly to maintain. Hence operational costs should be carefully evaluated upfront.

Efforts in regard of skilled personnel and spare parts require well-organized logistics and provision for a long period. Service personnel must also be provided for near-term trouble-shooting and repairing actions. Such deliberations must be considered at the moment of thinking about the kind of system architecture in this regard. The decision is not only influenced by cost-benefit analysis. Other resources like those mentioned above are also important.

Especially in countries with little qualified manpower available, the preference may tend to VLS-PV with fixed installed PV generators. Also the still ongoing decrease of PV module production costs and sales prices may in the mid-term endanger the economic viability.

(2) ELECTRICAL SYSTEM ARCHITECTURES

Two kinds of electrical system architectures for VLS-PV are currently applied (Figure 6.10). The kind of architecture is determined by the PV inverter. Since VLS-PV has a high

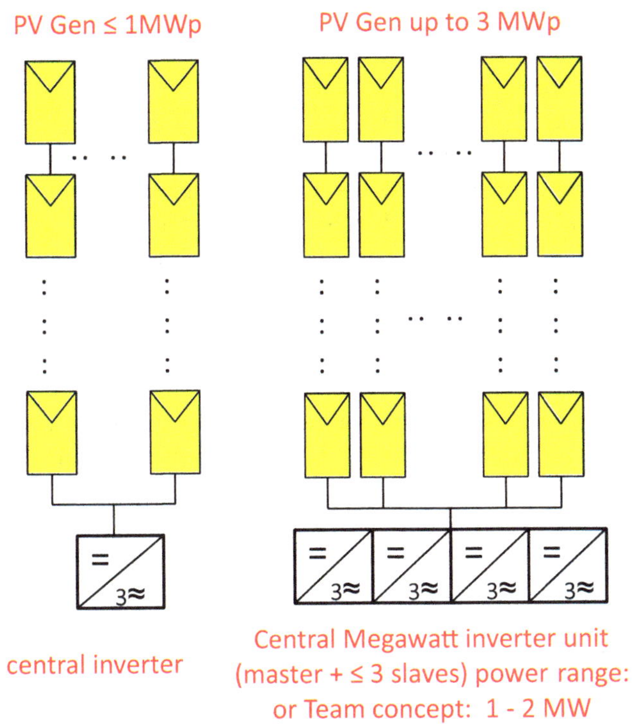

Figure 6.10 Central inverter and central inverter unit

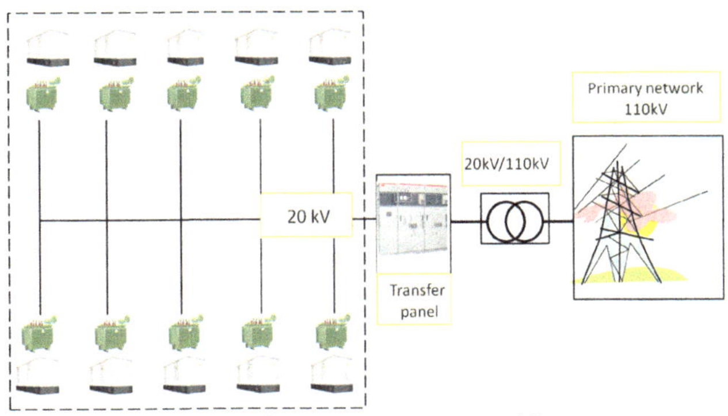

Figure 6.11 Example of a 30 MW VLS-PV PP bloc

repetition factor (a lot of sub-units / arrays) the essential architecture is determined by the choice of the kind of blocs or units.

The first architecture is the parallel connection of a lot of *central inverters*. Each inverter is linked to a PV sub-generator having a performance of about the same as that of the inverter. For example, in a 50 MW power plant 63 inverters of 800 kW each are applied. The 50 MW plant consists of 63 sub-systems accordingly. A failure of one of the inverters would then reduce the performance of the plant by only about 2%.

The inverter efficiency reaches its maximum usually at 20–30% of the rated power. The trend is to reach the rated efficiency earlier.

An alternative architecture uses the so-called *central inverter unit* constellation. Each unit consists of up to four inverters linked with each other in a so-called master-slave mode of operation (MSO). The rated power of the connected PV array / sub-generator is the sum of the entire capacity of the single inverters; commonly a design rule 1:1 is considered. The maximum capacity of a central inverter unit is currently 3–3,2 MW (4×800 kW). The entire plant consists of about 17 such units. In case of the failure of one 800 kW machine the overload capacity of the remaining machines could process the whole DC power.

The efficiency of a unit with four inverters reaches its maximum at about 3–4%. These very early and high values cannot be reached by a single inverter. This indicates one significant characteristic of MSO. While the third machine is only for peak power, a short-term failure of it may not significantly reduce the power. The overload capacity of the remaining other three inverters can cover most of the solar offer.

Sizes of VLS-PV can be varied to 100 MW rated power or more. It depends on the size of the available and suitable area for an installation. The example of a 30 MW bloc / module as shown in Figure 6.11 can be multiplied accordingly.

In the course of mid-term building up HVDC grids, AC transformers can be displaced by DC/DC converters allowing an increase of the plant efficiency since, among others, one conversion step will be avoided.

(3) POWER RELATION INVERTERS AND PV GENERATORS: DESIGN RULES

Two design rules can be considered.

DESIGN RULE 1 APPROX. 1:1 ($P_{inv} = P_{PVgen}$)

Inverter power at STC = rated power of the PV generator (P_{gen}), sometimes the rated power of the P_{gen} is divided through (P_{gen} × efficiency of inverter)

DESIGN RULE 2 1.X:1 ($P_{inv} > P_{PVgen}$)

Sometimes used for sites having

a. high ambient temperatures and/ or
b. remarkably high altitude.

In those cases the rated inverter power should preferably be slightly higher than the one of the PV generator although PV modules may reach higher temperatures. A simulation of the highest instantaneous values and their frequency of occurrence may help in this design decision. Generally the slight oversize has a positive effect on the technical lifetime since the thermal stress by operation is reduced. On the other hand it may be a matter of cost calculation whether an oversized inverter could be accepted.

By this constellation the inverters are oversized for securing no power limitation caused by the thermal limitation of the machines, respective compensation forthinner air at high altitudes. The reduced thermal capacity is considered by rule 2.

Detailed recommendations should in any case be provided upfront by the inverter supplier. They have to assure the qualified selection of the type and size of inverter and to safeguard the product warranty.

(4) CONFIGURATION OF THE INTERNAL AC GRID

As was said before, VLS-PV consist of a lot of sub-plants' respective sub-systems. An MV or HV substation is the link between the internal AC and the external grid.

Each sub-system is linked via an RMU (Ring Main Unit) to a common MV ring / busbar. A combined MV switch gear connects the VLS-PV to the external local MV grid. When an HV connection is required additionally a high voltage transformation together with an HV switch gear is the main connection.

Regarding feed-in at HV level, building up an internal HV RMU could be a solution too. A cost comparison should support the decision process.

Another solution to configure the internal AC power connection is the pure parallel connection of all sub-systems to one or some central MV substations hosting MV switch gears. The owners of the local grid often have requirements and conditions in regard of switch and other connection-related issues.

6.9.4 Protection systems

Three kinds of protection systems are to be considered in the plant design:

1 Overvoltage protection (obligatory)
2 Lightning protection
3 Theft protection.

(2) LIGHTNING PROTECTION

In VLS-PV a lot of metal, like module frames and support structure, is exposed to the environment. They are endangered by direct lightning strikes. Some regions are probably less endangered and some are more endangered. The owner has to decide whether a lightning protection system (LPS) is needed or not. It is a matter of individual risk assessment but usually the contract conditions of the insurance company influence the decision in the end.

LPS deals with the protection of the entire power plant against the effects of lightning electromagnetic impulses. It takes into consideration the effects of electrical and magnetic fields and induced voltages and currents caused by direct and indirect lightning strikes. It protects the investment against damage of such external and internal influences.

LPS includes so-called external LP and internal LP. External measures are realized by lightning conductors protecting against direct strikes. Internal LP protects against induced effects by use of different kinds of surge protection devices.

The application of air terminals placed in the PV field causes shading. Thus they cause power losses. The influence of the yield cannot really be assessed by a yield study. Therefore only lump-sum loss percentages are to be considered. Other technical protection solutions should be checked which could replace high air terminals with their negative shading effect.

The international standard IEC 62305 is a guideline for consequent and proper LSP application.

Last but not least, the conditions of insurance companies have to be analysed. They often contain a claim for an LSP.

(3) THEFT PROTECTION

The matter of theft is often not addressed but is a matter of fact which should gain attention accordingly. Theft in this regard is mainly theft of PV modules, sometimes of cabling material too.

Theft is an increasing problem for the owners. Beginning with the final planning, the design should broach the issue of protection of the VLS-PV. This issue concerns the care of the investment and an undisturbed operation.

Figure 6.12 shows an overview of protection systems.

Central questions have to be discussed at the beginning like:

- How endangered is the site?
- What kind of protection should be realiszd?
- What is the cost of different systems?
- What are the requirements of the insurance company?

Three main protection systems are addressed here:

a. Protection by preventive, pro-active and passive measures
b. Protection by separate active systems
c. Protection of the plant by external security service

PREVENTIVE PRO-ACTIVE AND PASSIVE MEASURES

The simplest passive measure is fencing the entire power plant. By choosing a stable high fence with barbed wire, climbing over the fence is hindered. A contact in the gates

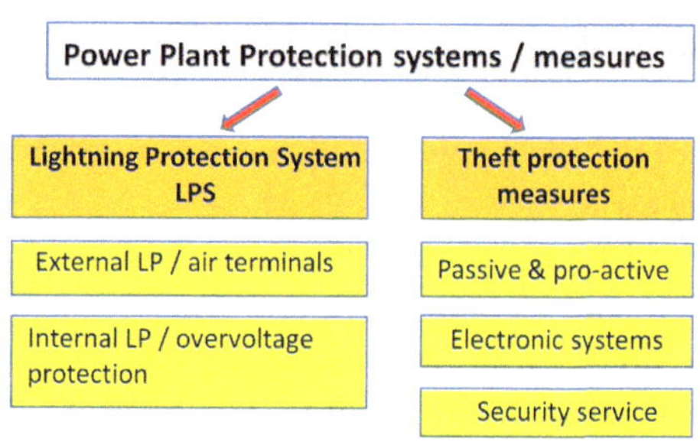

Figure 6.12 Overview of protection systems

can generate a visible and/or acoustic alarm when persons enter the site without permission. The alarm could then automatically transmit to a remote control centre as email, SMS message, fax etc.

Another passive measure is fixing PV modules with special devices to hinder removal of modules.

A further pro-active measure is called string monitoring. Monitoring software identifies an interruption in string cabling. Signalling could also be interpreted as unauthorised dismantling of a module. Most of the systems, however, are inactive after stopping operation in the evening or at night time.

SEPARATE ACTIVE PROTECTION SYSTEMS

There are a lot of solutions available e.g.:

- Power fence with integrated electrical sensors
- Infrared (IR) barrier system
- Video technique for visible monitoring and alarming
- Radar technique.

EXTERNAL SECURITY SERVICES

Under certain circumstances an external security service could be the most effective and reasonable protection. The service patrols round the site at several irregular times in the evening or at night. Sometimes patrols are necessary during the day time too.

6.9.5 Safety concepts

The planner faces a choice between two concepts for realizing electrical safety measures. Sometimes their choice depends on the type of PV module and its technology and/or local regulations and standards.

In principle two safety concepts are applicable:

1. Grounding, in Europe regulated by EN 50310, elsewhere local standards are valid,
2. Ground-free, a so-called double-isolated system (IT grid), see IEC 60364 standard. All components like PV modules, cabling and other electrical devices must then be certified as "Protection class II".

Ground-free operation is the optimal concept for VLS-PV power plants. It assures the highest availability. Should a ground fault occur in a PV generator the fault is detected by an earth-leakage monitor permanently measuring the ground resistance at the DC side. The safety of personnel is maintained and the plant operation continues. The fault then has to be locally identified and remedied in an appropriate time.

In a grounded system, first the relevant pole to connect to the ground has to be identified. A look at the manual of the PV module is sometimes helpful for orientation. Should a ground fault occur operation can continue. But if a person wants to enter the site the plant must shut down operation since it is dangerous to life. Or the design allows a switch from a grounded system to a ground-free operation for a certain period. In any case, only personnel who are trained, introduced to the operation and authorized are allowed to enter the site during operation.

Finally, the fencing must have sufficient number of easily readable warning signs showing that it is dangerous for life to enter the operational area without permission.

6.9.6 Auxiliary service power demand

In each power plant there are consumers / loads of electrical energy: beside inverters and fans, there are control units, monitoring system, air-conditioning and heater if needed. An assessment of the energy demand is obligatory for planning this kind of internal power supply. Some loads require a back-up, e.g. a monitoring system. The back-up time has to be assessed and accumulator capacity determined accordingly.

Where is the auxiliary service power AC energy to come from? Is there an LV grid with sufficient capacity nearby so that the AC power can be taken out of it? If not, a separate LV supply contract is necessary.

Since VLS-PV sites are often far away from a low voltage grid the power supply can be realized by an own MV connection. The size of the transformer has to be calculated. An additional contract with the MV grid owner must be negotiated.

In some countries it is likely that the regulatory or national programme requires that auxiliary service power must be generated by the PV plant itself.

6.10 SELECTION OF KEY COMPONENTS

The correct selection and combination of key components mainly contributes to the success of the realization of a VLS-PV.

Key components can be classified in two categories as shown in Figure 6.13:

1. Mechanical components
2. Electrical and electronic components.
3. Sensor technology.

6.10.1 Key mechanical components

Under this heading the following components are needed:

- Module support structures
- Foundations.

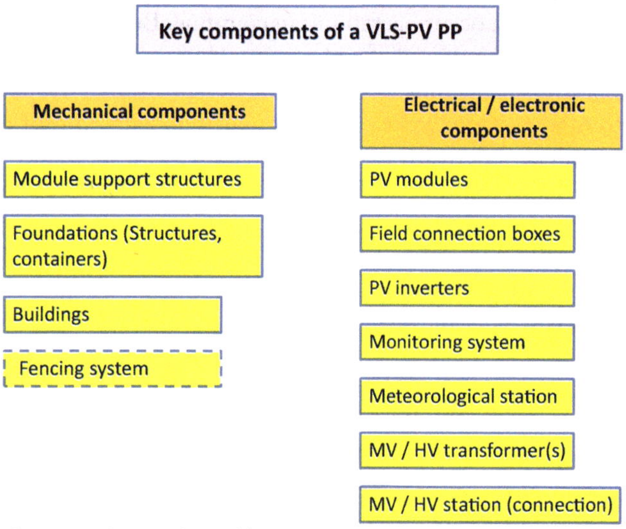

Figure 6.13 Overview of key components

(1) MODULE SUPPORT STRUCTURES

Selection and design of module support structures sometimes depend on the building permission which may include specific restrictions in regard to e.g. kind of foundations, absolute height of the structures and the material applied.

It is unusual if suppliers' standard support structures can be used. Normally the design is custom designed due to the fact that each project has its own special conditions. Conditions are site-related (e.g. wind, soil composite), PV module's mechanical structure and dimensions, restrictions of the module producer regarding installation, module string configuration, tilt angle, site profile, ground conditions, etc. Another criterion could be the requirement of local content forcing the structures to be bought within the country concerned.

In any case the PV planner should influence the design of the structures. With their contribution an optimal technical as well as cost solution can be found. The optimal technical solution considers the module string placement and fixation of the modules on the structures, reduction and fixing of string cabling, and ideas for theft prevention if necessary. LP issues could also be discussed with the supplier.

Module support structures for VLS-PV applications are made of aluminium (Al) or steel. Since Al is normally self-protecting, damage by corrosion is unlikely. The disadvantage of steel is less flexibility of profiles and corrosion issues. In the end, the current material prices could probably influence the purchase decision.

Under the assumption that the support structures will remain for many decades serving as supports for following generations of PV modules too, aluminium is the optimal material.

In regions with remarkable dirt content of the air, module cleaning and technical solutions are also an issue to be discussed and considered in the design of the structures.

Figs 6.14 and 6.15 show examples of support structures.

Figure 6.14 Monopole support structure (1×20)

Figure 6.15 Dual-pole support structure (4×18)

From the static point of view a dual-pole support structure is the more stable variant. A one-pole construction has significantly lower load limits than a dual-pole structure. Wind-induced oscillation is a risk to be considered.

VLS-PV static calculations of support structures may consider the so-called "wind-scale effect". The effect takes into consideration that only the outer support structures face the design wind speed. The majority of the inner support structures receive a smaller design wind speed and thus less material is required.

(2) FOUNDATIONS

Selection and design of foundations for module support structures sometimes depend on the building permission which may include specific restrictions. Restrictions could be in regard of material allowed, on-ground foundations only etc.

In any case the PV planner should influence the design of the foundations too. Several variants lend themselves to applications. Concrete foundations were often preferred in the past. They are still applied in countries where a local content is required. More and more common now are steel profiles hammered in the ground or earth screws. The latter are recommended by lower cost and much quicker installation helping to reduce the installation time and earlier feed-in. Results of soil investigations must however

Figure 6.16 Ramming of foundations

Figure 6.17 Earth screw foundations

allow this application. Figures 6.16 and 6.17 show ramming and screwing of foundations.

The involvement of engineers results in an optimal technical as well as cost solution. The optimal technical solution considers the module string placement and fixation of the modules on the structures, reduction and fixing of string cabling, and ideas of theft prevention if necessary. LP (lightning protection) issues could also be considered.

Foundations for other buildings like inverter containers and MV or HV stations must be considered in the context of the building permission, too.

(3) BUILDINGS

Normally a number of containers or onsite erected buildings protect the power electronics like inverters, transformers and MV switch gears from the environment. Not all VLS-PV are operated under remote control. Increasingly personnel in a separate control centre building are on site and monitor and control operations. Operation is then permanently controlled; a monitoring system provides online information on the status of the VLS-PV. The control centre is equipped among others with sanitary installations, too.

In the MV or HV substation a separate room should be reserved for the grid owner. Here their equipment for metering etc. is accommodated and locked. An own fenced entrance keeps them out of the power plant.

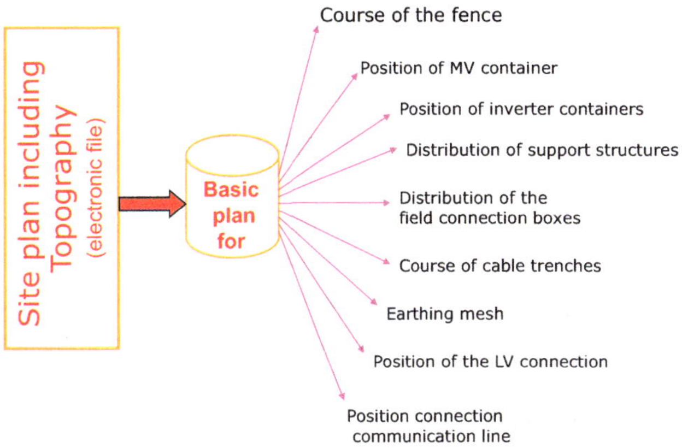

Figure 6.18 Importance of the site plan

The connection to HV mains requires an outdoor switchgear station. HV is a matter of definition and may differ from country to country.

6.11 FIELD LAYOUT

One of the most important planning steps is the field design. It comprises the distribution of support structures in a manner that shading row by row is always equal. For conducting this planning special know-how and professional software is required. Only an exact distance assures loss minimizing by shading. Not only module rows but also all container placements and defined distances from the fencing system and service lanes must be considered. The distance of structures from the module support structures must be defined before construction.

Field design draws on the exact border lines and topography of the site as basic plans. Thus the importance of the site plan is obvious (Figure 6.18). The plan must be available in an electronic format (for example, .dwg format). The figure shows the high importance of this file.

6.12 KEY ELECTRICAL AND ELECTRONIC COMPONENTS

The selection of the manufacturer and type of PV module is one of the most important decisions to be made. Next and also of high importance are the inverters to be selected. The majority of inverter suppliers offer a package of components directly coupled with the machines. This combination is very useful because of the importance of inverter communication and connection with the machines. On the other hand the planner is excused from market research to find suitable components matching to the special kind of inverter. And last but not least the guaranty and warranty issue is reduced to only one supplier.

The package inverter suppliers usually offer includes:

- Container for inverters
- Field connection boxes (AJB and GJB)
- Monitoring program and software licenses

- Meteorological station
- MW/HV switch gear.

6.13 PV MODULES

6.13.1 Technology overview

PV modules are beside the DC/AC PV inverters the main system components. They consist of solar cells connected in series (so-called strings) and in parallel connected strings. Solar cells directly convert solar irradiation into electricity.

The development of PV module technologies in recent decades has generated a lot of different types. For VLS-PV applications two main technologies are of relevance: crystalline technology (c-Si) and some thin-film technologies.

Currently the main categories and efficiencies are as follows:

- Crystalline or wafer-based technology (c-Si)
    - Mono crystalline (15–20% efficiency)
    - Multi crystalline (13–16% efficiency)
- Thin-film (TF) technologies
    - a-Si and a-Si combinations: amorphous / microcrystalline silicon (7–10 % efficiency)
    - CdTe: cadmium telluride (11,5–12,5 % efficiency)
    - CIS / CIGSS: $Cu(In,Ga)(S,Se)_2$ (11–14% efficiency)
- Hetero-junction technology
    - a combination of c-Si and a-Si.

Figure 6.19 shows a set of different types and sizes of PV modules for different kind of applications.

The majority of VLS-PV applications around the world by far consist of the crystalline silicon cell technology. It is currently the best understood and proven technology in the field with around 30 years of experience.

The share of TF technologies worldwide is increasing although the TF share for VLS-PV applications is still rather small. Its main advantage is still the low cost.

The power of a PV module is expressed by so-called Watt peak (Wp). The Wp indicates the rated power of PV modules to be found in module data sheets and on the label on the back of the module. This value is determined by the PV factory and measured and proven by an independent institute. Wp values in data sheets are lab values (so-called STC values, Standard Testing Conditions: air mass AM 1,5, 25°C cell temperature, simulated solar irradiation of 1000 W/m^2). Such data sheet information mainly allows customers/users to compare different types of modules. The values do not provide detailed information about the energetic behaviour of the modules in the field. Field conditions normally deviate a lot from STC conditions which may occur for only some hours in a year.

Some suppliers indicate within the delivery the real measured Wp on the modules' label.

Figure 6.19 Different PV modules

6.13.2 Crystalline or wafer-based technology (c-Si)

Crystalline silicon solar cells and modules have dominated the PV technology market since the beginning of terrestrial PV application. The most important reasons for crystalline technology diffusion are related to the success of well-known basic materials applied in the semiconductor industry, unlimited resources and last but not least the stable behaviour and impressive efficiency and potential of c-Si solar cells.

At its beginning the PV solar industry was dependent on semiconductor grade silicon. Special purified Si for PV application (solar grade material) has been produced for years now in factories.

Crystalline silicon technologies can be divided in two different groups (see Figure 6.20):

- Mono crystalline, where cells are manufactured from a single crystal ingot
- Multi crystalline, where cells are results of different silicon grain arrangements.

Mono crystalline cells are manufactured from a single crystal ingot, typically grown with a very high temperature and energetic process, called the "Czochralski process". Ingots are sliced and processed to make cells. In order to increase cell efficiency and reduce the reflected sunlight by the modules, a thin antireflection (AR) coating is applied.

Poly crystalline silicon cells, also known as multi crystalline silicon cells, have an important position in the PV technology portfolio. The starting material for the realization of multi crystalline cells is melted and cast in a cuboid form. After forming grains of different sizes, starting from a few millimetres up to a few centimetres, the ingot is cut and sliced into wafers used to build cells.

The grain boundaries for poly crystalline cells reduce efficiency with respect to mono crystalline technology but at the same time reduce the cost of production. Multi crystalline modules have a poorer quality with respect to mono crystalline technology because of crystalline defects, as grain boundaries and dislocations.

Figure 6.20 Example of appearance of mono- and multi-crystalline PV modules

Figure 6.21 a-Si (left) and micro-crystalline (right) module

Figure 6.22 Example of appearance of CdTe module

6.13.3 Thin film technologies

Thin film (TF) technologies are designed in order to reduce production costs by use of only a fraction of material compared with wafer-based solar cells. The deposition of a thin layer of photoactive material, less than 10 micro millimetres, leads to about a factor of 30–100 less material than crystalline wafer based cells require. On the other hand, some of the resources of processed material required in TF production are limited.

Photoactive TF material is deposited directly on low cost large area substrates. Flexible lightweight modules can also be produced for consumer applications using thin polymer or metal substrates and roll-coating techniques, although currently the majority of commercial products currently use glass as substrate.

There are currently two relevant thin-film technologies for VLS-PV applications:

- cadmium telluride CdTe (see Figure 6.22)
- Cu(In,Ga)$(S,Se)_2$ CIS / CIGSS.

Remarkable technical features of TF PV modules are:

- The power output behaviour of TF modules is semi-stable over time.
- The performance at low light is generally higher than that of c-Si modules.
- The temperature coefficients of TF modules (except thin-film Si) are lower than of crystalline modules. In consequence power losses caused by high cell temperatures are lower and energy yields are expected to be slightly higher.

The different slopes of curves in Figure 6.23 show how power temperature coefficients influence energy production of diverse technology, in particular in sites of high temperature stress condition for PV modules. The backsheet temperature represents in first approximation the cell temperature of the module. A low power temperature coefficient, as shown for TF technology, allows maintaining the efficiency nearer to the nominal one.

- Each TF technology has a specific spectrum density of the global irradiation. The air mass (AM) values influence the efficiency of the PV modules too, as shown in Figure 6.24. Some technologies have a better so-called low-light behaviour compared with c-Si modules.
- TF modules based on amorphous silicon (a-Si, Homo-junction) currently have an efficiency of between 7–10%. It is possible to improve the efficiency of a-Si based modules using micro crystalline silicon thin film (μ-Si). μ-Si is more stable and can convert a different portion of the solar spectrum, so double junctions made of a-Si and μ-Si, called tandem technologies, are more efficiency and stable than a pure Homo-junction type.

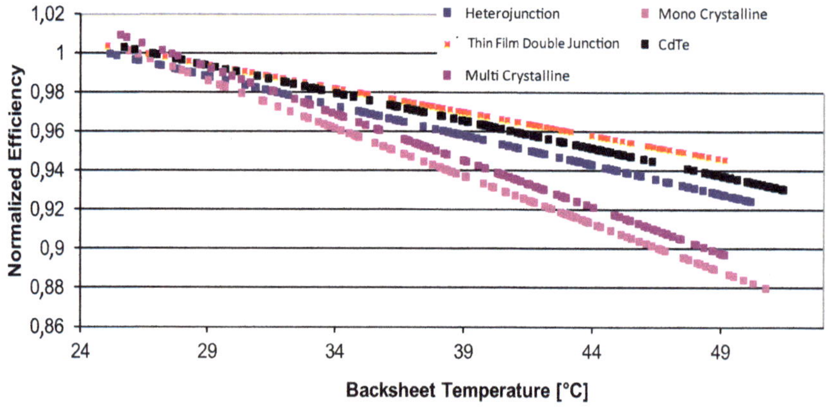

Figure 6.23 Influence of different coefficients

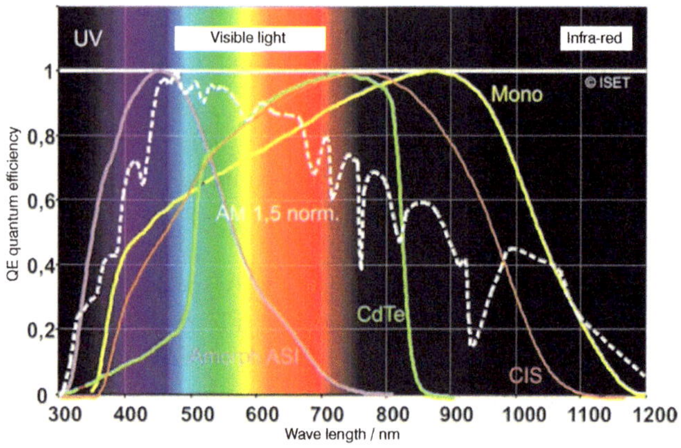

Figure 6.24 Spectrum density of different technologies at air mass 1,5

- Another TF technology is processing other photoactive materials like cadmium telluride (CdTe). The attractive features of CdTe are its chemical simplicity, relative stability, and low production cost. CdTe technology has a promising potential of a further increase in the existing efficiency of 11,5–12,5%.
- The material combination of Copper Indium / Gallium Diselenide / Disulphide and related I-II-VI compounds (CIGSS) technology currently exhibits the highest cell and module efficiencies of all thin-film flat plate technologies: 11–14%. Like CdTe the theoretical achievable efficiency is also higher and a challenge for future R&D.

6.13.4 Hetero-junction

In order to improve the efficiency of a-Si technology (Homo-junction) it is possible to combine thin film layers of different materials and different composition. Every layer is able to convert a different portion of solar spectrum, so it is possible to increase the conversion efficiency. These kinds of modules are called multi-junction thin film modules. Generally they have two or three layers. Another technical solution to optimise efficiency and energy production using a-Si is the Hetero-junction type, a junction made with amorphous silicon and crystalline silicon. The use of a Hetero-junction, instead of a Homo-junction, allows raising voltage and increasing output power using the same materials. Due to the presence of crystalline silicon, Hetero-junction modules are a combination of conventional silicon solar cells with a TF technology.

Module efficiency reached by Hetero-junction in labs is between 15% and 20%. Currently the cost of the Hetero-junction process is too expensive due to the fabrication process, but the energy produced by a Hetero-junction module can be up to 10% more than a mono-crystalline module of the same efficiency. In fact, as shown in Figure 6.23, losses due to temperature in Hetero-junction cells are lower than in the case of crystalline technology.

6.13.5 Valid international PV module standards

Standards from the International Electrical Committee (IEC) deal with basic issues of minimal requirements on PV modules. They list and describe the design qualification and type approval. For assuring the minimal quality of a PV module a certificate provides the proof. The European standard EN 50380:2003-09 describes datasheet and nameplate information for PV modules. It also gives minimum requirements as information for the user (see also Section. 6.22).

Relating to c-Si modules is the standard IEC 61215 (Figure 6.25), for TF modules IEC 61646. The standard IEC 61730 describes minimal requirements for module safety qualification, also three so-called protection classes. Class A (class II) is valid for VLS-PV applications.

6.13.6 Warranties and guaranties

(1) WARRANTIES

Mentioning warranties in combination with a VLS-PV, two kinds have to be stated. One is for special components e.g. like PV modules and inverters, and the other one is for the PV power plant which is rather similar to common conditions of other industrial complex contracts. An extension of warranty times may be negotiated.

Standard IEC 61215 (c-Si Modules) Entrance Test Procedure

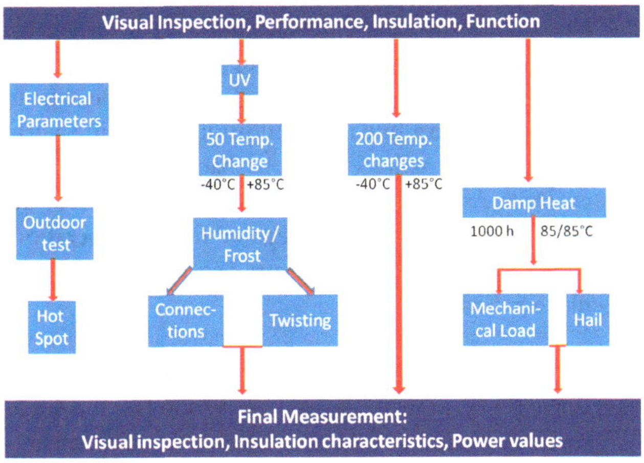

Figure 6.25 Environmental quality testing sequences for c-Si modules

Warranty conditions for PV modules currently sometimes vary from 5 year up to 15 years. The warranty conditions of the suppliers should be studied very carefully. For the EPC plant contract the same check must be done.

(2) GUARANTIED MODULE POWER AND PLANT EFFICIENCY

Guaranty conditions are at the choice of suppliers. Accordingly the power guaranty conditions for PV modules differ from supplier to supplier. To ensure that the delivered power of the PV modules ordered is within the contractual limits, the proof has to be provided by the manufacturer. The validation of the manufacturer's flash protocol values can be executed by sample testing in an independent and certified PV module test institute. It performs flash testing on samples from delivery lots of the project for comparison to the values provided by the manufacturer. This is done on the basis of mutually agreed testing procedures and tolerances.

The investor will probably demand a guarantied minimum plant efficiency of the VLS-PV. The efficiency mostly is described by the plant quality factor PR (see also Section 6.16.3). A performance ratio contract with the EPC is one tool serving to secure the investment.

(3) TYPE OF PV MODULE INFLUENCES PLANT DESIGN

A general decision has to be made on how energy intensive the selected site shall be. Should the VLS-PV site produce the highest feasible energy? Then the PV module chosen should have a high efficiency. Or should the price of the power plant be as low as possible? Then the efficiency is not the only dominating factor, the price of the module is then the main criterion. Low price modules are currently mostly TF modules.

General influencing issues:

- Module technology: c-Si modules are generally suitable for ground-free operation. Some TF modules require a grounded system.
- Module efficiency: the higher the efficiency is, the less are the relative costs of support structures, foundations, cabling and installation as well as the cost for the land required. For achieving the same power TF modules need significantly more area compared with high efficiency crystalline modules.
- Module size (mechanical dimension): an increase in the size normally corresponds to an increase of the rated power (Wp). The more powerful a module is, the fewer module devices need to be installed, the less strings are needed in the PV generator. The design is clearer and more lucid.
- Module power (electrical size): the higher it is, the lower the installation cost.
- Admissible system voltage: the higher it is, the lower the cost for DC cabling (cross-section and module string numbers) and inverters etc.
- Open circuit voltage V_{oc}: the number of PV modules in a string with a V_{oc} with a value of about 100 V_{DC} each (rather often common for thin-film modules) is much less than an applied c-Si module with around 36 V_{oc}. In consequence the number of parallel strings with such TF modules could be about 3 times higher compared with strings containing c-Si modules. More strings in a PV generator increase the number of connection boxes, costs of BOS (balance of system components) and reduces the clearness and lucidity of the structure design.
- Module with frame or without: modules without frames result in higher costs for assembling the modules and for the support structures and foundations but could have other advantages which have to be balanced (lower product price, power losses by dirt is less or will occur later etc.).

Beside a comparison of technical values, non-technical arguments may influence the purchase decision. New types of solar cells / modules have a lack of long-term field experience. Thus banks are cautious to provide financing for projects in which these modules are to be applied. Other arguments could be trust in the module producer, quality issues, power tolerance of the module, warranty and performance guarantee conditions etc.

6.14 PV INVERTERS

Beside PV modules the second key components in a VLS-PV are inverters. They convert the DC power produced by the PV generators into normally low voltage AC power. There are two different electrical inverter topologies: transformer-less inverters and inverters with electrical isolation / transformer. A transformer-less type has a higher efficiency. But the ones equipped with a transformer assure a proper electrical system isolation which sometimes is more advantageous.

A further distinction is in the size itself. It is expressed by so-called decentral or central application. Decentral means a large number of relatively low power inverters, e.g. 50–100 kW each. They each are connected with lower power PV generators (e.g. 50–100 kW); all sub-systems are connected in parallel.

The rated power of a central inverter is for example,in the range of 500 kW up to 1 MW or even higher.

The technical requirements from the grid owner's power system side need to be satisfied to ensure the local and national safety conditions as well as the reliability of the grid owner. Identifying the technical requirements for grid interconnection and solving the interconnect problems such as islanding detection, harmonic distortion requirements and electromagnetic interference are therefore very important issues for the official acceptance of VLS-PV systems.

The type of inverter has an influence on the system architecture (see also Section 6.9.3).

Inverter units compared with central inverters can operate with larger PV generators. The effects of inverter units operating in a master-slave mode are manifold:

- Less systems in parallel
- Reduction of complexity of a VLS-PV
- Highest overall efficiency in conversion along with optimised yields.

Depending on the PV module to be applied the kind of operation must be clarified: grounded operation or floating operation (IT).

The following selection of characteristics distinguishes a state-of-the-art product for VLS-PV:

- Very high reliability
- High rated power, e.g. 800–1 000 kW
- High conversion efficiency even at low DC power
- Highest rated efficiency even at 100% DC load
- Overload ability to a certain extent
- High MPP dynamics and accuracy
- Highest rated efficiency at operation temperatures from 0°C to 45°C (or higher)
- Wide operational voltage
- Rated / maximum DC system voltage up to 1 000 V
- Reactive power compensation
- Active THD filter
- Qualified and certified according to standards
- Safety integrity level (SIL)2 to IEC
- Humidity according to IEC 60068-2-56 (e.g. operation: 10–90% RH, 27°C)
- Appropriate and reliable cooling concept, especially valid for regions with high ambient temperatures
- EMC (electromagnetic compatibility) certificates
- Interference resistance IEC801-2/3/4/6; emitted interference EN 50091-2 tables 2 and 4
- In some countries PV power plants must be capable of a dynamic line backing / support operation. The generation plant is not allowed to disconnect from the line due to a line fault. Additionally during a grid fault the grid voltage must be backed by a reactive current. As soon as the fault has disappeared the inverter has to continue regular feed-in operation. This requirement will in future be valid for countries like e.g. Spain.

Advanced inverters take over powerful grid management functions. Relevant features are described in Section 6.8.3.

The final selection of the type of inverter takes other – non-technical – issues also into account and may influence the purchase decision too such as:

- Supplier with sufficient field experience
- As an operational issue: qualified information on internal values to be processed in the monitoring system
- Supplier with matching service grid (technical problem solving by short reaction time)
- Mid- and long-term service contracts available (implying among other things a guarantied access to spare parts over the service contract time).

Figure 6.26 shows a central inverter unit container planned and realized.

6.15 FIELD CONNECTION BOXES

Inverter suppliers additionally offer auxiliary packages matching their inverter concept. Field connections boxes are within such a package. Such an offer helps the planning engineers. It avoids own engineering and the boxes are within the warranty conditions of the supplier. In any case, it does not release the planner from checking the suitability of the boxes for his individual application.

Field connection boxes are the links between PV module strings and inlets of the inverters. Two kinds of field connection boxes exist:

- AJB (array junction boxes): a lot of them are connected in parallel with a GJB. By optimizing the cross-section of cabling the DC losses can be reduced.
- GJB (generator junction boxes): the number of GJB are determined by the number of DC inlets of the inverters.

In the course of cost optimization increasingly AJBs are replaced by GJBs. These GJB boxes have therefore many inlets for strings.

Field connection boxes are usually equipped with:

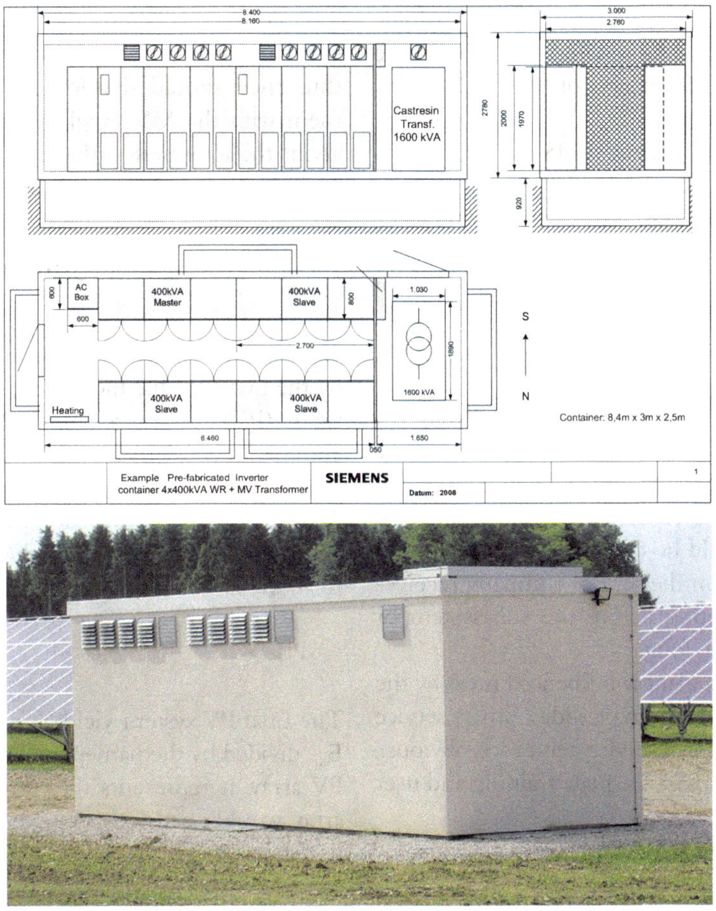

Figure 6.26 Example of a 1,6 MW central inverter unit: plans (top) and in situ (bottom)

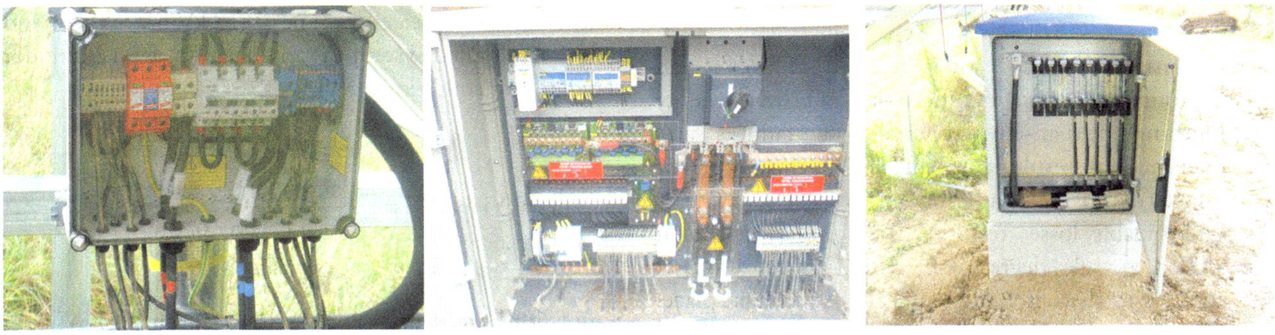

Figure 6.27 Types of field connection boxes

- String protection devices (fuses)
- Devices for surge protection
- String monitoring devices
- Power switches for disconnecting parts of the PV arrays e.g. for service works.

An impression of the variety of boxes offered is provided in Figure 6.27.

6.16 MONITORING SYSTEM AND PERFORMANCE RATIO

6.16.1 Monitoring system

A monitoring system (MS) is the main tool for operation; it is also called System Control and Data Acquisition (SCADA). It requires great attention to its features since the MS will be the tool for plant control over the entire operational period.

Operators have full responsibility for keeping the plant running regularly. Therefore it will have all the features that a professional control desk requires.

Generation of reports in particular is increasingly important. The MS provides the owner with up-to-date information about the performance of the VLS-PV. A yearly resumé is not sufficient.

Four main purposes of an advanced MS are:

1 Monitoring of the behaviour of the plant in operation
2 Identification, analysis and characterization of loss mechanisms for further reduction to the minimum level as possible
3 Generation of reports for the owner / investor
4 Archival storage of process data.

Professional monitoring has to ensure that problems like faults are identified and addressed quickly. Then the skilled operator can in a short time classify the faults and decide what remedial action should be taken and by whom. In a well-structured organization the service company receives an order for trouble-shooting before the sub-system or plant has been down for too long.

Since the availability of a plant is influenced most by the reliability of the inverter, in order to provide a timely service the importance of a professional service contract is obvious.

For conducting the planning adequate training and user knowledge is required.

6.16.2 Classification of failure modes

Three main categories of failure can be classified by their severity:

1 Requiring urgent intervention: failures and malfunctions causing significant loss of power / energy
2 Needing median-term intervention: failures and malfunctions resulting in smaller losses but in the mid-term causing significant cumulative losses
3 Unavoidable losses resulting from imperfect design and/or module degradation: After identification of such losses discussion of modifications such take place. Optional technical solutions should be developed and the consequences should be evaluated by a cost-benefit analysis.

The following types of communication are sent from the plant to the operator:

- Messages: Normally the operator receives messages containing status information which is induced by manual or automatic switching operations.
- Warnings: These are fault conditions in the plant involving no shut-off of the plant or partial shut-offs.
- Faults: Fault conditions involve a shut-off of the plant.

Most inverter suppliers have designed their own monitoring system according to their own needs. Apart from these internal standards users may define their own special needs and additional requirements to be tailored.

A customer currently cannot refer to an international standard dealing with general requirements for VLS-PV, thus they should develop their own ideas and discuss them with the MS supplier. They should think carefully about the long period that the MS will be in use and the personnel who will work with the system.

6.16.3 Performance ratio

How can the efficiency or quality of a VLS-PV be screened by the owner? One indicator is the so-called performance ratio (PR). A professional MS has software which automatically calculates this ratio.

PR has been established by the International Energy Agency (IEA) Photovoltaic Power Systems Program. It is described in the IEC standard 61724 (IEC, 1998); it is the relation

$$PR = \frac{Y_f}{Y_r}$$

The final PV system yield Y_f is the net AC energy output E_{AC} divided by the nameplate DC power P_0 of the installed PV array. It represents the number of hours that the PV array would need to operate at its rated power to provide the same energy. The reference yield Y_r is the total in-plane irradiance G_{Mod} divided by the PVs reference irradiance G_0. It represents an equivalent number of hours at the reference irradiance.

The annual yield of a solar system is defined as the ratio of generated electrical energy in kWh or MWh and the installed rated system capacity in kWp or MWp under standard test conditions (STC).

$$\text{Annual Yield } (Y_a) = \frac{\text{annual AC electric energy output of the system}}{\text{rated power of the installed system}}$$

Simulations are usually executed by a simulation / sizing program. It uses irradiation and temperature values from the irradiance data sources, including global and diffuse components of the solar irradiation. Further, it takes into account all system gains and losses.

Monitoring the PR is also important: by analysing the trend of the PR values over the years the degradation behaviour of PV modules can be studied and identified.

A performance ratio (PR) contract in place requires precise and true operational data for the calculation of the PR. In this context sensors in the meteorological station among others must deliver accurate values of solar irradiation and module temperature.

An increasing number of PV module producers now guaranty linear degradation limits. Should the owner identify significant deviations in power output they can take samples for testing in a certified institute. Getting a positive result (i.e. a decrease in module power) is a reason for checking again the power guaranty conditions. It may lead to a decision to lodge claims with the module manufacturer.

6.17 METEOROLOGICAL STATION

The meteorological station comprises different sensors of diverse meteorological data and converts them into electrical signals sent to the monitoring system.

The following sensors are commonly used (Figure 6.28):

- Pyranometer, measuring the global irradiation at the site at plane level
- Calibrated reference cells / modules, in-plane with the modules applied, for measuring the solar irradiation and measuring of cell / module temperature; the reference cells provide temperature-compensated solar irradiation values with an accuracy of about ±2% or better. It should have the same cell technology as the applied modules
- Measuring device for ambient temperature
- Anemometers (wind speed), 2D
- Sometimes relative humidity RH and rainfall is also measured.

Redundancies of reference cells should be taken into account.

Pyranometers measure at the horizontal level the total solar spectrum from 0.3 to 3 micrometers wavelength at the site and give an integrated measurement of the total short-wave solar energy available under all conditions.

Reference cells of the same cell technology as the applied modules are placed in-plane of the PV modules and measure

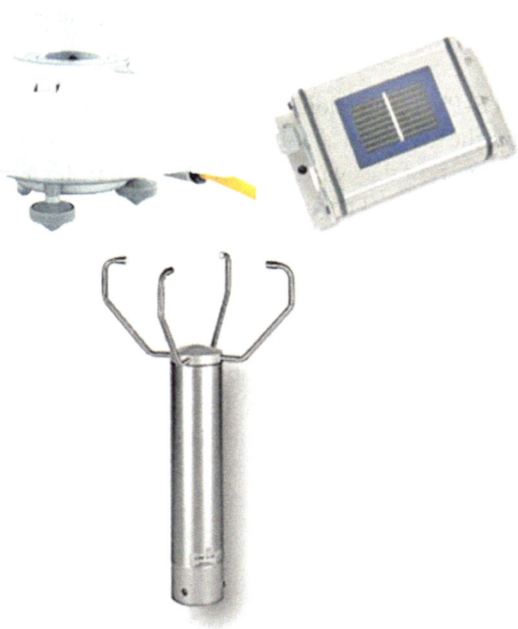

Figure 6.28 Example of pyranometer, reference cell and 2D anemometer

the cell-type-relevant solar irradiation all other power modules are facing. The calibration assures that the thermal mechanism of the module is eliminated so that the real irradiation values are provided for a PR calculation. A recalibration should be carried out every year (or second year).

The accuracy of values of the solar irradiation and module temperatures should be high, e.g. error margin ≤ ±4%. The higher the accuracy of sensors, the more qualified system information is generated, allowing better evaluation of the system and operation behaviour.

6.18 DC LOSSES AND DC/AC CABLING

6.18.1 DC losses

Among other factors influencing the overall plant efficiency (PR) DC losses play an important role. They can well be influenced and adjusted by cabling calculations. Other PR influencing loss factors can only be varied slightly by design like the tilt angle of the PV modules, or by shading angle determination between rows. Efficiencies of DC/AC conversion by the type of chosen inverter and transformer and their losses are the result of selection of suppliers and relevant types.

DC power losses are usually expressed by the voltage drop. It requires a consequent adjustment of the cross-section of the DC cabling. Values of 1% up to 1,5% calculated at STC are usually acceptable. Losses under operation are about 70% to 80% of the STC calculated percentage, depending on the particular site and design conditions.

Yield assessment studies are basic technical data sources for the economic calculation of the investment count with DC losses too. A negative deviation of the losses in reality compared with calculated DC losses would affect the entire economic calculation having then critical negative consequences.

An advanced calculation could compare the voltage drop with current copper or aluminium prices to find the optimal energy/cost relation. For a 10 MW PV generator the volume of copper cabling is around 150–160 tons.

6.18.2 AC cabling

A floating DC system requires special certified cabling material. It must be certified as protection class II; this means being double isolated (see Figure 6.29).

Attention should also be paid to power losses of the internal AC cabling. They among other factors influence the overall efficiency (PR) of the VLS-PV. Therefore in the course of the plant design the maximum voltage drop / power loss must be defined. A value of 0,2% to 0,3% at STC is usually acceptable.

Longer internal AC cabling lengths increasingly cause inductive or capacitive appearances in the internal

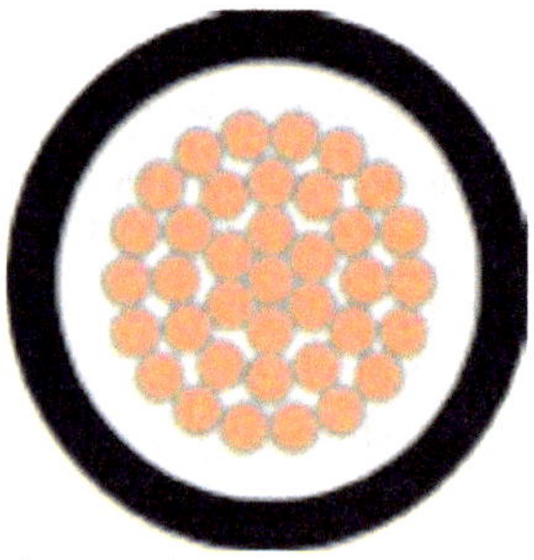

Figure 6.29 One-core double isolated cable

transmitting part. State-of-the-art inverters are able to compensate for these appearances.

Fibre optic cables should preferably be used for internal data communication cabling. They cannot be disturbed by atmospheric influences and can be put together with power cabling in the same cable ditches.

6.19 TRANSFORMERS

An international agreement of the definition of the range of medium voltage (MV) and high voltage (HV) does not exist. In several European countries a common definition is used as MV = 1 000–50 000 V, and HV ≥ 50 000 V.

A grid-connection to an MV grid requires one or more transformers, depending on the size of the VLS-PV and the MV architecture chosen. If the primary network is at HV level-two transformation steps are common (LV–MV, MV–HV).

The transformer to be applied is either an oil-filled transformer or a resin-encapsulated one. The first one is a cheaper solution since they are standard and less costly to produce. However, from the environmental view the oil-filled type requires maintenance like exchanging the oil. A leakage may cause contamination of the soil which must be avoided. A basin below the transformer is a safety measure.

Resin-encapsulated transformers are environmentally friendly types. Their overload behavior is significantly better; service actions are limited to measuring only. The disadvantage is the higher price. For VLS-PV far away from towns they may be preferred anyway because of the low maintenance costs.

6.20 YIELD ASSESSMENT STUDY / REPORT

The investment decision needs basic inputs for calculating the economic viability. One basic input is the expected annual yield in MWh of a VLS-PV. The yield can be calculated by using so-called simulation programs or sizing programs. The output of this study is the yield assessment report. There are a lot of simulation programs available.

In parallel to plant engineering and after the set up of the plant design, basic data values are available for the yield assessment study. All programs have a common need for project related detailed inputs like geographical site information, plant layout, data sheets of main components like PV modules, drawings of the module support structures, inverters, transformers, tilt angles of the modules, shading angles, DC and AC losses. Furthermore an assessment of the degradation of the modules (minimum values are mostly indicated in the module data sheet) and the availability is required for calculating the yield over the entire operation period.

The report results are then embedded in the economic project calculation. Figures of the electricity generated per year multiplied by the price for the electricity show the gross income from PV production. Direct production costs also have to be estimated by the planning engineers. These costs include assessed expenses for operation like maintenance costs, costs for regular service routines, estimated repairs and auxiliary power supply costs.

Depending on the kind of program used and the inputs provided by the planning engineers, the results of the calculations may differ. Reasons are e.g. use of different data sources (see also Section 6.8.3), program algorithms, different approaches to calculation etc.

Examples of some common simulation programs are given in Table 6.3.

Table 6.3 Some common simulation programs

Name of program	Company / developer
Meteonorm	Meteocontrol (GER)
PV*Sol Expert	Valentin Software (GER)
PVSYS	Fraunhofer Institute ISE (GER)
PVSYST	University of Geneve (CH)
SolEM	DGS / FH Munich (GER)
Greenius Expert	FHTW Berlin / DLR, PV and solar thermal (parabolic trough) (GER)

6.21 MV CONNECTION POINT

VLS-PV connection with the local MV grid is realized by the PV main substation being usually located on the property of the owner of the VLS-PV. Transfer terminals are the interface. The grid owner is normally given a reserved separate room for MV measuring devices and counters.

The physical connection point sometimes is outside and some distance away in the main station of the grid owner. In this case the MV cabling system from the PV site to the switch gear in the main station has to be executed and paid for by the PV plant owner. This causes higher cabling costs and AC losses, reducing the PR accordingly.

6.22 TECHNICAL DOCUMENTATION AND DESIGN-RELATED STANDARDS

6.22.1 Technical documentation

Part of the final planning is the compilation of the technical documentation. It should incorporate a well-structured identification system for all plans, drawings, files etc. in view of the long operational time of decades. It has to serve as a detailed information source for further works like repairs, trouble-shooting, diverse services and a maintenance plan. The latter has to list regular maintenance and pro-active measures.

In parallel to the planning process, a manual for the operator and a trouble-shooting guide must be prepared. Corresponding to them, a list of spare parts required must be developed.

6.22.2 Design-related specific standards

Development of PV related standardization is an ongoing process. Since around the year 2000 the market and products have developed rapidly and several international technical committees (TC) are increasingly engaged in setting up well balanced and coordinated standards. Apart from standards of the International Electrotechnical Commission (IEC), standardization in Europe was the precursor. Initiated by the inauguration of feed-in tariffs, Europe was the first booming region for grid-connected PV application and VLS-PV in particular. For that reason the European photovoltaic industry had a vital interest in backing the development by setting its own PV standards. Thus in parallel to IEC there are also EN (European Norm/Normes) standards.

Apart from PV standards other electrical, mechanical, environmental and other related standards must be considered.

In Table 6.4 a selection of current PV related standards at the international level is listed. Beside these standards for each project and PV design, valid national standards and local rules, standards and regulations have also to be evaluated and considered.

6.23 END OF TECHNICAL LIFETIME

A VLS-PV construction is not really designed for a certain technical lifetime. The primary time span commonly is the

Table 6.4 Selection of PV related standards

Standard	Object / Title
IEC 60269-1:2010	Low-voltage fuses – Part 6: Supplementary requirements for fuse-links for the protection of solar photovoltaic energy systems
IEC 60364 (2002-05)	Electrical installations of buildings – Part 712: Requirements for special installations or locations – Solar photovoltaic power supply systems
IEC 60904-2:2007	Photovoltaic devices – Part 2: Requirements for reference solar devices
IEC 60904-12:2007	Photovoltaic devices – Part 3: Measurement principles for terrestrial photovoltaic (PV) solar devices with reference spectral irradiance data (STC definition)
IEC 61215:2005	Crystalline silicon terrestrial PV modules: Design qualification and type approval
IEC 61557:2007	Electrical safety in low voltage distribution systems up to 1 000 V a.c. and 1 500 V d.c. Equipment for testing, measuring or monitoring of protective measures. General requirements.
IEC 61646:2008	Thin-film terrestrial PV modules: Design qualification and type approval
IEC 61727:2004-12	PV systems – Characteristics of the utility interface
IEC 61730-2:(2004-10)	PV modules – Safety qualification (protection classes)
IEC 62109-1:2010-04	Safety of power converters for use in photovoltaic power systems
IEC 62446:2009	Grid-connected photovoltaic systems – Minimum requirements for system documentation, commissioning tests and inspection
IEC 62305 1-4:2010	Protection against lightning
EN 50539-11:2010	Low-voltage surge protection devices – Surge protective devices for specific application including d.c. – Part 11: Requirements and tests for SPDs in photovoltaic applications
EN 50380:2003	Datasheet and nameplate information for photovoltaic modules
EN 50521:2009-10	Connectors for photovoltaic systems – security requirements and approval (German version EN 50521:2008)
EN 50530:2010	Overall efficiency of grid connected photovoltaic inverters
EN 50548:2009-10	Junction boxes for PV modules
IEC 62446:2009	Grid-connected photovoltaic systems – Minimum requirements for system documentation, commissioning tests and inspection
Draft EN 50548:2009-12	Junction boxes for PV modules

project period depending on the valid period of the feed-in tariff. Financing contracts and long-term service contracts for inverters, maintenance contracts, and power guaranties for PV modules are backing a secure and reliable operation. What could be the situation after the phase-out of the project period? What can be expected from the technical point of view? When should a plant be dismantled? No clear answers are possible.

First considering the PV modules: studies have shown that the technical lifetime can be 30 years and longer. Then looking at large central inverters, their technical lifetime can be much longer than the project period. A complete exchange of an entire machine is an exception. What sometimes happens is, e.g., an exchange of control boards only. Finally the definite end of availability of spare parts limits their operation. Transformers are least effected by ageing, as practice proves.

Module support structures are also long lasting. Except in regions with an aggressive environment, aluminum structures with foundations, for example, are able to last for many decades. After removing modules that have reached their end of technical lifetime, the next generation of PV modules could be put on.

Should the dismantling of the plant be decided on, usually all materials can be recycled. For PV modules recycling systems are initiated by the PV industry. All other materials are non-critical too. Dismantling costs will be much below the price of the materials themselves like aluminum, copper and steel.

6.24 CONCLUSIONS

The main factors regarding design and engineering which influence the success of planning results essentially are as follows:

- Identification of all relevant site special features
- Experience in planning VLS-PV
- Planners being able identifying the optimal system architecture
- State-of-the-art market information on key products like PV modules, inverters etc.
- Selection of high quality products
- Incorporating pro-active measures in the design phase
- Design of a quality strategy for all phases of the project
- Strict enforcement of the quality concept
- Proper safety and security concepts
- Qualified and experienced subcontractors in the realization phase.

The quality of execution of the planning is mainly dependent on the qualification and experience of the engineers and site managers involved. Beside electrical know-how, basic knowledge of neighboring fields is required, such as testing procedures, mechanical and static as well as soil science and meteorology.

Inherent to VLS-PV are a very large number of conversion devices (PV modules) connected, a lot of power electronics (inverters), measuring systems etc. The larger the VLS-PV, the higher the complexity. Depending on the region, more and more VLS-PV are being situated far away from living environments. Therefore the issues of safe, secure and trouble-free operation, maintenance and servicing should increasingly be considered in the course of plant design.

Several issues could be realized by reducing the complexity of the architectural design. This can go along with application of rugged key components like specially adapted inverter design and even the selection of qualified PV modules with features promising long-term stable behaviour. The selection of components requiring less maintenance is another contribution in this direction. Pro-active maintenance efforts could also prevent and reduce technical problems in the medium and the long term. Last, but not least, well-developed technical documentation is proof of a professional engineering job.

CHAPTER SEVEN

Financial guidelines for VLS-PV

7.1 INTRODUCTION

A project is considered as "successfully developed" once the financial closing stage is reached. Indeed, at that point onward, the investors and the lenders are committed to disburse their funds for financing the plant's construction, which will start shortly after the financial closing.

To reach this stage where the project is developed successfully, the project has to be approved by credit committee of the banks and investment committee of the investors concerned. In the case of approval, technical and legal due diligences will be conducted by independent auditors mandated by the financiers. The result of such audits will condition the availability of funds for later stages of the project.

The feasibility study should anticipate the requirements of the credit and investment committees and the contents of due diligences. A high quality level, pluri-disciplinary feasibility study is therefore necessary to secure the financing. It should address, together with the technical aspects, all the key financial and legal issues. Figure 7.1 attempts to describe the key milestones in the development of a VLS-PV transaction.

This chapter is an overview of what is required to achieve the financial closing stage and build an acceptable financial model for a VLS-PV transaction.

7.2 ELABORATION OF THE PROJECT

There are two main legal environments in which a promoter can develop a project: either the State has fixed a rule to define the electricity selling price for all eligible projects (namely, a *feed-in-tariff* mechanism, called "REFIT"), or there is no rule and each project has to demonstrate its added value to the country and the selling price is negotiated on an individual case basis.

We will not deeply discuss the different REFIT mechanisms in this chapter. Some issues are addressed in the case study for West Africa (see Section 11.1). Nonetheless, some clarifications are necessary in order to identify the key elements of the feasibility study which should be addressed, whose contents depend highly on the legal environment of the project.

A REFIT is defined by the following rules:

- Obligation for the Utility to buy the electricity at a fixed price for a certain period of time;
- Priority access to the grid (no or limited disconnection);
- Transparent eligibility criteria to benefit from the fixed tariff.

There are several ways to define the electricity price in a REFIT system. The price could be set through a tender

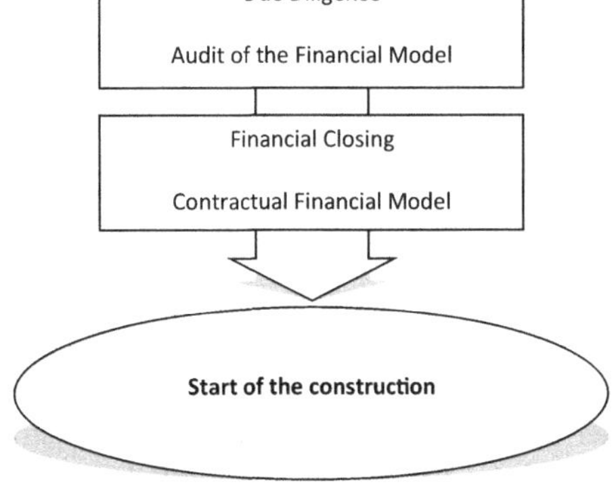

Figure 7.1 Key project development milestones – evolution of the business model [1]

and be different for each project, or it could be set by law, therefore relevant to all projects.

Given the size of a VLS-PV program, we believe that it will have in any case to demonstrate its added value to the country. The electricity generated should perfectly fit in their energy mix. Therefore, our key assumption is that the huge amount of investment required will not be financed unless the project makes sense in the short, medium and long term, from energy and economic perspectives. The services provided by the electricity produced with a VLS-PV system should be clearly identified.

In this case, the development process of the project will be comparable to a typical energy transaction in the power sector. This leads in particular to the following considerations during the feasibility study:

- Technical aspects: for the grid operator, PV is a solution, not a problem. This should be clearly demonstrated (positive impact on the grid, management of intermittency, etc.);
- Economic aspects: PV price should be competitive and fit in the energy mix of the country. The assessment methodology is described later in this chapter;
- Financial aspects: in emerging markets, access to debt and equity is a problem and the risk profile of the country is usually too high for traditional, long-term investors. Therefore the structure of the transaction often requires a sophisticated guarantee package.

Based on the technical, economic and financial studies, the promoters will then design a project proposal detailing, in particular, the capacity of the plant, the site, the technology involved, the price of electricity and the overall guarantee framework.

Setting up the technical design is the first step of the development. On the technical aspects, a special focus should be made on:

- The grid stability studies and intermittency regulations;
- The maintenance, logistic supply chain and the operating guarantees;
- The radiation and climate assessment;
- The technological choices and their bankability in the context of a VLS-PV program.

The financial and economic aspects may also influence the technical design. For instance, financing of new technologies may be possible, but the size of the transaction may be limited by several MW. Or, in order to reach a certain kWh price, it may be necessary to benefit from a specific source of financing, which is only available if the project has the required profile (origin of the equipment, size, end-users profile).

Once the technical design is made (details were given in Chapter 6), it is possible to evaluate the investment and operating budget and all the technical data (production, degradation, etc.) that will be the basis of the financial model.

7.3 ASSESSMENT OF THE LEGAL AND TAX ENVIRONMENT

7.3.1 The legal and tax study

(1) CHOOSING THE MOST EFFICIENT LEGAL SCHEME

Having made the technical design of the plant, the first question to tackle is the legal environment in which the project will be developed. A first legal assessment should go through the following key questions:

- Does the legal framework of the country allow Independent Power Producers (IPP) to sell electricity to the Utility?
- What is the process, when IPP status exists, to obtain the two key authorizations, namely the Concession (authorization signed with the State that allows the promoter to produce electricity) and the Power Purchase Agreement (PPA)?
- Could an energy transaction in the country benefit from a state guarantee and what could be its form?

Depending on the answers to these questions, the promoters will be able to choose the appropriate legal design. For example, if an IPP is not allowed to sell electricity to the Utility or the State, then the only financial strategy is to have the State or the Utility buy the VLS-PV system and support the financing of the transaction. What's more, if the PPA and/or the Concession are submitted to tender, with few exception rules, the promoters may be willing, in order to secure their investments during the development phase, to design a project that will fall under the exception rules. More details are given in Section 7.3.2 on the two most common legal schemes, the BOOT mode and the public market.

(2) ANALYSIS OF THE TAX ENVIRONMENT

Once the legal framework is set, it is then necessary to analyze the tax scheme relevant to the transaction. Given the fact that corporate tax is in most emerging markets above 25%, the relationship between the tax environment and the electricity generation costs is significant.

The promoter will have to carefully understand the tax environment based on the following elements, among others:

- Customs: import taxes for all equipment during construction and operation;
- Corporate: level, special regimes, retained losses;
- VAT;
- Local taxes: roads, land, others;
- Social taxes: operational staff for instance;

- Financial taxes (loan, dividends, etc.): taxable interests, tax on dividends, bilateral country agreements, tax perception system, etc.

Setting up the legal and tax environment is the second step in the development. The next section gives details concerning the most likely legal framework that is offered to a promoter.

7.3.2 Two main project structures: BOOT vs public transaction

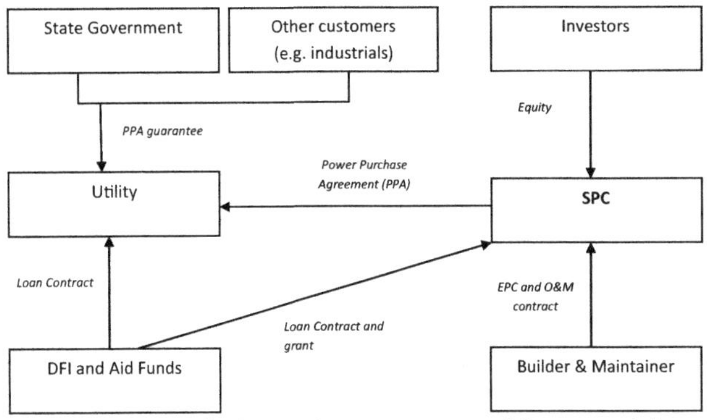

Figure 7.2 A typical BOOT scheme contractual framework for sub-Saharan Africa [1]

The promoters will have to choose mainly between two different models:

- The BOOT mode, where the Utility buy the electricity produced by the operator;
- The public mode, where the Utility or the State acquire a system of their own.

(1) THE BOOT SCHEME

The BOOT mode is based on the creation of a project company (a Special Purpose Company or SPC) which will be the operator of the plant. The SPC is a legal entity which will bear all the responsibilities and obligations on the behalf of the project promoter. In particular, the SPC will be responsible for the construction, the operation and the plant maintenance, and it will sell the electricity to the Utility. The SPC will also be the direct borrower from the banks, and it will be owned by the promoter (along with other investors, if any). Figure 7.2 describes an example of the contractual framework.

In a BOOT mode, the financiers expect to be paid by the revenues generated by the electricity produced, and only by that. The risk they take is a "project risk", which means that if the plant does not operate properly they will lose their money. That is why, in such a scheme, each party will strongly negotiate their obligations and the guarantees they should obtain from or give to the other parties. As a consequence, all the financiers are bound to each other during the accomplishment of their contractual obligations.

Usually, many parties are implicated in a BOOT mode and the cost of the transaction is high. Not to mention all the independent experts required to validate all aspects of the transaction, the banks, the investors and the promoter/developer, a typical BOOT scheme implicates dozens of entities and lots of development work.

The role of the developer in this respect should be emphasized. In a BOOT scheme, there should be a leading player, which can be called either the promoter or the developer, that will manage the elaboration of the project during all the development phases and coordinate the work of all the parties. Its role is key, since it is this player which will make sure that everyone is going in the right direction and that, in the end, the project will reach financial closing. The developer is the only party in the transaction that has access to all the negotiations between all the players, and its expertise should cover all the topics of the transaction. Sometimes, the promoter could be the government itself or the constructor/operator of the plant. Even though these cases may happen, it is important to remark that the development work is a specific task that requires a strong team with great experience, with good technical and financial skills. In this sense, the promoter ought to be multiple-task and investment ready since project development consumes a fair deal of time and money.

There are two main types of BOOT scheme:

- If the loan or the revenues are guaranteed by the State, the BOOT scheme is a Public-Private Partnership.
- If there are neither guaranteed revenues nor a fully State guaranteed loan, then the BOOT scheme is a purely private transaction.

Depending on the capacity of the government to provide strong financial guarantees, the financial situation of the Utility and the size of the transaction, a promoter can choose the most appropriate scheme. Needless to say, the PPP scheme is the most secure one, but it requires a strong implication of the country.

(2) THE PUBLIC MARKET SCHEME

The public market mode is based on the acquisition by the Utility or the State of the VLS-PV plant. In this case, a public entity will directly finance, construct and operate this asset. The utility can subcontract the construction and maintenance of the plant to a private company, or manage the plant directly. Figure 7.3 describes the contractual framework.

Given the size of a VLS-PV system, the relative simplicity and short time of the construction, we believe that this scheme is not appropriate to this kind of transaction. The public scheme is indeed more appropriate to small or standard transactions (small or large) well known by the public entities. Also, the utility should have sufficient

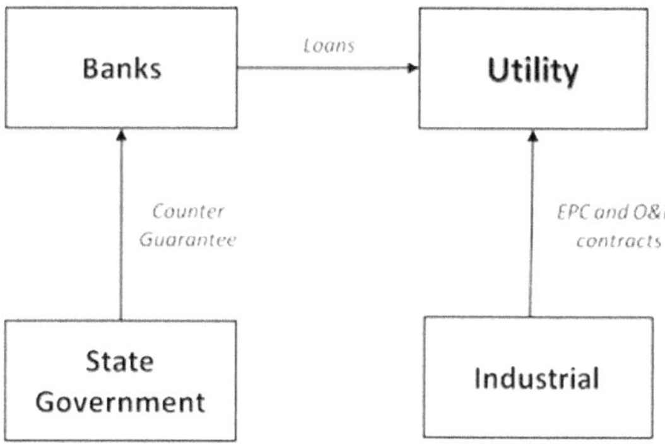

Figure 7.3 A simplified public scheme contractual framework

knowledge to manage the overall process, which is not the case in many emerging markets.

Even in this case, the feasibility study should be conducted with the same quality level and, even though the development of the project is less complex since the number of players and contracts is fewer than with a BOOT scheme, the promoter of the project should address the same questions as the ones mentioned in the previous section. Relying on the strong public guarantee of this scheme, this point is often forgotten and, in some cases, project development is of lesser quality than in a BOOT scheme – which leads to either a failure to reach the financial closing or to problems during the operational phase.

(3) COMPARISON BETWEEN THE TWO SCHEMES

Choosing one scheme or another is the promoter's decision and there is, in many cases, no obligation to favor one scheme over another. Everything relies on the promoter's strategy and on the outcomes of the discussions with the state and the financiers.

Needless to say, the public scheme is almost impossible to implement in the poorest countries for a VLS-PV system, since the experience required for managing the transaction and the financial commitments may represent way too heavy a burden for the State. But the BOOT scheme is also sometimes impossible to implement, especially in countries where utilities face financial difficulties and hold no experience in dealing with IPPs. The BOOT scheme requires a lot more expertise in the development phase and the transaction costs are higher, so many promoters will try to advocate, when possible, for a public scheme, especially in countries where it is difficult to attract investors.

Our recommendation is that, whenever there is a private consortium that is willing and capable to implement a BOOT scheme in a poor country, all the public institutions should support such projects, since it represents a major development of the private sector of the country and contributes to setting up the institutional tools of the development of private investment in the country.

Providing that the development of the private sector is one of the fundamentals for the economic and social development of the poorest countries, helping such a private initiative may be the best use of international aid. Unfortunately, this basic approach is not yet fully understood by the international agencies, since almost all their tools are constrained by procurement rules which make it difficult to support bilateral agreements between a private consortium and a public institution. By imposing their procurement rules and developing public schemes in countries where some private initiatives are emerging, the international institutions in some cases prevent the very countries they want to help from achieving sustainable development, forcing them to live in a subsidized environment, contributing to local corruption, destabilizing the private sector and stopping the flow of international private investment.

7.3.3 The guarantee package

An important aspect of the legal scheme is the guarantee package that should be put in place to secure the financing. Among all the key guarantee components, the promoter should devote particular attention to the following aspects:

- The payment guarantees form brought by the Utility. In most cases, the obligations of the Utility are also counter-guaranteed by the State, in which case the form of this counter-guarantee should also be analyzed;
- The responsibilities of the promoter, the Utility and the State with regard to the Concession and PPA. In particular, the *force majeur* clauses, the breach of contract and their consequences and all the technical services that should be brought either by the promoter or the Utility – quality of the electricity produced, availability/minimum production guarantees from the producer, grid availability guarantee, take-or-pay clauses and exceptions list, etc.;
- The construction and maintenance guarantees, which should depend on what the financiers request and the obligations included in the Concession and the PPA.
- The insurance scheme.

The definition of the appropriate guarantee package may influence the technical design of the plant. The obligations in the concession and PPA will clearly have a strong impact on the definition of the monitoring system and on the management of the plant. Also, the choice between one technology and another could also be related to the guarantees provided by the equipment suppliers such as degradation, lifetime, performance, etc. Last but not least, some equipment is difficult to insure in specific environments. Therefore, depending on the situation it could restrict the technical choices. For instance, in sites close to the seashore, or subject to typhoons or earthquakes.

7.4 ELABORATION OF THE DEVELOPMENT BUDGET

As previously mentioned, it is during the development phase that all the pieces required for reaching financial closing are gathered. Regardless of the legal and financial structure that has been selected, the most important aspect here is the quality of the development, the best chances the promoter has to take the project to completion.

In this sense, the third step of the development of the project is to elaborate the detailed organizational chart of the transaction, with a timeline and a budget, and to find the financial resources that will support all the development costs. This step will help the promoter to:

- Identify the resources needed to reach the financial closing;
- Identify the financial strategy to support the costs of the development.

Depending on the resources required and on the total development costs, the promoter will adapt its partnership strategy. At this stage, the technical design of the project may evolve again, since the partners that support the development costs may have some technical requirements: technology, origin of the equipment, etc. It may also influence the choice of the legal scheme, since again the financial partners may have a preference for a specific scheme.

7.5 IDENTIFICATION OF THE FINANCIAL PARTNERS

7.5.1 Role of the financial partners

The fourth step of the development is the selection of the key financiers of the project. The first characteristic of the financial partners is commonly considered to be the financing costs. Any promoter will be willing to lower the WACC (Weighted Average Cost of Capital) in order to achieve a competitive tariff. Table 7.1 shows the impact of the WACC on the electricity selling price necessary to achieve an acceptable IRR for investors.

Even though the WACC has a considerable impact on the final selling price and therefore on the competitiveness of the project, it is actually only one aspect of the financiers' profile. For example, an equipment provider that would act as a lead investor may expect a lower return as an investor if he can sell his equipment at a higher price. Or a bilateral institution may offer attractive financial conditions but insist in the use of specific equipment, which may negatively impact the technical design. Other services, such as political support, capacity to secure grants to finance part of the project or development costs, or specific skills (e.g. asset management in emerging markets, experience in international power transactions), are also elements to be considered in the package the promoter will gather for investors and banks in order to support their decision making process.

7.5.2 Choosing the lead investor

The lead investor, when it is not the promoter itself, should be selected carefully. There are indeed several types of investors, such as PV developers, PV industries, operators or investment funds, and they all have a different approach in their transactions, particularly with respect to their technical and guarantee requirements as well as expected returns.

In a BOOT mode, the lead investor is in most cases a private entity, while there may be some cases in a PPP scheme where the state may act as the majority shareholder of the SPC, therefore being the lead investor. In a public scheme, the State or the Utility is the lead investor.

In a nutshell, the lead investor is the key shareholder of the SPC. Its responsibilities are the following:

- Reassure the banks;
- Gather other investors;
- Lead the negotiation of the key contracts;
- Provide several guarantees likely to be required by banks and/or the Utility.

Again, at this stage of the project, the technical design may be influenced by the selection of the lead investor – technological choices, origin of the equipment. Therefore, it is important to carefully review the full package offered by each investor before selecting the lead investor.

7.5.3 Selecting the banks

It is necessary to have a preliminary idea of the financial conditions expected to achieve a certain electricity selling price before approaching the banks. Indeed, given these expectations, it is easier to shortlist several banks and optimize the process. For example, if the financing requires a loan whose maturity is more than 15 years, then in most emerging markets it is almost impossible to work with commercial banks. Also, depending on the country risk, some banks simply cannot make any offer. Additionally,

Table7.1 kWh price in Euro cents giving an average 15,5% investors' IRR for various WACC values [1]

WACC (%)	4,9	5,5	6,0	6,5	7,0	7,5	8,0	8,5	9,0
kWh price (Euro cents)	16,5	17	17,5	18	18,5	19	19,7	20,3	20,8

other banks have a regional expertise and are more competitive in specific markets than others.

It is almost certain that development institutions, international, regional, or bilateral – such as the World Bank, EIB, AfDB, AsDB, IADB or the IDB – will be major players in the financing of VLS-PV systems. Their services will range from direct private loans to concessional financing, as well as grants, technical and political support. Even though there currently are no PV specific funds, one could expect that within the next five years such funds will be implemented by these institutions, given the recent drop in module prices and the subsequent low price of electricity achieved by the latest technologies.

Here again, the promoter should carefully evaluate the full package offered by each of these institutions, since they all have their own way of structuring a transaction. Even though they are all used to working together on transactions, they act as MLA (Mandated Lead Arranger) very differently from one institution to another and may have different expectations, in terms of guarantees, technology, etc.

7.6 CALCULATION OF THE SALE ELECTRICITY PRICE AND ELECTRICITY COST OF PRODUCTION

The last step of the development consists in the financial closing, which comprises the following activities:

- The due diligences: their scopes vary from one transaction to another. For a VLS-PV transaction, one should expect at least five independent audits: the technical review of the project, the legal review of the financial structure and all the contracts, the audit of the insurance package, an economic assessment of the impact of the transaction on the national/regional level, and an audit of the financial model;
- The signing of all the contracts.

Only at the financial closing stage is the electricity generation price firm and definite. There are several approaches to fixing it and we will mention the two main strategies, which should theoretically converge in the end: the cost approach and the value approach.

7.6.1 The cost approach

The calculation of the electricity generation costs is quite simple for a promoter: it is the minimum selling price required to cover all the expenses during the operation phase. Namely:

- The operating costs, such as the maintenance (preventive and corrective), spare parts, all taxes, facility management, administration, etc.;

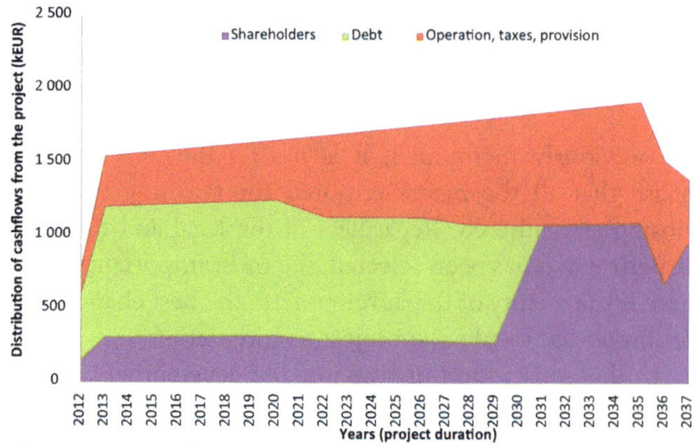

Figure 7.4 Use of revenue for a standard VLS-PV project in sub-Saharan Africa [1]

- The financial costs, namely the debt service and the dividends;
- The provisions.

Figure 7.4 gives an overview of the uses of the revenues during the operating phase.

The revenue distribution profile featured in Figure 7.4 is drawn from a financial model that will evolve during all the different steps of the development, and the assumptions will progressively go from preliminary expectations to contractual figures or audited estimations, finally reaching the final electricity generation cost, which corresponds to the minimum electricity selling price that could be offered by the promoter.

7.6.2 The value approach

Since in the case of a VLS-PV transaction the project should "make sense", from an economic, social and energy perspective, the cost approach may not be the appropriate way to calculate the selling price of the electricity. Indeed, the grid operator may proceed to decision making on PPA with little regard to the promoter costs: only the selling price and the service the producer can offer will be considered. In this respect, the value of PV is the maximum price that the grid operator can pay for a certain volume of PV electricity.

For example, since PV electricity is not available at night, the management of a PV plant on the grid is drastically different from a gas-fired plant or other "base load" systems. Considering the fact that a kWh has not the same value for the operator, depending on the period it is produced, the PV electricity will have different values depending on the demand profile and on the national/regional energy mix.

Assessing the value of PV requires a lot of information and a complex model with multiple variables, indeed a complex task. Among the main parameters, one could mention the following:

- The decentralized value: even though its intermittency may generate grid management costs and strong

investments in regulation devices, PV can increase the quality of the electricity and lower the overall transportation costs, if properly located;
- The production curve profile: production of PV electricity is directly dependent on solar irradiation as well as local climate. If the demand is also influenced by these parameters – consumption of some industries, air cooling systems, etc. – then the PV electricity has a greater value by competing with the peak load production systems. On the contrary, the value decreases;
- The absence of GHG emissions: PV is an environmentally friendly technology and this feature increases the value of its electricity. In particular, when looking at all the logistic chains of thermal plants – oil and gas production, transportation and storage – the environmental value of PV may be greater than the avoided GHG emissions. For instance, in sub-Saharan Africa where trucks are mostly used for oil transportation, the environmental savings consist also in reduction of the degradation of the roads and a possible reduction in fatal accidents;
- The endogen value: PV contributes to energy independency. Given the geopolitical aspects of the energy sector, this value is in some cases rather important;
- The fixed selling price over a long-term period (more than 25 years): this aspect is fundamental for the grid operator. Indeed, PV can contribute to stabilizing the average price of electricity in an energy mix which relies even partially on oil and gas, whose price are volatile, to produce electricity. In this respect, PV acts as a natural hedge on energy prices, and if we apply the concepts founding the "portfolio theory" to PV [1], it might even be possible to determine the optimal PV penetration rate given a specific energy mix;
- The low marginal production costs and the short installation time: capacities of PV systems are very modular and it is possible to increase the capacity of any plant very easily, as long as space is not an issue, and that regardless of the size of the capacity expansion: from several kW to tens of MW. As a consequence, when power generation development time is an issue for a country to match its increasing demand, the value of quickly having more energy available may be sometimes tremendous, since emerging systems (mostly diesel generators) are very expensive and the cost of not having electricity is also significant.

In many emerging markets, it is difficult to get the required information in order to assess precisely the value of PV. Information systems are in many cases awkward: data is manually collected, monitoring systems are poorly maintained, etc. Nonetheless, it is still possible to conduct studies to produce initial value assessments and the results are often of great use for the State and the Utility.

7.6.3 How to reconcile the cost and value approaches

If the cost of electricity is the minimum selling price offered by the producer and the value of PV is the maximum buying price proposed by the operator, what happens if these prices are too different to converge? Different scenarios are possible:

1 If costs equal value, then the situation is optimized;
2 If costs are inferior to value, then the producer is in a position to have greater margins and the grid operator may be able to generate savings;
3 If costs are superior to value, then the situation will potentially lead to the cancellation of the transaction.

Scenarios 1 and 2 are favorable for the development of the PV sector. Scenario 3 is more complex:

- If, in the short, medium or long term, the costs of production will decrease so that it becomes inferior to the value, then it is possible to subsidize the first extra costs in order to reach scenarios 1 or 2;
- If this situation will never happen, then PV is not a viable option.

This situation should be looked at on an individual case basis, considering a specific plant, in a specific location. For example, even a country with an energy mix fully based on hydro may have an interest in developing PV systems for certain usages, such as electrifying an isolated village.

7.7 CONCLUSION: THE LONG WAY TO REACH THE FINANCIAL CLOSING

The development of a VLS-PV system is a long process where almost all parameters will evolve until the financial closing stage is reached. The preliminary technical design will continuously evolve during the development process, taking into account the economic, financial and legal issues that are raised all along the way.

In a VLS-PV transaction, the development costs and risks are quite high and the studies are complex. In this respect, the support of governments and development agencies should be as great as possible directed toward removing the most difficult barriers. Complete studies should point out such barriers and suggest a pluri-disciplinary approach and appropriate management to guide these players.

Unfortunately, such studies are not yet available because, for most national support mechanisms, the value of PV is not at stake. The promoters that develop in current markets projects with the greatest value will not be awarded any extra benefit. On the contrary, as developing such projects is more expensive and time-consuming, promoters will tend to completely avoid developing high value projects, searching only for the cost reduction.

We believe that, given the great cost reduction of recent years, future markets, based on value, will rapidly emerge and that PV will definitely combine with the conventional energy sector before 2020 in order to take its natural and specific place in the energy mix of all countries. At this point, developing VLS-PV systems will be as common as developing combined cycle gas-fired plants.

REFERENCE

1 Awerbuch, S., and Berger, M. (2003) 'Applying portfolio theory to EU electricity planning and policy-making', IEA/EET Working Paper.

PART III

VLS-PV technical options and applications

CHAPTER EIGHT

Grid matching issues

8.1 INTRODUCTION

The intermittency in output from a VLS-PV plant renders its matching to the requirements of a conventional electricity grid an issue of quite complex proportions. In the first place, the grid has its own intermittency problems, associated with customer requirements. These problems are handled via a combination of economic considerations and the *controlled input* from three basic kinds of power plant. Base-load plants (typically coal-fired or nuclear) have limited ramping capabilities. They operate on a continuous basis for extended periods of time (days or weeks) with relatively infrequent shut-downs for maintenance. Intermediate-load plants (typically gas-fired or oil-fired) have improved ramping capabilities. They can be started up in the early morning and shut down in the evening. Peak-load plants (typically fast-reacting gas turbines) are capable of ramping from zero to full power in typically 10 minutes and can undergo smaller ramping steps with extreme rapidity. They are used to follow the moment-to-moment requirements of the grid. Each of these plant types has its own economic considerations based on: fuel cost, frequency of maintenance, age of plant, cost of replacement, and moment-to-moment selling price of the electricity.

When it comes to attempting to input the power from a VLS-PV plant, a number of additional problems arise that result in new grid matching challenges. First, on clear days, a VLS-PV plant performs something like an intermediate-load plant in that it starts up in the morning and shuts down at night. However, unlike fossil-fuelled plants, a VLS-PV plant cannot be ramped up and down at will. Moreover, its peak output, which occurs at noon, will generally not coincide with the time of day at which grid requirements reach their peak. Thus, for example, if a VLS-PV plant is sized to provide the grid's peak load on a given summer day, it will produce more power at noon than can be used.

Secondly, even on clear days, the peak output of a VLS-PV plant depends upon the time of year and its specific design. For example, a fixed-tilt plant will typically exhibit its peak output in the spring season, when noontime solar angles are close to normal incidence and ambient temperatures have not yet reached their summer highs. Now a consequence of the relatively benign springtime ambient temperatures is that air-conditioning loads have not yet reached their summer maxima. Therefore, if a VLS-PV plant is sized to meet the peak demand in summer, it will produce more power in spring than can be used.

Yet a third challenge is the intermittency of clear days, a solution to which requires a combination of some form of weather forecasting, a facility for storage, and an appropriate grid operation strategy. Otherwise, precisely the same amount of backup conventional generating capacity will have to be maintained for the provision of power when solar input is not available.

Finally, the effect of clouds on less than perfectly clear days can lead to grid instabilities unless the output of the VLS-PV plant is suitably smoothed.

A pioneering study on grid matching VLS-PV systems was carried out by Denholm and Margolis [1, 2] in reference to the giant ERCOT electricity grid in Texas, USA. Those authors defined a specific mix of various PV technology types (fixed, tracking, etc.), geographically distributed over the ERCOT domain. From their study they concluded that VLS-PV plants would generate increasingly large amounts of unusable energy once annual penetration levels of 10–20% were reached.

The ERCOT study did not distinguish between the contributions from the individual technology types, neither did it attempt to quantify the effect of geographical distribution of the various VLS-PV plants. And of course, being a study of a specific grid system there was no way of knowing how general their results were. In an attempt to gain more insight into these issues, they were systematically addressed in a series of papers that analyzed the hourly requirements of the Israel electricity grid during the year 2006 [3–8]. The latter study simulated the outputs of various kinds of VLS-PV plant; alone, in combination, in combination with wind energy generation systems (WEGS); localized, and distributed over the Negev Desert. It led to new insights and possibilities beyond those envisaged in [1, 2]. A summary of the Negev results and their wider implications is accordingly presented here.

8.2 THE ISRAEL ELECTRICITY GRID IN 2006

During the year 2006, the Israel Electric Corporation (IEC) operated 10,5 GW of generating capacity, producing a total of 50,35 TWh. The various fuels involved were: coal (35,65 TWh), gas (9,06 TWh), fuel oil (2,87 TWh) and gas oil (2,77 TWh) [9]. This total averaged out at 6 055 kWh per capita, a figure comparable to the production of several European countries.

Although the Israeli grid is of the "island" kind (load-sharing with neighboring states not being feasible the way it is in Europe or the USA), the results of this study are of more general relevance for a number of reasons. Firstly the initial desert-based VLS-PV plants and their local customers will, in all likelihood, constitute island grids. At a later stage, mutual interconnection will render the combined system closer in type to the European or US electricity grids. Secondly, to a certain extent, an island grid is a microcosm of a continental inter-connected grid, so certain results should be capable of being generalized. Thirdly, precisely because it is an island-type grid, it is useful to be able to compare the results of its grid matching considerations with those of the more continental-type ERCOT grid.

8.3 GRID FLEXIBILITY

From the above introduction the degree to which VLS-PV power can penetrate the grid depends upon how flexible the grid is vis-à-vis its ability to rapidly change the amount of conventional generating capacity that is operating at any given time. A highly flexible grid would be one consisting entirely of fast-ramping gas turbines. At the other extreme, an all-nuclear powered grid would be extremely inflexible to the intermittent input from a VLS-PV plant. The term flexibility was quantified in [1, 3] as:

$$ff = 1 - L_{min}/L_{max}$$

where L_{min} and L_{max} are respectively the minimum and maximum total hourly load of the grid. Assessment of the characteristics of the power plants in the IEC's 2006 fleet revealed that, technically, the grid could have operated at $ff = 0{,}80$, but that for economic reasons, it was operated closer to the flexibility $ff = 0{,}65$.

8.4 A NO-DUMP VLS-PV PLANT

The first question of interest is: in the absence of storage, what is the largest VLS-PV plant that the IEC grid could have operated in 2006 that would not have produced any surplus output during any single hour of the year? Naturally, any surplus would have needed to be dumped as it could not be stored for later use. The largest such plant that would not have produced any surplus is accordingly referred to as a *no-dump* plant [10, 3].

There are two answers to this question. The first is a purely theoretical answer in that it assumes the unrealistic possibility that during one hour of the year the VLS-PV plant would have provided 100% of the load. Such a situation is unrealistic because during that hour the IEC could not have shut down its base-load plants, i.e., its flexibility was not $ff = 1$. Nevertheless, the answer is instructive. It was found to be 5,4 GWp [3]. Such a theoretical plant would have provided 17,4% of the stated 50,35 TWh annual load requirements for the year 2006.

A more realistic answer must allow for the fact that the VLS-PV plant can at best only provide the *difference* between the total load and the base-load. Approximating the IEC grid in 2006 as having been one with a flexibility of $ff = 0{,}65$, the no-dump size of a VLS-PV plant would have been a mere 827 MWp – which could have provided only 2,7% of the annual load. On the other hand, had the IEC grid been operated at its technically feasible flexibility of $ff = 0{,}80$, the no-dump size of a VLS-PV plant would have been 3 046 MWp and its annual grid penetration would have been 9,8%.

One thus sees that without storage to take up surplus solar generation, or a different grid operation strategy, any VLS-PV plant larger than 827 MWp, if connected to the IEC grid during 2006, would have needed to dump some of its generated energy.

8.5 ENERGY DUMPING

In the absence of storage, an immediate question that arises from the no-dump calculation is: to what extent might it be worthwhile to slightly oversize the VLS-PV system and, thereby, increase grid penetration for only a modest dumping of solar-generated energy? This is obviously an economic question the answer to which depends upon VLS-PV construction costs and electricity prices. However, it is easy to calculate a technical answer by considering a number of successively oversized VLS-PV plants and asking how much annual penetration each would achieve. Table 8.1 lists the results for the 2006 IEC grid for the two flexibilities $ff = 0{,}65$ and $ff = 0{,}80$.

From Table 8.1, it can be seen that there is an enormous improvement in grid-penetration if one constructs a VLS-PV plant that dumps 5% of its annual generation. But for larger plants, and their consequential larger amount of energy dumping, the improvement rapidly becomes

Table 8.1 Percentage of grid penetration achieved by successively over-sizing a no-dump VLS-PV plant and dumping its surplus energy, for grid flexibilities *ff* = 0,65 and *ff* = 0,80

Dumped PV energy [% of total PV generation]	Grid penetration [% of annual demand]	
	ff = 0,65	*ff* = 0,80
0	2,7	9,8
5	10,4	18,7
10	12,5	21,3
15	13,8	22,9
20	15,2	24,0
25	16,5	25,0

marginal. It has already been seen that a no-dump system could theoretically provide 17,4% annual penetration for a perfectly flexible grid (i.e. with *ff* = 1). It is accordingly worthwhile pointing out that for such an ideal grid, an oversized VLS-PV plant that dumped 5% of its annual generation could have achieved 29,0% penetration.

These considerations enable one to conclude that a modest amount of energy dumping – typically 5% – can achieve a significant improvement in grid penetration irrespective of grid flexibility. Conversely, too much energy dumping rapidly becomes wasteful in that it requires an increase in VLS-PV plant cost for insignificant increase in grid penetration.

8.6 VARIATIONS IN TECHNOLOGY TYPE

The numbers that have been presented thus far all refer to a simulated static, tilt = latitude VLS-PV plant at a single location (Sede Boqer). In order to investigate the improvements that might be possible from other kinds of technology at that site, the authors also simulated [4]: (a) a single (horizontal) axis system that tracks the sun from east to west; (b) a dual-axis sun-tracking system; (c) a CPV system. Table 8.2 shows the no-dump size associated with each technology type and its annual grid penetration for the year 2006. Results are given for the two grid flexibilities mentioned above: *ff* = 0,65 and *ff* = 0,8.

Table 8.2 reveals two extremely important results. First, if the IEC grid had been operated at its *economically optimum* flexibility of *ff* = 0,65 (i.e. optimum without taking into consideration the possibility of solar generation), then nothing would have been gained by employing sun-tracking technologies: a static no-dump system would have achieved the highest annual penetration. However, (not shown in Table 8.2) with 5% energy dumping, even at *ff* = 0,65, sun-tracking would have resulted in some improved penetration [4].

On the other hand, if the grid had been operated at the *technically possible* flexibility of *ff* = 0,80, the annual penetration of a no-dump static system would have more than tripled, i.e. rising from 2,7% to 9,8%. Furthermore, the employment of any kind of sun-tracking would have increased the annual penetration by a further 20% above that of the static system, and resulted in a smaller no-dump power rating than for the static system.

Although CPV results are included in Table 8.2, they involve a different set of assumptions from the simulations of the other three technology types. Therefore, they are not directly comparable to the latter.

The lesson to be learned from Table 8.2 is that even though it is obvious that the "squarer" daily output profile of a sun-tracking system should present a better fit to the daily load requirements than does the more "peaked" profile of a static system, one really needs to operate the grid at as high a flexibility as possible in order to gain maximum benefit from this effect.

8.7 GEOGRAPHICAL DISTRIBUTION

Compared to Texas, the Negev Desert is minuscule in size. However, it contains eight meteorological stations that have provided up to approximately 20 years of hourly data that may be used for simulating the performance of a wide variety of renewable energy systems [11]. The greatest distance between two such stations located relative to one another along an approximately north-south line is 195 km (from Arad in the northern Negev to Eilat on its Red Sea coast). In the east–west direction, the greatest distance is 88 km (from Sedom near the border with Jordan, to Besor Farm near the Mediterranean coast).

The authors of ref. [4] performed three sets of simulations to investigate the cloud-smoothing effect of geographical distribution, and the extent to which this might improve grid penetration. The first set of simulations studied a maximally separated north-south pair of VLS-PV plants, located at Arad and Eilat respectively. A range of relative

Table 8.2 No-dump size and annual percentage grid penetration for various VLS-PV technology types and grid flexibilities *ff* = 0,65 and *ff* = 0,80

Technology type	No-dump size [MWp]		Annual penetration [%]	
	ff = 0,65	*ff* = 0,80	*ff* = 0,65	*ff* = 0,80
Static	827	3 046	2,7	9,8
1-axis	457	3 022	1,8	12,1
2-axis	464	2 826	2,1	12,5
CPV	591	3 072	2,2	11,5

Table 8.2 No-dump size and annual percentage grid penetration for various VLS-PV technology types and grid flexibilities *ff* = 0,65 and *ff* = 0,80

Technology type	No-dump size [MWp]		Annual penetration [%]	
	ff = 0,65	*ff* = 0,80	*ff* = 0,65	*ff* = 0,80
Static	827	3 046	2,7	9,8
1-axis	457	3 022	1,8	12,1
2-axis	464	2 826	2,1	12,5
CPV	591	3 072	2,2	11,5

Table 8.3 Highest achievable improvement in grid-penetration brought about by distributing a VLS-PV no-dump combination among two maximally located north-south sites in the Negev Desert; for static PV and 2-axis sun-tracking CPV; *ff* = 1

	Static no-dump rating [GWp]	Annual grid penetration [%]	CPV no-dump rating [GWp]	Annual grid penetration [%]
Arad	5,40	17,6	5,70	21,2
Eilat	5,48	17,6	5,88	20,9
Optimal combination	5,55	17,8	5,95	21,7

Table 8.4 Highest achievable improvement in grid-penetration brought about by distributing a VLS-PV no-dump combination among two maximally located east-west sites in the Negev Desert; for static PV and 2-axis sun-tracking CPV; *ff* = 1

	Static no-dump rating [GWp]	Annual grid penetration [%]	CPV no-dump rating [GWp]	Annual grid penetration [%]
Besor Farm	5,52	16,8	5,97	19,4
Sedom	5,62	16,8	6,28	19,8
Optimal combination	5,71	17,2	6,29	19,9

Table 8.5 Maximum, minimum and mean improvement in grid-penetration brought about by distributing a VLS-PV no-dump combination among 8 locations in the Negev Desert, for 400 random combinations of relative sizes; for static PV and 2-axis sun-tracking CPV; *ff* = 1

	Static no-dump rating [GWp]	Annual grid penetration [%]	CPV no-dump rating [GWp]	Annual grid penetration [%]
Maximum	5,71	18,1	6,27	22,2
Minimum	5,44	17,5	5,80	20,5
Mean ± sigma	6,59 ± 0,04	17,8 ± 0,1	6,04 ± 0,07	20,7 ± 0,3

sizes was computed so that the aggregate had no-dump status. The aim was to see whether a combination could be found that would have higher annual grid-penetration than a no-dump plant at either site alone. These calculations were performed separately for fixed flat-panel PV, and 2-axis sun-tracking CPV systems, the former representing the currently most popular technology type, and the latter representing the technology which is most sensitive to cloud movement. These calculations were performed assuming grid flexibility *ff* = 1 in order not to complicate the results still further. Table 8.3 summarizes the results.

The second set of simulations was similar in kind, but aimed at investigating the possible improvement in penetration that could be effected by spreading out the no-dump combination in maximal manner along an east-west line: i.e. by locating the two VLS-PV plants in Besor Farm and Sedom, respectively. The results of these simulations are given in Table 8.4.

The third set of simulations consisted of locating parts of a distributed VLS-PV plant among all eight Negev sites, in 400 random combinations. As was the case for the pairs of plants, the present combinations were normalized so that the aggregate was no-dump. For these calculations the Monte Carlo method was employed for generating and analyzing the results. The latter are given in Table 8.5.

Perusal of Tables 8.3, 8.4 and 8.5 reveals that very little improvement in grid penetration may be expected by distributing VLS-PV plants around the Negev. It is not clear whether this is a general result, or one that is specific to small geographical regions like the Negev. However, a positive result of this study is that geographical distribution of VLS-PV plants significantly reduces the additional ramping requirements that a single plant would impose upon the grid. This has important implications regarding the lifetime of the non-solar plants in the grid and on their frequency of maintenance.

8.8 COMBINING VLS-PV WITH WECS

In ref. [5] the authors studied the possible improvements in annual grid penetration that might be brought about by combining VLS-PV plants with wind energy conversion systems (WECS). The motivation here is that instantaneous fluctuations in these two resources are uncorrelated, and the latter even continues at night time. A slight complication in providing a quantitative comparison between the wind study and the earlier work in refs. [4] and [5] is that Sede Boqer is not a very good wind site. Therefore the authors selected the two best wind sites among the eight: Mitzpe Ramon (best wind – also the best solar site) and Arad (second best wind). Once again the two VLS-PV technology types, fixed flat-panel and 2-axis CPV, were studied from the viewpoint of investigating the extent to which the addition of WECS could improve grid-penetration.

For reasons that were given in [5], but which need not concern us here, each of the technology types under consideration was considered to be distributed equally (in power rating) between the two sites, such that the combination was no-dump. Table 8.6 compares the annual grid penetration of these "50-50" technology distributions for WECS alone, VLS-PV alone (both technology types) and the best hybrid combination. Although several grid flexibilities were investigated, in the present summary we address only the ideal *ff* = 1 case.

Table 8.6 shows that in the case of fixed flat-panel VLS-PV, the addition of a suitably sized WECS plant can increase the annual grid penetration from a pure solar value of 17,6% to a combined value of 20,0%. In the case of VLS-CPV, annual penetration rises from 21,8% without wind, to a combined vale of 23,9%. It also shows that the penetration of VLS-PV alone (irrespective of technology type) is greater than for WECS alone.

The study in [5] also considered the effect of some modest energy dumping.[1] If, as was considered above, 5% of the renewable energy is allowed to be dumped, penetration for a hybrid WECS+VLS-PV (flat panel) system reaches close to 45% and that of a WECS+VLS-CPV hybrid reaches 47%.

However, returning to the more realistic flexibility values of the IEC grid in 2006: for a WECS+VLS-PV (flat panel) hybrid, 5% energy dumping would result in 15% penetration at *ff* = 0,65 or 28,9% penetration at *ff* = 0,80.

Table 8.6 Comparison of annual grid penetration of "50-50" no-dump distributions between Mitzpe Ramon and Arad of: pure VLS-PV (both flat-panel and CPV), pure WECS, and optimal blends of WECS and VLS-PV; for grid flexibility *ff* = 1

Technology type	Annual penetration [%]
VLS-PV (flat panel)	17,6
VLS-PV (CPV)	21,8
WECS	15,7
WECS + flat panel	19,8
WECS + CPV	23,9

For a WECS+VLS-CPV hybrid, 5% energy dumping would result in 16,9% penetration at *ff* = 0,65 or 30,8% penetration at *ff* = 0,80.

From this wind study one may infer that an appropriate combination of VLS-PV and WECS could have enabled renewables to provide approximately 20% of the IEC's 2006 load requirements without the need to alter their operating strategy. On the other hand, had they been willing to operate the grid at the technically possible flexibility of *ff* = 0,80 then a WECS+VLS-PV hybrid could have provided approximately 30% of the annual requirements. This appears to be the best that one can do without storage.

8.9 STORAGE

All of the discussion thus far has addressed the question of how to achieve maximum grid penetration for a minimum loss of (expensive) PV-generated energy. In particular, no-dump systems were discussed, and 5% energy losses were considered to be a not unreasonable sacrifice, given the fact that this is actually an economic question which has deliberately not been addressed in the studies reported in [3–8].

The potentially attractive feature of storage is that, in principle, it allows one to construct VLS-PV plants that are greatly in excess of no-dump size, and to store the excess energy for such times as it may be needed. However, the dual issues of sizing and operation of storage are extremely complex.

First, one must realize that no storage system has 100% efficiency. Thus, not all of the stored energy from the VLS-PV plant will be available for later use. Following the earlier work in [2], the authors of [6–8] employed 75% as representative of the typical round-trip efficiency for presently-available storage systems. But even this is a gross simplification because, as discussed in [6], many storage devices have efficiencies that vary with their state of charge, rate of charge/discharge, and useful lifetime.

Second, one must take into consideration both the energy capacity and the *power capacity* of storage. Power capacity relates to the maximum power that the storage device can receive within a given time (in our case, one hour) and, equally important, to the maximum power that the device can deliver within an hour. This is a non-trivial issue because, as shown in [6], there is a four-way linkage among: VLS-PV power rating, storage energy and power capacity, and the grid energy-with-time profile. Since, for any given electricity grid, the amount of VLS-PV will grow with time (probably over tens of years) the ratio of energy capacity to power capacity of the storage system may change. Therefore the storage system must be designed in a manner that can accommodate such change.

Third, one must take into consideration the fact that there can be very large differences in the amount of storage that is required at different times of the year. For example, in desert countries such as Israel, a VLS-PV system can

Table 8.7 Grid components that would have been needed in 2006 for VLS-PV to provide approximately 90% of the IEC's annual load with no electricity brown-outs, assuming grid flexibility *ff* = 1.

Power rating of VLS-PV [GWp]	24,3
Energy capacity of storage [GWh]	94,5
Power capacity of storage [GW]	17,4
Power rating of backup [GW]	6,6

produce large amounts of power in the spring season, before air-conditioning loads become of appreciable importance. This means that a storage system sized to capture the spring excess may remain almost empty for the rest of the year. Thus, it is necessary to determine criteria for the optimal sizing of storage. This is, of course, ultimately an economic question, but the authors of [7] showed that there are also physical constraints that limit the range of the eventual decision-making process.

Fourth, one must take into consideration the grid operating strategy. The situation was already mentioned above whereby, for economic reasons, the IEC preferred to operate the grid in 2006 at a flexibility of *ff* = 0,65 rather than at the technically feasible flexibility of *ff* = 0,80. With the availability of large storage systems – as will be needed for the massive penetration of VLS-PV – there will be an economic incentive to use the storage for transferring nighttime base-load electricity to times of day when it can be marketed in a more cost-effective manner.

Finally, there is the important question of how much shadow fossil-fueled backup must be maintained in order to ensure that the grid will always be able to provide for customer requirements. This question was addressed in [8].

Because the next chapter will discuss practical storage systems in greater detail, we conclude the present chapter with an illustration from [9] of one VLS-PV+storage configuration that could have provided approximately 90% of the IEC's 2006 grid requirements, and the amount of shadow backup (in the form of fast-ramping, quick-starting gas turbines) that could have ensured that no electricity "brown-outs" would have occurred. The system configuration is given in Table 8.7.

A number of comments are of relevance to Table 8.7. First, the VLS-PV simulations pertain to a fixed flat-panel system at a single location (Sede Boqer). In practice, such a system with a 24,3 GWp power rating would need to occupy approximately 900 km^2 of land. Although the Negev Desert occupies approximately 10 000 km^2, such a large VLS-PV plant would presumably need to be distributed throughout the Negev, where the climatic parameters would be comparable or even superior to those at Sede Boqer [11].

Secondly, it will be noted that the 6,6 GW of backup fossil-fueled generating capacity is actually less than the 10,5 GW of capacity that constituted the IEC fleet in 2006.

Thirdly, the calculations in Table 8.7 were performed at a grid flexibility *ff* = 1. This would not be an unrealistic idealization, given the fast-response kind of technology employed for the backup, and given the fact that what had previously been base-load generation would have been largely replaced by storage and a new grid-operating strategy. The latter would be informed by short-range weather forecasting (hours and days) and would involve operating the gas turbines in a manner that would tend to prolong their lifetime more than does current operating practice.

As an example, Figure 8.1 displays the hourly contributions from the three components in Table 8.7 during the "worst" (from the IEC view-point) week of 2006. A number of important features are evident from Fig 8.1. Firstly, it shows hours for which the fossil backup (in black) is completely switched off. One can thus see that the grid flexibility really is *ff* = 1. Secondly, the fast-response backup turbines are never called upon to provide peak power as is their normal mode of employment. On the contrary, under the illustrated strategy of use, when called upon, they operate at almost constant power output. This mode of operation should considerably extend their service life compared with their conventional usage. Thirdly, one sees how the backup is often employed at nighttime to ensure adequate energy in storage for the next day. Naturally the precise time at which the backup is actuated will be determined by weather forecasting.)

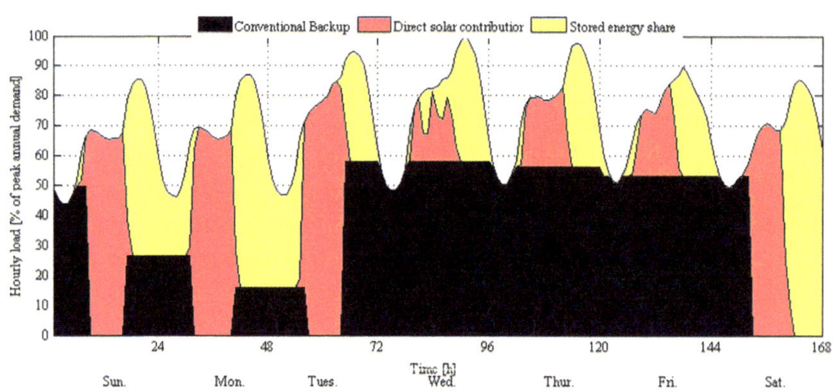

Figure 8.1 One possible employment of the VLS-PV (pink), storage (yellow) and backup (black) power system itemized in Table 8.7, during December 22–28, 2006. (Note that Dec 25 constituted the IEC's highest hourly load for that year)

8.10 CONCLUSIONS

This chapter summarizes the principal findings of a detailed study [3–8] of the extent to which VLS-PV could provide substantial amounts (up to 90%) of Israel's electricity requirements. Although these results may not be quantitatively applicable to other regions, many of the results should still be approximately correct in other desert areas. Furthermore the overall methodology will certainly be of wide applicability and the various details discussed here will all need to be taken into account for optimizing VLS-PV plants wherever they will be.

NOTE

1 Unlike the calculations in Table 8.6, for which different percentages of WECS and VLS-PV were found to be optimum for the flat-panel and CPV situations, all energy dumping calculations were performed for a standard system configuration defined as being both "50-50" in distribution, as above, *and* "50-50" in power rating.

REFERENCES

1 Denholm, P. and R.M. Margolis (2007) Evaluating the limits of solar photovoltaics (PV) in traditional electric power systems. *Energy Policy* 35 (2007) 2852–61.

2 Denholm, P. and R.M. Margolis (2007) Evaluating the limits of solar photovoltaics (PV) in electric power systems utilizing energy storage and other enabling technologies. *Energy Policy* 35 (2007) 4424–4433.

3 Solomon, A.A., D. Faiman and G. Meron (2010) An energy-based evaluation of the matching possibilities of very large photovoltaic plants to the electricity grid: Israel as a case study. *Energy Policy* 38 5457–68.

4 Solomon, A.A., D. Faiman and G. Meron (2010) The effects on grid matching and ramping requirements, of single and distributed PV systems employing various fixed and sun-tracking technologies. *Energy Policy* 38 5469–81.

5 Solomon, A.A., D. Faiman and G. Meron (2010) Grid matching of large-scale wind energy conversion systems, alone and in tandem with large-scale photovoltaic systems: An Israeli case study. *Energy Policy* 38 7070–81.

6 Solomon, A.A., D. Faiman and G. Meron (2010) Properties and uses of storage for enhancing the grid penetration of very large photovoltaic systems. *Energy Policy* 38 5208–22.

7 Solomon, A.A., D. Faiman and G. Meron (2010) Appropriate storage for high-penetration grid-connected photovoltaic plants. *Energy Policy* 40 335–44.

8 Solomon, A.A., D. Faiman and G. Meron (forthcoming) The role of conventional power plants in a grid fed mainly by PV and storage, and the largest shadow capacity requirement. *Energy Policy*, DOI: 10.1016/j.enpol.2012.05.050

9 Israel Electric Corporation statistical report for 2006. <www.israel-electric.co.il/Static/WorkFolder/IRR/2006%20Stat%20ENG.pdf>

10 Rabl, Ari (1985) *Active Solar Collectors and Their Applications* (Oxford University Press, New York, Oxford), 340 *et seq*.

11 Faiman, D., D. Feuermann, P. Ibbetson, B. Medwed, A. Zemel, A. Ianetz, V. Liubansky, I. Seter and S. Suraqui (2004) The Negev radiation survey. *Jour. of Solar Energy Engineering* 126: 906–14.

CHAPTER NINE

VLS-PV intermittence and stationary storage for VLS-PV

Matching grid requirements implies that plant operators are aware of VLS-PV systems intermittence in order to conduct an informed comparison of available solutions. Therefore before analyzing the range of available stationary storage solutions (with regard to the VLS-PV plant size), power output intermittence and power output prediction strategies have to addressed.

9.1 VLS-PV SYSTEMS POWER OUTPUT INTERMITTENCE AND PREDICTION

9.1.1 Definition

Intermittence is a term used to indicate the fact that the solar resource is not constant at the photovoltaic module level, making the power output of a photovoltaic plant vary with time. Indeed, in-plane solar irradiance (in kW·m^{-2}) fluctuates with sun activity and position, with atmospheric and oceanic circulation, and with meteorological conditions.

Photovoltaic power output prediction, and thus intermittence characterization, encompasses planning, forecasting and nowcasting. Nowcasting anticipates intermittence from the present time up to few hours ahead, forecasting covers up to 10 days, and planning refers to longer time periods [1].

Conventional power plant operation is predicted to anticipate unavailability, load peaks, reserves, maintenance, etc., typically for three time scales: one month, one week and one day before operation. Planning infringement or infra-day unavailability is considered fortuitous and is therefore subject to penalties from the utility.

When coupling VLS-PV with other systems, including storage, to get systems managed as firm power sources, the operator, the utility and the automaton deciding the electricity injection, storage or dumping will require prediction from planning to nowcasting. This allows to predict power output and to anticipate the best operating strategy, in conjunction with what has been contracted.

Predicting photovoltaic power output typically requires a two-stage method:

1 The first stage consists in predicting the main meteorological parameters influencing the photovoltaic electricity production, i.e. the Global Horizontal Irradiance (GHI), the temperature and the wind speed. The direct component (Direct Normal Irradiation, DNI) of GHI is highly influenced by cloud cover.
2 The second stage uses these parameters to predict the DC power production of the photovoltaic system.

IEA Solar Heating & Cooling programme Task 36 focused on solar resource knowledge management and initiated reviews and benchmarks on solar radiation planning and forecasting. Intermittence is defined for the time scales of interest, covering planning to nowcasting, as prediction models, which basically are split into two model categories: numerical weather prediction and cloud motion (satellite-based and ground-based).

9.1.2 Planning

(1) YEARLY TO SEASONAL INTERMITTENCE

Yearly solar irradiation received (in kWh·m^{-2}) is typically in a range of 1–2% around a statistical mean, based on historical data. While this is true for up-to-now data, expected climate change and mutations in climate indices frequency (such as successions of El Niño / El Niña events) could slightly modify this [2]. This accuracy is part of the determinant information for warranties in PV project EPC (Engineering, Procurement and Construction) contracts.

Satellite-based methods give annual production expected for the PV plant lifetime (typically 25 years) with an uncertainty of ±5–10%. Results cover large areas and are fast to obtain, requiring mainly historical databases to produce accurate statistics. For small PV plants, the resolution may not be enough (typically 1 km^2 pixels), but it is sufficient for LS- and VLS-PV.

Ground-based methods have in general the best accuracy, but only for a specific area [3]. They often act as a correction factor to compensate the satellite resolution. They are then adapted to small to LS-PV plants; several stations are needed to capture the whole variability at the scale of a VLS-PV plant [4].

The MotherPV method [5, 6], for example, proposes a simple approach, based on the site-specific distribution function of solar illumination and on the thermal characterization of the module.

Infra-yearly intermittence is rarely assessed for PV systems; today grid-connected PV systems do not require such a level of detail because of the 30% variable renewable penetration limit that is not reached. However, this limit will be rapidly exceeded (and is already in some insular grids), above all with integration of VLS-PV. Therefore, to compensate impacts on grids and to be comparable to firm power sources, infra-yearly intermittence should be defined.

Seasonal variations in irradiation are of importance when considering implementation of VLS-PV on grids: if conceived as bulk power, it can limit penetration with less production during winter or a rainy season [7]. Therefore, perspectives of on-site seasonal use of electricity produced, or storage, could enhance the penetration capacity of grids, compensating the seasonal intermittence due to the position of the Sun / Earth system.

(2) NUMERICAL WEATHER PREDICTION MODELS (NWP)

NWP models are used in planning to predict the value of a large range of parameters, with the use of computer simulations for various purposes (agriculture, tourism, safety, etc.) including solar radiation applications. A large number of NWP models exist, which differ from each other by the area covered, resolution, updates, dissemination, etc.

At the synoptic scale, GFS (Global Forecast System) run by NOAA (National Oceanic and Atmospheric Administration), IFS (Integrated Forecast System) run by ECMWF (European Center for Medium-Range Weather Forecasts) and GEM (Global Environmental Multiscale) model run by CMC (Canadian Meteorological Centre) are the three major mathematical models covering respectively the USA, eastern and central Europe, and Canada. They typically provide irradiation forecasts every 3 hours to 72 hours (3 days) and every 6 hours to 168 hours (7 days).

Regional and local models are more numerous: Mesoscale Atmospheric Simulation System (MASS), National Digital Forecast Database (NDFD), Weather Research & Forecasting Model (WRF) and Advanced Regional Prediction System (ARPS) are the main ones. They tend to higher spatial resolution, from 4 to 25 km, and to higher temporal resolution, especially for Rapid Update Cycle (RUC) models, allowing hourly predictions.

Model Output Statistics (MOS) techniques use NWP model outputs to obtain GHI or other parameters, or to directly predict the PV power output [8]. MOS techniques can be based on (artificial) neural networks, fuzzy logic, radial basis functions, multi-layer perceptron models, Markov chains, or a combination of some of these [9].

NWP models are usually combined to multiply the sources of uncertainty, and to provide a higher degree of confidence on the forecast boundaries [10]. Because NWP models are probabilistic, they do not capture the exact position of clouds, limiting their use for short-term application.

9.1.3 Forecasting and nowcasting

(1) DAILY TO VERY SHORT-TERM INTERMITTENCE

The alternation of day and night conditions is well known and only depends on a few meteorological parameters (elevation and annual average local pressure and temperature), and on the location. The Solar Position Algorithm [11] gives today the best accuracy for sunrise and sunset times.

Infra-daily variations of irradiation depend on meteorological parameter fluctuations: mainly temperature, pressure, wind speed and direction, aerosol content and water vapor content.

These variations are not used in the most common representation, i.e. the classification of daily solar radiation distributions in typical days, which is based mainly on the distribution of the clearness index, i.e. the ratio between global horizontal irradiation and extraterrestrial irradiation (depending on solar zenith angle, earth radius vector and solar constant values). Solar zenith angle and earth radius vector are calculated with the SPA algorithm [11]. Since February 2003, the solar constant has been measured with a 6 hour time-step by the SORCE (SOlar Radiation & Climate Experiment).

Various classification methods are based on the clearness index: Muselli *et al.* [12] use a Ward aggregation method to perform the clustering with 4 years of hourly sampled data, Maafi and Harrouni [13] use fractals with one year of 10 minutes sampled data, Soubdhan *et al.* [14] use a mixture of finite Dirichlet distributions with one year of one second sampled data, and Gastón-Romeo *et al.* [15] use morphological clustering with around 2 years of 10 minutes sampled data.

The study [15] tends to show better classification results than the others, in particular for the overcast days classification between intermittent clear sky days and intermittent cloudy sky days. Two interesting conclusions also came from [14]:

1 The occurrence of totally cloudy sky days, when almost zero PV production could occur, is very low (only 23 days – 6% – for the year studied).
2 Three consecutive cloudy sky days were not observed.

Therefore daily solar radiation distribution classification on typical days is important to identify days of cloudy sky on which maintenance would be preferrable. It requires at least one year of data measurement with a sufficient time step, typically less than 10 minutes, to perform the above cited methodologies of daily solar radiation distribution classification, and to make cross-comparisons.

Nevertheless, these studies are based on data provided by a single station, leading to the interpretation that rapid and large variations of the power output can be expected, essentially from the cloud cover variation. While this can be true for small PV installations, where daily distributions of solar irradiation measured by a single station will be closely linked to power output distributions [13], in larger PV plants, there will be a smoothing effect on the power output fluctuations.

Hoff and Perez [16] introduced a simple approach and a general method to predict the maximum power output variability of a PV fleet, based on size, orientation and distribution of the PV fleet, cloud transit speed and the time-scale of interest.

They illustrate LS-PV power output variation with a 4,59 MW plant in Springerville (Arizona), where the maximum relative power output can reach 60% in 60 seconds. This determination is in agreement with Apt and Curtright [17] which determined the spectrum of power for the same PV plant and three PV tracking systems separated by 100 km. They found it follows a Kolmogorov spectrum ($f^{1/3}$), differing from a wind turbine $f^{5/3}$ spectrum, and leading to the same conclusions.

But, Kawasaki *et al.* [4] demonstrated that a more intensive PV plant would decrease this relative power output variation, in particular for the most concerning classes, i.e. overcast days. Nine irradiance stations with 1 minute time step were implemented in an approximately 14 km^2 square area (Tsukuba, 100 km from Tokyo), representative of a 500 MW plant area.

Getting the nine sites average reduces (i) the average spectrum, i.e. the average of the power spectrum more than 1 mHz; this low limit for averaging corresponds to the lowest frequency limit where load frequency control systems can operate to regulate power flow between different areas; the larger the average spectrum is, the larger the fluctuations in power output of the considered power plant will be., and (ii) the magnitude of fluctuation, i.e. the maximum power output variability that can occur in 1 minute.

This smoothing effect is more effective when individual site fluctuations are important, i.e. during intermittent cloudy days, which is the most problematic type of day.

Taking back Hoff and Perez's general method [16], assume a 1 GWc VLS-PV plant with a 15 km length and a maximum cloud transit rate of 10 m·s^{-1}. Then, in a 10 minute interval, the maximum power output variability will be around 55%. In a 5 minute interval, it drops to less than 30%, 10% in a 1 minute interval.

Therefore, the VLS-PV scale will decrease the short-term intermittence, limiting the relative power output fluctuations that can occur, from 60% for a 4,59 MW system to 10% for a 1 GW system.

It is identified in [16] for Springerville, Arizona, that maximum relative power output (MRPO) can reach 60% in 60 seconds. Because of the VLS-PV plant size, the MRPO is much lower (10%). If 60% is considered problematic as in [16], this will be measured for a 1 GWp VLS-PV plant only on a time-scale of 10 min.

And this time-scale is not so much problematic, being the time usually needed to start compensating devices.

(2) CLOUD MOTION MODELS

Cloud motion models are deterministic, i.e. they predict cloud cover advection based on inputs such as wind speed and direction, topography, atmospheric composition on two consecutive images.

A satellite-imagery based model has been recently set up [19] for very short-term forecasting, on the basis of above one hour products, as SolarAnywhere, SolarGIS or NSRDB [1]. SolarAnywhere covers the USA and is run by Clean Power Research. SolarGIS covers approximately from 60° W to 120° E in latitude and is run by GeoModel Solar. National Solar Radiation Data Base covers the USA and is run by the Renewable Resource Data Center (RReDC) from the National Renewable Energy Laboratory (NREL). Cloud advection is determined by defining for each image pixel a GHI index (ratio between GHI and clear-sky GHI [18]), and assigning a motion vector. Future cloud cover is then simulated and the GHI index is deduced back.

A satellite-based model has recently been set up [19] for very short-term forecasting, on the basis of hourly above developed one hour products. Intermittence within a 1 km^2 pixel resolution is predicted with 20 seconds, 1, 5 and 15 minutes time-steps through variable estimation capturing the mean, the maximum and the standard deviation of the GHI index.

Variables are estimated through a three-parameter calibrated crosstab, parameters being GHI and DNI (Direct Normal Irradiance) indexes and variability of GHI index in the eight surrounding image pixels. Calibration was performed using 17 stations with a 20 second sampling period and 7 stations with a 1 minute sampling period.

Ground-based models use measurements generally performed on site with a Total Sky Imager (TSI). Cross-correlation of consecutive images allows the cloud motion vectors to be detected, and the cloud cover advection to be deduced. A ceilometer can be coupled to imagery to detect cloud base height, and helps to optimize the cloud shadow mapping. GHI is retrieved through comparison with a clear-sky model [18].

Imagery techniques are the most promising as regards very short-term forecasting, being able to deliver infra-hour and sub-kilometer forecasts, with a mean error below 30% for 5 minutes-ahead forecasts [20]. Yet, optimization perspectives are important: image processing, cloud evolution between captures, cloud base height and composition definition, limited horizon, etc.

9.1.4 Conclusion

In conclusion, VLS-PV power output intermittence can be quantified at various time scales, from the yearly yield to the short-time variations of the power output. It requires (i) detailed and important databases and (ii) mathematical and statistical tools to predict the power output variation range at the desired time-scale. Solar resource experts today have the expertise to meet such challenges.

Even if short-term variations of irradiance can be expected, the smoothing effect will tend to limit the power output variations of a whole VLS-PV system, pushing the fluctuation preoccupation limit to around 10 minutes, which is more consistent with already existing ramping capacities.

Prediction can also be achieved at almost all time-scales of interest for VLS-PV systems, with the use of NWP models and cloud motion models.

Cloud motion models tend to present more accurate forecasts below 5 hours [1], and sky imagery systems are the most suitable models for infra-hour nowcasts [20]. Above 10 days, the limit of NWP models, predicting solar irradiation relies on climatology and climate indices, and is beyond the scope of the present review.

Persistence models are based on the simple assumption that the forecast is equal to the instantaneous measured parameters, with adjustment of solar position [11]. They usually are used in benchmarking NWP or cloud motion models, i.e. they can be more accurate than these models for very short time periods, where variations are more unlikely, and for periods above 10 days, where NWP-based forecast availability is limited.

Figure 9.1 summarizes the temporal chain of VLS-PV systems power output intermittence and prediction; models are in blue boxes.

Long-term intermittence characterization and power output planning is based on climatology and climate indices.

Climatology and climate indices are also part of NWP model inputs, which mainly use statistics and mathematical computing to determine either the main meteorological parameters or the power output directly. NWP models can also help with seasonal definition, mainly based on historical data banks.

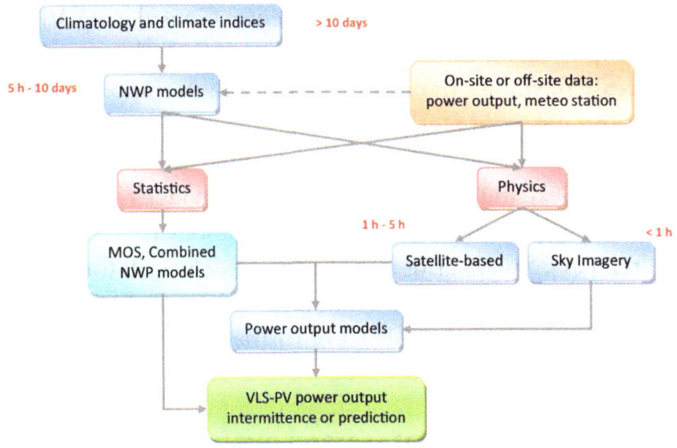

Figure 9.1 PV systems power output intermittence and prediction at various time scales

Physics-based techniques use NWP models as inputs to find the accuracy of the cloud motion advection, and to determine wind pattern data. They do not directly define VLS-PV power output but the main meteorological parameters that are then used in power output models.

On-site or remote data can enhance the accuracy of the models, except for NWP models, where stations have to be recognized as standards, and therefore may not be considered as inputs. Still, such data are essential for learning the real behavior of the photovoltaic systems in different meteorological environments, and for studying the aging of modules.

9.2 STATIONARY STORAGE

9.2.1 Introduction

Stationary storage is divided into three sectors: chemical, mechanical (including kinetic, magnetic and electrostatic) and thermal storage.[1] Their grid applications can be of two types: grid services and energy management. While power and capacity can be uncorrelated in some storage sizing, they are linked in the use of storage by the capacity ratio, i.e. the time step of use; we therefore prefer to refer to capacity instead of power, except when it is explicitly cited.

Grid services include power quality control (frequency and voltage), through the fast delivery of mainly large amounts of power, and bridging power applications, where storage is used by the dispatching grid to switch from one source (for example, VLS-PV) to another. Grid services are meant to deliver to the utility active and reactive power for grid balancing purposes.

Energy management, or load leveling, consists of storing electricity when the kWh price is low or when unused (in case of variable renewable plants) and injecting back when the kWh price is high or to provide guaranteed electricity during identified periods.

Stationary storage can be either centralized or decentralized, at the grid or plant scale, to fit utility, operator and consumer requirements. Considering VLS-PV deployment, stationary storage is envisaged on-grid to increase flexibility, or on site to limit the intermittence and/or to provide grid regulation reserves.

A brief storage overview is presented below, centered on chemical and mechanical storage. A wider review and details on each stationary storage technology can be found in ref. [21]. Smart-grids and electric vehicles will form part of solutions to raise grid flexibility, but are not discussed here ; they will be the subject of further publications.

9.2.2 Brief storage overview

(1) CHEMICAL STORAGE

BATTERIES

These types of electrochemical storage all have a fast response time, i.e. they deliver a large amount of energy and/or power in less than one minute. They are summarized below:

- Sodium-sulfur (NaS): encapsulated molten metal cells (50 Ah) assembled in modules, and then in the battery, which can be designed either for grid services or for energy management. Used at medium scale, from 500 kW to around 5 MW. Around 15 years lifetime (3 000 cycles). Maintained at 350°C for – and by – the process. Replacement represents 75% of the installation (cells mainly). Already industrialized.
  - Advantages: Good efficiency (~70%) and energy density (~150 kWh·m^{-3}).
  - Drawbacks: Explosion risk and corrosion (limited with encapsulation). Important self-discharge rate (> 2%/day).
- Redox-flow (RFB) [22]: one reactor containing cells (power), producing energy through the circulation of electrolyte contained in two separate tanks (capacity). Three types: vanadium, zinc-bromide and polysulphide-bromide. Dedicated for grid services and energy management. Around 10 years lifetime (2 500 cycles). Replacement represents a third of the installation. Industrialized for infra-MW power.
  - Advantages: Good efficiency (~70%). Independent power and capacity. No self-discharge.
  - Drawbacks: Low energy density (~25 kWh·m^{-3}). Corrosion. Limitation on vanadium resources.
- Nickel-cadmium (NiCd) and nickel-metal hybrid (NiMH): popularized through pocket batteries, vented batteries with nickel oxyhydroxide as positive active material, and respectively metallic cadmium and hydrogen absorbing alloy as the negative electrode. Dedicated to grid services. Multi-MW scale applications in USA. Industrialized for infra-MW power.
  - Advantages: Good efficiency (~70%) and energy density (~150 kWh·m^{-3}).
  - Drawbacks: Memory effect. Hazardous cadmium. Important self-discharge rate (> 2%/day).
- SoNick ($NaNiCl_2$): sodium nickel chloride molten battery are available today in high power and high capacity versions. Under 10 years lifetime (2 000 cycles). Maintained at 270°C for – and by – the process. Industrialized for infra-MW power.
  - Advantages: Good efficiency (~70%) and energy density (~140 kWh·m^{-3}).
  - Drawbacks: Limitation on nickel resources. High self-discharge rate (> 10%/day).
- Valve-regulated lead-acid (VRLA): most developed storage made of lead plates immersed in sulfuric acid and water. Important work performed to rethink the battery in the last ten years. Still correct energy density (~80 kWh·m^{-3}). Applied for large range of storage, from some kWh to multi-MWh. Correct self-discharge (< 0.25%/day). Replacement concerns 75% of the installation (cells mainly). Industrialized.
  - Advantages: Scalability. Proven experience. Good efficiency (~80%).
  - Drawbacks: Low lifetime (5 to 7 years, 1 000 cycles). Requires specific recycling process. Low mass energy density (40 Wh·kg^{-1}).
- Lithium-ion (Li-ion): developed for electric vehicle applications, as for MW-scale. Based on movement of lithium ions between positive (metal oxide with lithium) and negative (graphite) electrodes. Major use in grid services, but can also be applied to energy management. Lifetime from 7 to 10 years (2 500 to 3 000 cycles). Replacement concerns 75% of the battery (cells mainly). Already industrialized.
  - Advantages: High energy density (~300 kWh·m^{-3}) and good efficiency (~75%).
  - Drawbacks: Limitations on lithium resources.

In conclusion, batteries tend to be considered only for grid service applications, but they can also be designed for energy management purposes. When considering energy management the following four problems arise, because of the induced MW-scale.

- Self-discharge and depth of discharge (DOD): If self-discharge is too high (i.e. SoNick), varying electricity production will add losses to the usable capacity, reducing the overall round-trip efficiency.

  High-DOD operation can decrease the battery lifetime (i.e. Li-ion). Thus oversizing is practiced to optimize the balance between additional cost due to oversizing and longer life.
- Energy density: Low energy density (i.e. RFB) can make storage volume important. To get a 5 MWh usable capacity for RFB, with an energy density of 25 kWh·m^{-3}, requires 200 m^3 of electrolyte, plus the volume required for the reactor and the electronics.

This may not be problematic for LS-PV, which already occupies large areas, but it also increases the operation risks.

- Environment: Large installations induce large requirements, and therefore significant pressure on resources (lithium, vanadium). As seen above, 200 m^3 of electrolyte would require approximately 20 tons of vanadium, assuming vanadium molarity of 2 $mol \cdot l^{-1}$ (maximum concentration limit for vanadium redox-flow batteries – VRB), with a molar mass of 50,941 5 $g \cdot mol^{-1}$. In 2010, the vanadium used for non-metallurgical purposes was only 1,680 × 10^3 tons [23].

  Large installations also concentrate potential pollutants (cadmium, lead). 5 MWh of lead-acid batteries represents around 75 tons of lead.[2] Yet, it can help to develop a local recycling industry if large volumes are to be produced by a VLS-PV plant.

  Hot batteries need to be heated (this can be achieved partially by the process), but backup support for this heating is also necessary, limiting efficiency by burdening storage requirements.

  Other batteries generally require venting or cooling, especially in desert areas. Indeed, out of the optimal temperature range battery performance can decrease greatly because of irreversible induced deposition or crystallization [22].

- Maintenance, dismantling and recycling: Batteries require specific maintenance and waste treatment industries, at least for replacements over intermittent renewables power plant lifetime (typically 25 years), and above all when dismantling. Maintenance can be diminished by, for example, using a smaller DOD, but this increases the required capacity. Replacement frequency can equally be decreased with a smaller DOD or with efficient heating/cooling; it also requires greater oversizing and/or consumption.

  Component recycling is achievable for almost all batteries technologies, and should not be neglected when selecting one system over another.

FUEL CELLS

Fuel cells gather all the systems designed to produce electricity through the reaction in a cell of a fuel and an oxidant, brought by a flow, in the presence of an electrolyte. They are either in development or demonstration for the MW scale requested in stationary storage. Indeed, they present an important interest, in the way of several similarities with oil and gas (essentially high energy density and scalability), allowing thinking they could benefit the development achieved for such resources (pipes, facilities, transportation, etc.).

Hydrogen is of constant interest due to its high energy density (above 1 000 $kWh \cdot m^{-3}$) and to the water it produces when reacting with oxygen in the fuel cell. But still round-trip efficiency is low – essentially due to the energy produced by the hydrogen process – and storage is difficult (high pressure and/or low temperature).

In the Direct Methanol Fuel Cell (DMFC), methanol is fed with water to get H^+ ions that react at the cathode with air (O_2) to produce water. Methanol is less complex to store than hydrogen, but is toxic and flammable. Also, the process produces CO_2, which is difficult to manage. Development is mainly oriented to military applications.

Molten carbonate fuel cells (MCFC) and solid oxide fuel cells (SOFC) are high temperature fuel cells. MCFC operates at 650°C where lithium-based salts melt to form carbonate ions that produce water and CO_2 when combined with hydrogen. SOFC work at higher temperatures (up to 1 000°C), and therefore are able to directly oxidize fuel without requiring a reforming process. SOFCs are targeted to operate in cogeneration units.

A metal-air battery uses metal (zinc or aluminum mainly) as fuel, and air as oxidant. Such metals confer to the battery a high energy density (above 2 000 $kWh \cdot m^{-3}$), while being commonly available. However, the recharging phase is still limited, forming dendrites in the electrolyte; the lifetime is mainly under 200 cycles and this battery must be mechanically recharged.

To conclude, fuel cells are bringing promising opportunities to energy (electricity and heat) production. They are very similar to oil in terms of energy density and scalability, and therefore could be suitable for progressive substitution. Nevertheless, their development could be restrained by the lack of demonstration sites and by the exploitation risks still attached to the processes.

KINETIC, MAGNETIC AND ELECTROSTATIC STORAGE

- Flywheel: spinning machine delivering energy through system inertia. Dedicated to short-term grid services, generally power quality. Fast response time. Achieved at small and MW-scales (grid stabilization). 20 years lifetime (> 10 000 cycles). Already industrialized.
  - Advantages: Fast discharge start (< 1 second). High efficiency (~90%).
  - Drawbacks: Cannot store for important duration, rapid self-discharge.
- Superconducting magnetic energy storage (SMES): energy carried through a refrigerated superconducting coil in a vacuum. Dedicated to grid services because of the fast discharge rate. Applications up to multi-MW scale. 30 years lifetime (> 10 000 cycles). Already industrialized.
  - Advantages: Instantaneous discharge. High efficiency (~90%).
  - Drawbacks: Disturbing magnetic fields. Requires intensive cooling.
- Supercapacitor energy storage (SCES): parallel plates separated by a dielectric insulator; capacity is linked to the exchange surface. Designed for grid services (fast discharge of power). 20 years lifetime (> 10 000 cycles). Very low energy density (< 0,025 $kWh \cdot m^{-3}$)

and very high mass power density (~2 000 W·kg^{-1}). Already industrialized.

- Advantages: Fast charge and discharge. High efficiency (~90%).
- Drawbacks: Low energy density.

These storages are like batteries designed for fast response time ability, except they are today only dedicated to that, and cannot be envisaged for energy management purposes. They are more suitable for uninterruptible power supply (UPS) applications, because of their specifications and their implementation.

For grid purposes, they are adapted to specific services, such as maintaining power quality for crucial applications (defense, health, internet, etc.).

OTHER POTENTIAL ENERGY STORAGE

- Pumped-hydro energy storage (PHES): extension of the reversible dam principle; excess energy is used to pump water from a lower into an upper reservoir (capacity) and energy is recovered from inverse flow through reversible – or dedicated – turbines (power). Power and capacity are then decoupled. Most developed stationary storage after lead-acid batteries. Requires height between reservoir and turbines. Dedicated to energy management, except for systems with adjustable speed that can change their power output in 10 to 30 ms. 30 years lifetime. Already industrialized.
  - Advantages: Good efficiency (~80%). Can store large amounts of energy (> 1 GWh). Independent power and capacity.
  - Drawbacks: Site-specific storage. Evaporation issues. Long development (typically > 5 years). Very low energy density (0,218 kWh·m^{-3} with 100 m height and 80% efficiency).
- Compressed air energy storage (CAES): excess energy is used to compress air – 40 to 50 bars – and energy is recovered through heated expansion in a turbine (power). Large-scale units require underground cavities (salt mines, depleted gas cavities) for storing air (capacity). Heated expansion is achieved with gas firing, except for adiabatic systems, recovering, storing and delivering heat. Dedicated to energy management. 30 years lifetime. Already industrialized except for small scale (< 5 MW) and for adiabatic (R&D in process on small and large scales).
  - Advantages: No hazardous compounds. Can store large amounts of energy (> 1 GWh). Independent power and capacity.
  - Drawbacks: Site-specific storage. Low efficiency (45%), except for adiabatic (65–70%). Low energy density (~15 kWh·m^{-3}).

Potential energy storages are the most important installations that can be implemented on a site, i.e. it is centralized storage. They are today reserved for energy management, as in 15 minutes they can provide large amount of energy for a long period (typically several hours).

Interesting perspectives are formulated for small-scale applications, keeping their advantages while getting rid of the most problematic drawback: a specific site.

For PHES, the small scale is targeted to reach specific markets, as insular grids, benefiting from sea proximity; use of salted water has to be taken into account, even if already achieved (Okinawa, Japan). It is also viewed as a way to limit development time.

For CAES, adiabatic operation is targeted to get rid of gas for expansion, and then increase the round-trip efficiency while decreasing environmental impacts. Small scale is also envisaged to avoid specific site requirements, i.e. a suitable geological cavern for air storage and proximity to high voltage lines.

9.2.3 Cost comparison between RFB, NaS, LI-ion and VRLA

Cost comparison between technologies will vary with the project location and the operation conditions. It also varies from one technology to another, because they are based on the same principle but constructed with different architectures and in different locations. It is typically the case for VRLA and Li-ion batteries, where actors are numerous, making costs double between the cheapest and the most expensive model.

Only RFB, NaS, Li-ion and VRLA are studied here, because they are already available for MW-MWh applications, and are the main technologies under consideration in current renewable projects coupled to storage. The analysis is focused on a 3 MWh capacity to reflect the current looked-up capacity in the small storage market coupled to renewables.

Table 9.1 presents the results, split between CAPEX and OPEX-related costs. Green cells are assumptions; such values will vary regarding the project location and the battery conditioning, being more or less exposed to changes of temperature and corrosion.

Lifetime (in years) is based on number of cycles and annual operability (in days). The yield (total output / total input) is important, having an impact on the electricity sold to the utility.

Investments refer to construction costs, and include all the operations required to install the system on site. Installation and commissioning will vary essentially according to the location. Investments projected in 2015 and 2020 are similar for the all technologies explored.

However, replacement costs that can be secured through provisions make the total CAPEX over 25 years of operation highly variable. Replacement costs are linked to the lifetime, i.e. the number of times the system needs to be changed. Replacement essentially includes refurbishment costs, varying regarding the amount of the system that needs to be changed. Indeed, for RFB, the cells only need

Table 9.1 Technology cost comparisons between present day and projections for 2020; based on a 25 years operation of a 3 MWh capacity (green cells are assumptions)

Technology			Redox-flow (RFB)			Sodium-sulfur (NaS)			Lithium-ion (Li-ion)			Valve-regulated lead-acid (VRLA)		
			2011	2015	2020	2011	2015	2020	2011	2015	2020	2011	2015	2020
Life	Annual operability	days	330	330	330	330	330	330	330	330	330	330	330	330
	Cycles	cycles	5 000	5 000	5 000	2 800	2 800	2 800	3 000	3 000	3 000	1 400	1 400	1 400
	Lifetime	yr	15	15	15	8	8	8	9	9	9	4	4	4
Yield				0.7			0.7			0.85			0.8	
Investment	Procurement, assembly & tests	kEUR	3 090	1 500	1 050	1 850	1 650	1 350	3 350	1 500	1 200	1 750	1 750	1 750
	Installation & commissioning	kEUR	150	150	150	150	150	150	150	150	150	150	150	150
	Sub-total 1	kEUR	***3 240***	***1 650***	***1 200***	***2 000***	***1 800***	***1 500***	***3 500***	***1 650***	***1 350***	***1 900***	***1 900***	***1 900***
Replacement	Replacement (y/n)		y	y	y	y	y	y	y	y	y	y	y	y
	Number over lifetime		1	1	1	2	2	2	2	2	2	5	5	5
	Proportion	%		33%			75%			75%			75%	
	Refurbishment cost	kEUR	1 080	550	400	1 500	1 350	1 125	2 625	1 238	1 013	1 425	1 425	1 425
	Sub-total 2	kEUR	***1 080***	***550***	***400***	***3 000***	***2 700***	***2 250***	***5 250***	***2 475***	***2 025***	***7 125***	***7 125***	***7 125***
Total CAPEX over 25 yr (sub-total 1 + 2)		kEUR	**4 320**	**2 200**	**1 600**	**5 000**	**4 500**	**3 750**	**8 750**	**4 125**	**3 375**	**9 025**	**9 025**	**9 025**
O&M	Spare parts	kEUR/year	15	10	10	15	10	10	10	10	10	10	10	10
	Man-hours	kEUR/year	30	20	20	25	20	20	10	10	10	5	5	5
	Consumables	kEUR/year	5	5	5	6	6	6	3	3	3	3	3	3
Total OPEX per yr		**kEUR/year**	**50**	**35**	**35**	**46**	**36**	**36**	**23**	**23**	**23**	**18**	**18**	**18**

to be changed, and they represent only one third of the total investment costs.

The OPEX variation between the studied technologies is influenced not by the spare parts but by the man-hours needed to perform the battery maintenance: already mature and widely distributed technologies such as VRLA are easier to maintain than emerging technologies such as NaS for example.

This analysis accentuates the necessity of performing detailed cost comparison, on the basis of the criteria developed in Table 9.1. This cost comparison analysis will help operators see which technologies require higher CAPEX and which require higher OPEX. Depending on countries where it will be implemented, operators could prefer having a technology with higher CAPEX, but with lower OPEX, or conversely. Choice will be influenced by the local representation, the spare parts availability, etc.

9.2.4 Storage for VLS-PV systems

(1) GENERAL CONSIDERATIONS

Considering VLS-PV plants as firm power sources requires the intermittence to be compensated or used for other purposes than grid injection.

As seen in intermittence analysis, (i) seasonal intermittence can be problematic, but resource production could provide alternatives to this problem, (ii) day and night alternation can be partially compensated by combination with VLS-wind, and (iii) as regards infra-minute intermittence, the smoothing effect minimizes the power output fluctuations to below 10%.

It will sometimes happen that VLS-PV production will be greater than the demand. Instead of conceiving complex solutions that will not increase the value of the whole system, dumping of production can be envisaged. This approach helps to understand that VLS-PV production penetration on grids can be important when part of the production is neglected [24].

At the opposite end of the spectrum, lack of production can occur: resources may be insufficient to absorb the whole production during daytime, VLS-wind may not be capable of providing enough energy for the morning and evening load peaks [26, 27], or an installation may be asked to participate in grid balancing.

All these reasons can lead to the implementation of storage, for grid services or for energy management purposes. Such storage can be implemented by the utility or by private entities on the grid, if they project benefits on local production and load instability management. If VLS-PV plant operator targets profitability on grid-balance services or agreed with utility on grid-balance obligations, this could be more convenient to have the storage implemented on the production site instead of being remotely installed on the grid.

(2) STORAGE ON GRID

Implementation of variable renewable generation, as VLS-PV, can be drastically limited by the grid infrastructure itself. Indeed, if flexibility is too low, and grid reinforcement is needed to allow accepting the whole capacity of a variable renewable plant, it could jeopardize the project, by the duration of the process (five to ten years), not to mention the local acceptance.

To improve flexibility faster than line creation utilities can implement storage on the grid. It gives them time to anticipate grid developments while not restraining the development of production plants and while being scalable. It also gives policy makers opportunities to support the development of intermittent renewables by adding further flexibility capacities on the grid.

Storage can improve the grid stability, by virtually aggregating produced and/or consumed electricity on specific grid spots (homes, small plants, industries, etc.) [25].

It can also be part of grid strategy to trade electricity with interconnected regions or countries, giving the utility extra capacity that can be employed on demand.

Private entities already have invested in this sector, projecting profitability on production and demand aggregation, and on bridging power services for the grid. The main difference is that utility-owned storages are usually dedicated to the grid flexibility. In case of private ownership, the stored capacity will be traded as usable capacity or services, bringing a constraint to the storage availability and the associated flexibility.

Knowing the power, capacity and time response requirements, the optimum technical solution can be found from the list of storages and their specifications.

Targeting MW-scale grid services, batteries are the optimum choice, essentially because of their fast response ability with large power/energy capacities. If there are critical installations, such as administrations, hospitals, data centers, etc., flywheel, SMES or SCES should be privileged, without neglecting their limited duration, and therefore a combination with batteries for backup could be necessary.

If energy management is needed, some batteries can ensure this role, but careful attention must be paid to the limitations exposed above when considering the multi-MW scale. For important power and/or capacity, PHES or CAES installations are more suitable; this can be envisaged in major grid reinforcements or in interconnection stabilizations.

(3) ONSITE STORAGE

Onsite storage reflects (i) the capacity to modify the VLS-PV plant injection profile to fit to grid requirements and maximize the benefits of VLS-PV electricity produced and (ii) the provision of power and energy units for grid balance regulation.

Considering modifying the plant injection profile, it is needed because of the mismatch between VLS-PV production and load, especially with one of the major daily peaks, i.e. the evening peak [26, 27], requiring a large amount of energy when the VLS-PV plant starts decreasing its injection on the grid.

Load leveling is typically illustrated through such applications. Considering the plant size, a large amount of energy will be shifted, and therefore, large energy management capacities, such as PHES or CAES plants, would be employed on site. Their sizing depends not only on the grid flexibility, but also on the ability for the operator to value the electricity shifted to the evening peak, when storage brings adequate service to the utility.

Consider storing 10% of a 1 GW VLS-PV plant daily production, with a 1 900 kWh·kWp^{-1} yearly yield and with storage in operation 330 days per year (includes maintenance). This represents a yearly stored capacity of 190 GWh, and therefore, an approximate daily engaged capacity of 576 MWh. Such a size, with attention on the fact that it is purely virtual, clearly shows the range of capacity required, and therefore the need for massive storages, such as PHES or CAES.

To be considered as firm power sources, VLS-PV systems will be asked to participate in grid balancing with sufficient reserves for frequency and voltage regulation.

Even if illustrations of grid requirements were performed with some continental European utilities, VLS-PV systems could be asked to apply such rules, except this cannot be done directly by the PV plant itself. Thus it requires dedicated storage capacities, providing reserves for such applications.

Again, to apprehend the capacity range, virtual assumptions can be examined. Consider a maximum power injected of half the peak power installed (500 MW). VLS-PV plant could typically be asked for 25% of this maximum power as a reserve for 30 minutes operation. Thus, it requires 125 MW for 30 minutes: 62,5 MWh. If the plant is conceived with ten 100 MW sub-fields, it requires an installed capacity of 6,25 MWh per sub-field.

Flywheels, SMES and SCES are at first sight not adapted to such requirements, because it cannot be considered as UPS demand.

Batteries and fuel cells have specifications that fit to frequency and voltage requirements and such virtual sizing. Attention must always be paid to the volume of the installation it represents.

Potential energy storage can participate to such reserves, in particular for secondary frequency regulation, because they need to be already in operation for delivering active and reactive power for primary frequency regulation. Yet, if capacities are already scheduled for load leveling purposes, they can also contribute as frequency and voltage regulation reserves.

(4) CONCLUSION

To conclude, VLS-PV will have to be coupled to storage systems in order to participate in grid balancing, as do other firm power sources. This will require a combination of fast response time storages, such as batteries, and energy management storages, such as PHES or CAES, for secondary reserves.

Such massive storages will doubtless also be required for load leveling purposes, in order to shift electricity produced during daytime to evening and/or early morning times, when the load is typically higher in most countries.

Massive energy management storages represent the limit between storages that could be implemented on site, and storages that could be implemented on the grid, to reinforce it, to secure energy procurement or to serve utility trading strategies.

Economic considerations were not developed in this section, essentially because of their variability from one project to another. Every storage review [21] presents a wide range of prices, essentially based on estimations or in specific past projects.

The fact that the price can double in this range emphasizes the necessity of conducting a storage review not only in the sense of economics. Lifetime, OPEX, environmental impacts (LCA, recycling, hazards, etc.) and exploitation risks have to be carefully pointed out in storage benchmarking and also in policy making processes. For example, choosing storage with specific maintenance or with specific cooling requirements in desert areas will certainly lead to extra costs that could be prevented with the choice of another technology.

Storage for VLS-PV systems does not rely on a unique technological solution, but rather on a combination of at least two systems, one for energy management purposes, and one for grid services.

NOTES

1 Thermal storage will not be reviewed here as it is more in the scope of CSP power plants, which attain higher levels of temperature than PV plants could achieve.
2 Mass energy density of 40 Wh.kg^{-1}. For cars, 60% of the battery mass is lead. Therefore, 5 MWh requires 125 tons of battery, and represents 75 tons of lead.

REFERENCES

1 Perez, R., Kivalov, S., Schlemmer, J., Hemker Jr., K., Renné, D., and Hoff, T.E. (2010) 'Validation of short and medium term operational solar radiation forecasts in the US', *Solar Energy* 84: 2161–72.
2 Rossi, A., Massei, N., and Laignel, B., (2011) 'A synthesis of the time-scale variability of commonly used climate indices using continuous wavelet transform', *Global and Planetary Change* 78: 1–13.
3 Stein, J.S., Perez, R., and Parkins, A., (2010) 'Validation of PV performance models using satellite-based irradiance

measurements: A case study', *Proc ASES Annual Conference*, Phoenix, Arizona, 2010.

4 Kawasaki, N., Oozeki, T., Otania, K., and Kurokawa, K., (2006) 'An evaluation method of the fluctuation characteristics of photovoltaic systems by using frequency analysis', *Solar Energy Materials & Solar Cells* 90: 3356–63.

5 Guérin de Montgareuil, A. (2007) 'Description of MotherPV, the New Method Developed at INES/CEA for the Assessment of the Energy Production of Photovoltaic Modules', *Proceedings of the 22nd EU-PVSEC Conference*, Milan, 2608–12.

6 Williams, S.R. *et al.*, (1006) 'Accuracy of energy prediction methodologies', *Conference Record of the 4th IEEE World Conference on Photovoltaic Energy Conversion*, vol 2, Hawaii, USA, 7–12 May, 2206–9.

7 Solomon, A.A., Faiman, D., and Meron, G. (2010) 'An energy-based evaluation of the matching possibilities of very large photovoltaic plants to the electricity grid: Israel as a case study', *Energy Policy* 38: 5457–68.

8 Chen, C., Duan, S., Cai, T., and Liu, B., (2011) 'Online 24-h solar power forecasting based on weather type classification using artificial neural network', *Solar Energy* 85: 2856–70

9 Mellit, A., and Massi Pavan, A. (2010) 'A 24-h forecast of solar irradiance using artificial neural network: Application for performance prediction of a grid-connected PV plant at Trieste, Italy', *Solar Energy* 84: 807–821, 2010.

10 Pelland, S., Galanis, G., and Kallos, G. (2011) 'Solar and photovoltaic forecasting through post-processing of the Global Environmental Multiscale numerical weather prediction model', *Progress in Photovoltaics: Research and Applications*, doi: 10.1002/pip.1180.

11 Reda, I., and Andreas, A. (2003) 'Solar Position Algorithm for Solar Radiation Applications', NREL Report No. TP-560-34302, 2003 (revised January 2008).

12 Muselli, M., Poggi, P., Notton, G., and Louche, A.,(2000) 'Classification of typical meteorological days from global irradiation records and comparison between two Mediterranean coastal sites in Corsica island', *Energy Conversion and Management* 41: 1043–63.

13 Maafi, A., and Harrouni, S., (2003) 'Preliminary results of the fractal classification of daily solar irradiances', *Solar Energy* 75: 53–61.

14 Soubdhan, T., Emilion, R., and Calif, R. (2009) 'Classification of daily solar radiation distributions using a mixture of Dirichlet distributions', *Solar Energy* 83: 1056–63.

15 Gastón-Romeo, M. *et al.*, (2011) 'A Morphological Clustering Method for daily solar radiation curves', *Solar Energy* 85: 1824–36.

16 Hoff, T.E., and Perez, R. (2010) 'Quantifying PV power Output Variability', *Solar Energy* 84: 1782–93.

17 Apt, J., and Curtright, A. (2008) 'The Spectrum of Power from Utility-Scale Wind Farms and Solar Photovoltaic Arrays', Carnegie Mellon Electricity Industry Center.

18 Ineichen, P., (2006) 'Comparison of eight clear sky broadband models against 16 independent data banks', *Solar Energy* 80: 468–478.

19 Perez, R., Kivalov, S., Schlemmer, J., Hemker Jr., K., and Hoff, T.E. (2011) 'Parameterization of site-specific short-term irradiance variability', *Solar Energy* 85: 1343–53.

20 Chow, C.W., Urquhart, B., Lave, M., Dominguez, A., Kleissl, J., Shields, J., and Washom, B. (2011) 'Intra-hour forecasting with a total sky imager at the UC San Diego solar energy testbed', *Solar Energy* 85: 2881–93.

21 Haisheng, C., Ngoc Cong, T., Yang, W., Tan, C., Li, Y., and Ding, Y. (2009) 'Progress in electrical energy storage system: A critical review', *Progress in Natural Science* 19: 291–312.

22 Weber, A.Z., Mench, M.M., Meyers, J.P., Ross, P.N., Gostick, J.T., and Liu, Q. (2011) 'Redox flow batteries: a review', *Journal of Applied Electrochemistry* 41: 1137–64.

23 U.S. Geological Survey, (2011) Mineral Commodities Summaries, January

24 Solomon, A.A., Faiman, D., Meron, G. (2010) 'Properties and uses of storage for enhancing the grid penetration of very large photovoltaic systems', *Energy Policy* 38: 5208–22.

25 ENEL (2011) 'ENEL Distribuzione: Italy's first smart grid in Isernia', Press Release, November.

26 Martin-Rodriguez, G., and Caceres-Hernandez, J.J. (2005) 'Modeling the hourly Spanish electricity demand', *Economic Modeling* 22: 551–569.

27 Psiloglou, B.E., Giannakopoulos, C., Majithia, S., and Petrakis, M. (2009) 'Factors affecting electricity demand in Athens, Greece and London, UK: A comparative assessment', *Energy* 34: 1855–1863.

CHAPTER TEN

PV and wind-based renewable power methane

Photovoltaic (PV) is the fastest growing electricity generation technology in the world [1]. The second fastest growing electricity option is wind power [2]. Global solar and wind resource assessment clearly documents sustainable and by far the highest resource potentials for PV and wind of all power technologies available to the market. However the fluctuating solar and wind resources make it necessary to use fossil fuel power plants or storage capacities for balancing reasons. An emerging storage option is the renewable power methane (RPM) storage technology.

Research on hybrid PV power plants results in quite similar economics for PV and wind power plants by the end of the 2010s for many sites in the world [3]. The complementary characteristics of PV and wind power availability [4] generate the insight of firstly, taking into account both major new renewable power technologies and secondly, integrating the seasonal RPM storage technology for a 100% renewable power solution. This conceptual approach ends the need for fossil fuel based power plants in power systems.

This chapter introduces the concept of renewable power methane and presents an analysis of the global economic impact potential of RPM storage by the end of the 2010s which would enable hybrid PV-Wind-RPM power plants to establish a fully dispatchable power supply based on fluctuating solar and wind resources.

10.1 RENEWABLE POWER METHANE STORAGE

The current mix of risks, i.e. impact of climate change, diminishing fossil fuels and nuclear hazards, induces enormous pressure for restructuring the global energy supply, which is dependent to about 87% on exactly these risk-creating energy sources [5]. The only sustainable energy pathway follows the various renewable energies. A stable power supply needs to be based on full daily and especially seasonal adaption of renewable power supply to the load demand in the grids. The balance of seasonal renewable power supply and load demand is challenging, since both hydro storage dams and biomass power supply are very limited due to geographic and resource competition constraints. Additionally, the major renewable power supply options are represented by solar PV and wind power, which are both fluctuating. Hence, flexible power generation units are needed for balancing resource availability and load demand for a stable electricity supply.

Oil, natural gas and coal fired power plants can serve as flexible power generation units. However, the associated greenhouse gas emissions and diminishing fossil fuel resources allow these conventional power technologies only for a limited function in a transition phase prior to a fully renewable power supply. Beyond that, fossil natural gas fired power plants, technically better called methane power plants, can also be fired by RPM, whereas RPM can be produced by renewable power, air and water as input sources. The required seasonal storage of methane is already applied today. As a consequence, the power grid and natural gas grid become connected, enabling an energy flow in both directions, i.e. storage of electricity in the natural gas infrastructure and conversion of RPM into electricity, as shown in Figure 10.1.

Three elementary core processes are needed for RPM: electrolysis (conversion of electricity and water into hydrogen and oxygen), CO_2 supply (e.g. extraction from ambient air via dialysis process, by-product of biogas plants, fossil fuel power plants or industrial processes) and methanation (conversion of hydrogen, carbon dioxide and electricity into methane and water) [7]. A good overview of the RPM concept, the relevant components and the energy system integration is given by Sterner [7], Specht *et al.* [8, 9] and Sterner *et al.* [10]. The RPM technology will be scaled up to an industrial size in the next few years, i.e. the existing 25 kW_{el} demonstration facility is currently being

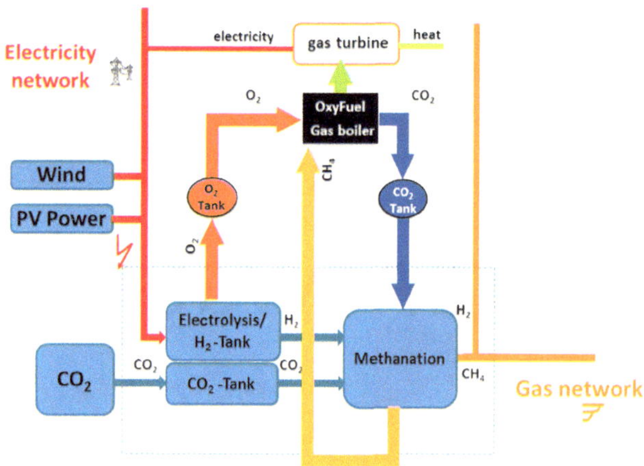

Figure 10.1 Hybrid PV-Wind-RPM power plant embedded in the power grid and the natural gas grid [6]

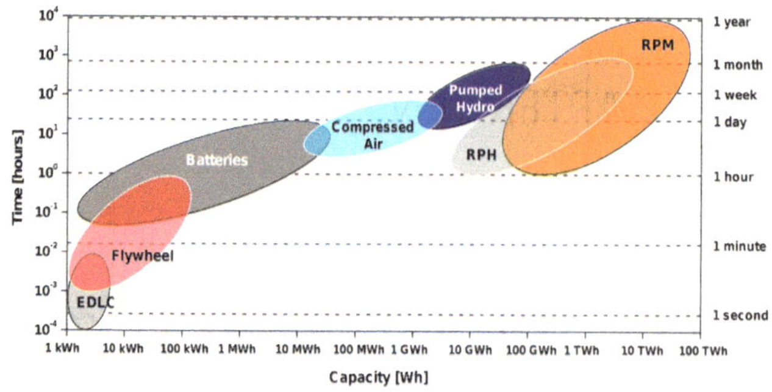

Figure 10.2 Overview of major electricity storage technologies in dependence on energetic storage capacity and charge cycling. RPM is one of only a few seasonal storage options for large-scale energy storage

expanded to 6.3 MW_{el} (in 2012) and will be further scaled to 20 MW_{el} on a modular basis by 2014 [11, 12].

The first core process is the electrolysis converting renewable electricity and water into hydrogen and oxygen. Several electrolysis technologies are available, enabling energy conversion efficiencies of up to 80%. The technology has been used for decades, can be operated at various pressures and temperatures, and is scalable for industrial applications in the range of some kW to MW [7].

The hydrogen is used in the methanation (Sabatier process), the second core process, to convert hydrogen and carbon dioxide to methane and water. The energy conversion can reach efficiencies up to 85% in a catalytic exothermal process on a temperature level of 180–350 °C and a pressure of 1–100 bars [7].

Several CO_2 sourcing routes are available, e.g. by-product of biogas plants, fossil fuel power plants or industrial processes; however the most elegant way is the extraction of CO_2 from ambient air. Several processes are known for extracting CO_2 from ambient air [7], whereas in the following the focus is put on the energy efficient ZSW process based on absorption and electrodialysis [13].

For producing 10 MJ_{th} RPM, 16 MJ_{el} are required for the electrolysis process including the methanation process plus a further 4,8 MJ_{el} in case of extraction of CO_2 from ambient air. This translates into a renewable electricity to RPM conversion efficiency of about 63% excluding energy for CO_2 sourcing. Specific energy consumption of the dialysis process extracting CO_2 from ambient air corresponds to a virtual efficiency of about 77%. Thus, RPM based on CO_2 extraction from ambient air can be generated on an efficiency level of about 49%, whereas the exothermic methanation process supplies high temperature process heat that might be used for other purposes. About 50% of the electric energy can be chemically stored in RPM and may afterwards be used for all purposes that fossil natural gas is used for. In the case of burning RPM in modern gas power plants, i.e. in 58% efficient combined cycle gas turbines (CCGT), the full cycle efficiency would be about 37% (CO_2 available on site) and 29% (CO_2 extraction from ambient air). This full cycle storage efficiency is rather low and induces high specific cost for such stored electricity; however an easy to use seasonal energy storage would be enabled and the entire natural gas infrastructure could be used.

Looking for alternative solutions, there are few and very limited seasonal electricity storage options, in particular for large scale energy storage (Figure 10.2).

Pumped hydro storage is the most preferred energy storage on the power plant and power grid level due to relatively low cost, large energy storage reservoirs and a high flexibility in switching from charge to discharge operation mode. However, geographic prerequisites of substantial difference in altitude and available topographic sites limit this large scale energy storage. The example of Germany emphasizes the already existing methane storage capacity, since 0,04 TWh_{el} pumped hydro storage and 217 TWh_{th} (equivalent to 125 TWh_{el}) natural gas storage capacities are currently available. A seasonal energy storage would also induce high storage costs due to a respective low cycle frequency. The remaining two seasonal storage solutions are hydrogen and methane. Seasonal storage on an RPM basis might be preferred due to higher storage efficiency and already existing transport infrastructure in most regions in the world plus available energy converting units, like power plants, heating for houses, powering the transport sector and using RPM in the chemical industry.

In case of available local hydrogen storage capacities and no need for balancing the stored hydrogen over regions via an infrastructure, it might be more preferable to opt for renewable power hydrogen (RPH). Using RPH would have the advantage of better economics due to less efficiency losses and lower capital expenditures (no methanation and dialysis process) but no physical exchange of the stored electricity might be an enormous obstacle due to no available hydrogen infrastructure. Hydrogen based CCGT are currently under development for similar technical and economic parameters like methane based CCGT. For the further discussion the RPM pathway is taken into account as a result of the infrastructure constraint.

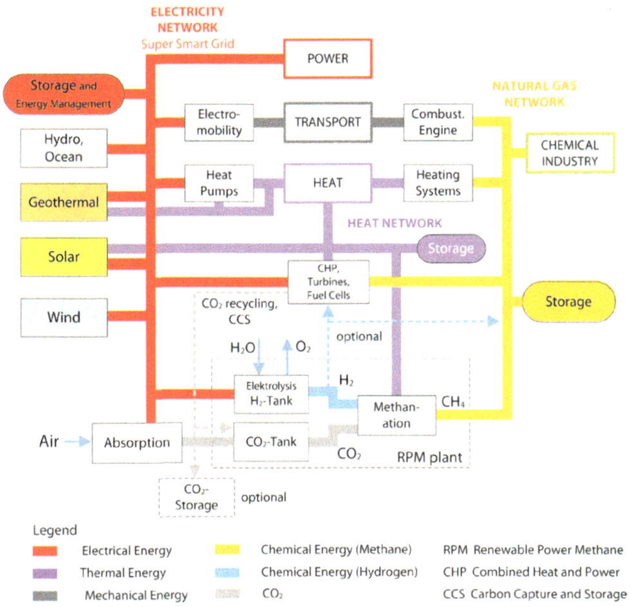

Figure 10.3 Hybrid PV-Wind-RPM plant as the integral centrepiece of a future sustainable energy supply system [6]

Based on the new link power-to-gas, 100% renewable energy supply systems are designed as shown in Figure 10.3. The four main energy systems, i.e. power network, natural gas network, heat network and transportation network, are integrated and positively influenced by renewable power methane. The key elements are direct renewable power generation (main "primary energy source"), renewable electromobility [14], heat pumps, RPM and overcoming traditional biomass. By integrating smart power networks, heat networks and natural gas networks, a full renewable energy supply is enabled. Several 100% renewable energy systems based on hybrid PV-Wind-RPM plants are presented, reducing global energy-related CO_2 emissions by 95%. One new key element is the mutual linking of power and gas networks.

Several integrated concepts with CO_2 from air, biomass, industrial processes and fossil fuels are designed. In this way, renewable power can be stored in the natural gas network and used temporarily and spatially flexible for balancing power and long-term power storage, for process heat and for (long-distance) transportation with a high-energy density CO_2-neutral energy carrier. The major advantage versus hydrogen is the use of the existing infrastructure. Hydrogen is "stored" in CO_2 and thus made available as a natural gas substitute. RPM can be produced basically anywhere where water, air (CO_2) and renewable power (wind, solar, hydro) are available and thus decreases dependence on fossil fuel imports and the need for new transmission lines by using existing gas grids. It can recycle CO_2 in the energy system by CO_2 capture from combustion or by the use of the generated O_2 for combustion of RPM in the oxyfuel process in combined cycle plants. This new approach can even act as a carbon sink in combination with CO_2 storage and thus create a "carbon sink energy system".

10.2 SOLAR AND WIND RESOURCE AVAILABILITY

The total global primary energy demand was about 151 200 TWh_{th} in the year 2008, i.e. about 17 TW of continuous energy flow [5]. However, a substantial amount of this primary energy demand is wasted in inefficient energy use based on burning fuels, i.e. the direct use of valuable electricity would reduce the aforementioned energy flow to about 11,5 TW provided for instance by solar PV or wind power [15].

The global energy supply potential of PV and wind power exceeds by far this energy demand of mankind. The technical energy potential of solar PV and wind is assessed differently by various authors but always by factors or orders higher than total global energy demand. In 1978, Weingart estimated the solar PV potential energy flow usable for mankind was higher than 100 TW [16]. In 2003, the German Advisory Council on Global Change derived a harvestable energy flow potential for wind power of about 90 TW and a practically unlimited potential for PV [17]. However, these numbers were adjusted in 2011 to a technical potential of about 54 TW for wind power and about 8 900 TW for solar energy and, thus, also for PV [18]. In the 2008 study 'energy [r]evolution' by Greenpeace, the utilizable energy flow was estimated as about 35 TW for wind power and 150 TW for PV [19]. Also in 2008, Sawin and Moomaw estimated the energy flow potential as about 145 TW for PV and about 55 TW for wind power [20]. In 2009, Lu *et al.* estimated the energy flow for wind power as about 80–150 TW [21]. Jacobson and Delucchi derived an energy flow of 40–85 TW for wind and 580 TW for PV [15]. In 2011, the IPCC derived a theoretically utilizable energy flow of about 190 TW for wind power and about 120 000 TW for PV [22]. The estimates for the technical potential of PV and wind power are summarized in Table 10.1. Other authors have clearly pointed out that wind and solar energy will become the backbone of the global energy supply and that this could happen before 2030 [23]. The awareness of the necessity to establish a solar powered society dates back many decades and was emphasized for instance by Hubbert in 1949 [24].

The three major power technologies in relation to minimized fully loaded social cost are solar PV, wind power and hydro power (see Section 10.3). But only solar PV and wind power have access to abundant resources, whereas the solar resource is the most homogeneous distributed energy source in the world (see Section 3.1). It might be considered as a godsend that the two least cost energy options for the twenty-firs st century are fully complementary to each other [4].

The first insights in the analysis of the complementarity of PV and wind power are depicted in Figure 10.4. This first global analysis gives plenty of indications that these two major renewable power technologies complement each other to a very large extent and show nearly no competition due to the fundamental underlying solar and wind resources [4]. Assumptions are PV fixed optimally

Table 10.1 Technical potential of PV and wind power versus the current global energy demand

Technical potential		Reference	PV [TW]	Wind [TW]
Weingart	1978	[16]	> 100	–
WBGU	2003	[17]	infinite	90
Greenpeace	2008	[19]	150	35
Sawin and Moomaw	2008	[20]	145	55
Lu *et al.*	2009	[21]	–	80–150
Jacobson and Delucchi	2009	[15]	580	40–85
WBGU	2011	[18]	8 900	54
IPCC SRREN	2011	[22]	120 000	190
Current global energy demand				
Including waste of heat	[TW]		17,0	
Direct energy demand	[TW]		11,5	

tilted power plants and wind power plants with 150 m hub height. Calculations are performed on a mesh of 1° × 1° latitude and longitude and a one hour time interval for the year 2005. Power capacities of PV and wind power are set to an equal value.

The degree of complementarity is measured by overlap full load hours (FLh), i.e. the amount of power provided by PV and wind power plants adjacent to each other in the same time interval. Typically, the overlap FLh indicates a good complementarity due to part load conditions of the respective power plants. For extracting the amount of power being problematic, the critical overlap FLh are defined, i.e. the amount of power being above the rated capacity of PV or wind power per geographic unit and time interval. In these cases the renewable power might be lost due to limited power line, balancing power plant or storage capacities. The overlap and critical overlap FLh are defined formally in Equation 10.1. Abbreviations stand for: full load hours (*FLh*), overlap (*OL*), critical overlap (*COL*), hours of the year (*i*), power (*P*), photovoltaic (*PV*) and wind power (*wind*).

$$FLh_{OL} = \left(\sum_{i=1}^{8760} \min\left(P_{PV,i}, P_{Wind,i}\right) \right) \cdot 1GW^{-1} \qquad \text{(Eq. 10.1a)}$$

$$FLh_{COL} = \left(\sum_{\forall i \in \{(P_{PV,i}+P_{Wind,i})>1GW\}} P_{PV,i} + P_{Wind,i} - 1GW \right) \cdot 1GW^{-1} \qquad \text{(Eq. 10.1b)}$$

Equation 10.1 Definition of overlap and critical overlap full load hours

Global average of the relative total overlap full load hours is about 15%, whereas maximum overlap is 25%. Critical overlap is significantly lower, i.e. available power per coordinate higher than rated power capacity of one power technology. Critical overlap FLh are worldwide below 9% and at most places even below 3% to 4%. Consequently, PV and wind power plants are no competition to each other.

Moreover, it can be expected that the complementarity of PV and wind power leads to a significant reduction in the amount of investment which is needed for the energy system and will enable the power sector to offer highly competitive solutions for the heat and transportation sector and maybe even for the chemical industry via renewable power generated methane.

In order to achieve a secure and stable power supply, balancing power plants are still needed. Natural gas (NG) power plant capacity of more than 1 100 GW is installed worldwide [5] and is perfectly suited for power balancing purposes. Hybrid PV-Wind-NG power plants are an excellent power plant option in the years to come [3, 25], but this technological approach still depends on fossil fuels. Greenhouse gas emissions of NG fired CCGT are in the range of 400–500 gCO_2/kWh considering the full life cycle: too much in a climate change constrained world. Moreover, a peak in conventional oil production is a matter of fact and it is only a question of time when a peak in NG production will occur. Nevertheless, the global installed NG infrastructure might be of utmost relevance for fighting climate change and diminishing fossil fuel resources – as basis for the RPM diffusion.

10.3 ECONOMICS OF SYSTEM COMPONENTS

The economics of the core system components related to hybrid PV-Wind-RPM-CCGT power plants under the technological frame conditions needed for RPM storage is focused on here. Relevant system components to be analyzed are 1-axis horizontal north-south continuous tracking PV power plants (PV 1N), wind power plants with 150 metre hub height (Wind 150m), combined cycle natural gas power plants in the conventional and carbon capture and storage (CCS) version (CCGT and CCGT-CCS), hard coal power plants in the conventional and CCS version (coal and coal-CCS) and RPM storage composed by dialysis, electrolysis and methanation units.

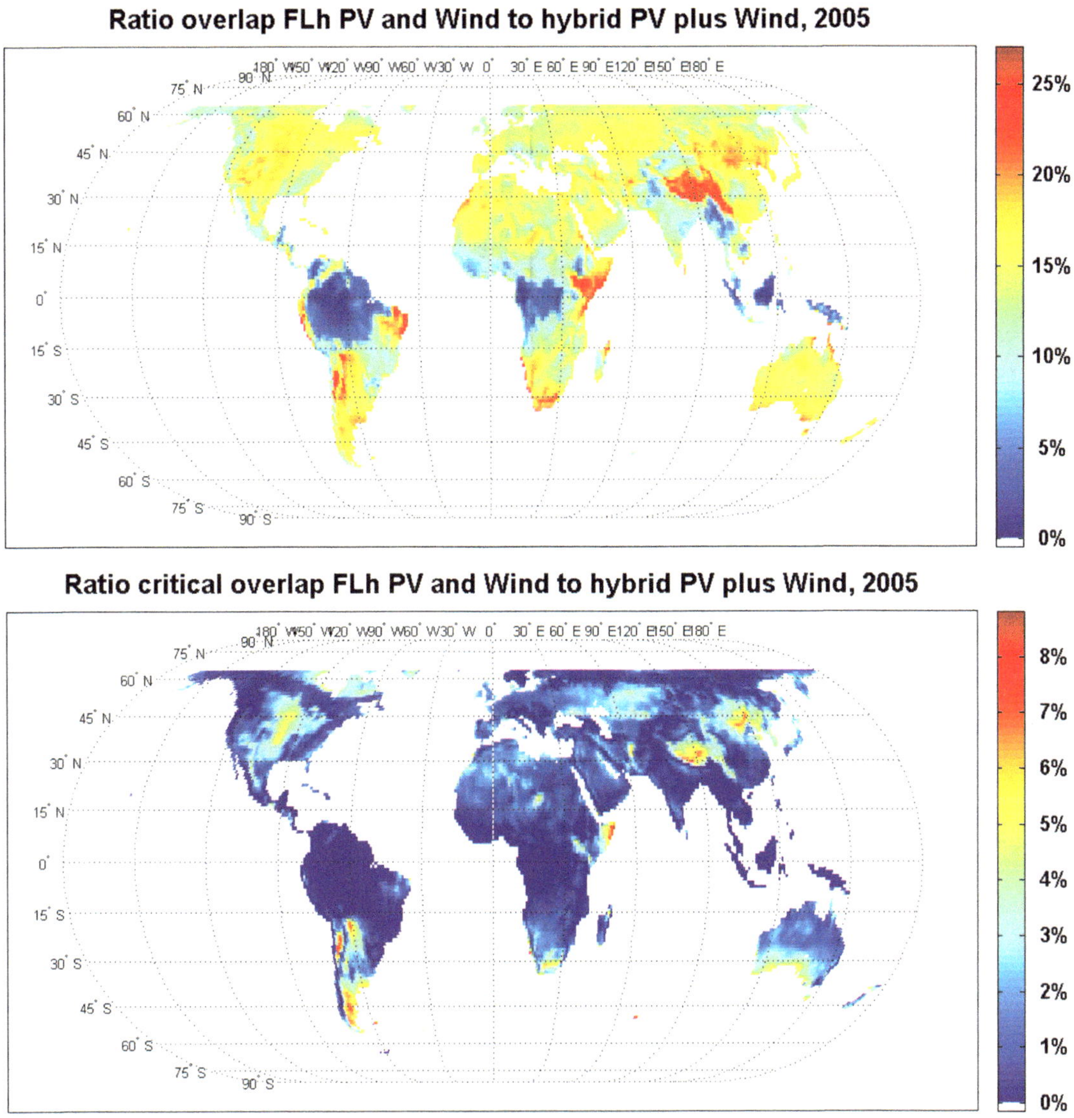

Figure 10.4 Ratio of annual total (top) and critical (bottom) overlap full load hours of PV and wind power to added up full load hours of both power technologies [4]

The preconditions for a successful hybridization are given for the hybrid power plants discussed in this chapter but are discussed elsewhere in detail [3, 25].

A cost model for all components of the hybrid PV-Wind-RPM-CCGT power plant is used for the calculation of the levelized cost of electricity (LCOE) [26] for all coordinates by applying local FLh for the PV and wind component in combination with FLh assumptions for the entire hybrid power plant. RPM storage is expected to be available on a large scale by the end of the 2010s, whereas the first CCS power plants, as a potential CO_2 source, might be in the demonstration phase at that point in time. Based on the availability of these two major relevant components for a broad hybrid PV-Wind-RPM-CCGT power plant analysis, the year 2020 is chosen for scenario evaluation. Besides typical capital expenditures (Capex) and operational expenditures (Opex), major cost positions are expenditures for fuel but not for related carbon emissions due to either CO_2 free power supply or application of CCS techniques. Such derived LCOE make it possible to compare them to LCOE of other renewable and conventional power plant technologies.

The methodology of calculating hybrid PV-Wind-RPM LCOE is summarized in Equation 10.2. Abbreviations stand for: capital expenditures (*Capex*), annuity factor (*crf*), annual operation and maintenance expenditures (*Opex*), annual fixed Opex ($Opex_{fix}$), variable Opex ($Opex_{var}$), annual full load hours of component *i* ($FLh_{i,el}$), fuel cost of component *i* ($fuel_i$), thermal energy conversion factor of component *i* ($PE_{th,i}$), primary to electric energy conversion efficiency of component *i* ($\eta_{i,el}$), weighted average cost of capital (*WACC*), lifetime of component *i* (N_i), equity (*E*), debt (*D*), return on equity (k_E), cost of debt (k_D), reference yield for a specific PV system at a specific site (Y_{ref}), PV performance ratio (*PR*), renewable power methane components (*RPM*), fuel cost of crude oil ($fuel_{crude\ oil}$), ratio of fossil fuel *i* to crude oil as coupling factor (cf_i) and primary to electric energy conversion efficiency of component *i* ($\eta_{i,el}$). Components *i* of the hybrid PV-Wind-RPM-CCGT power plant are: PV fixed optimally tilted,

Table 10.2 Key economic assumptions of hybrid PV-Wind-RPM power plants and competing power plants projected for the year 2020 [6]

In year 2020		PV 1-axis N-S horiz.	Wind 150m hub height	CCGT		Coal		Renewable power methane		
				Conventional	CCS	Conventional	CCS	Dial.	Electrol.	Methanation
Capex	[EUR/kW]	1 130	800	750	2 100	1 500	2 800	500	300	400
$Opex_{fix}$	[EUR/kW/y]	17	15	15	40	20	70	10	6	8
$Opex_{var}$	[EUR/MWhel]	–	–	1	2	1	3	1	1	1
Plant lifetime	[y]	30	25	30	30	40	40	30	30	30
Plant efficiency (PR)	[%]	80%	95%	58%	48%	44%	34%	78%	78%	82%
Fuel price coupling	[fuel/oil]	–	–	80%	80%	30%	30%	–	–	–

			Remarks
WACC	[%]	6,0%	despite a higher fossil risk profile identical for comparison reasons
Exchange rate	[USD/EUR]	1,40	
FLh methanation	[h/y]	8 000	baseload operation
FLh min CCGT	[h/y]	500	min FLh of CCGT component of hybrid RPM-CCGT plants
FLh min NG-CCS	[h/y]	500	min FLh of NG-CCS component of hybrid PV-Wind-NG-CCS plants
FLh min overlap	[h/y]	500	assumed time resolved FLh overlap of PV and Wind (conservative)

Abbreviations: 1-axis horizontal north-south continuous tracking (1-axis N-S horiz), combined cycle gas turbine (CCGT), conventional (conv), carbon capture and storage (CCS), capital expenditure (Capex), operational expenditure (Opex), performance ratio (PR), weighted average cost of capital (WACC), full load hours (FLh), renewable power methane (RPM) and natural gas (NG). Data are taken from various sources and in case of the non-RPM components given elsewhere [3].

PV 1-axis horizontal north-south continuous tracking, Wind 150 metre hub height, RPM storage units (dialysis, electrolysis, methanation), combined cycle gas turbine (CCGT) and carbon capture and sequestration (CCS). These components are compared partly or fully to natural gas (NG) fired CCGT-CCS and coal fired coal-CCS power plants.

$$LCOE = \sum_i LCOE_i \quad \text{(Eq. 10.2a)}$$

$$LCOE_i = \frac{Capex_i \cdot crf + Opex_{i,fix}}{FLh_{i,el}} + Opex_{i,var} + \frac{fuel_i}{PE_{th,i} \cdot \eta_{i,el}} + \frac{carbon \cdot GHG_i}{\eta_{i,el}} \quad \text{(Eq. 10.2b)}$$

$$crf_i = \frac{WACC \cdot (1+WACC)^{N_i}}{(1+WACC)^{N_i} - 1} \quad \text{(Eq. 10.2c)}$$

$$WACC = \frac{E}{E+D} \cdot k_E + \frac{D}{E+D} \cdot k_D \quad \text{(Eq. 10.2d)}$$

$$FLh_{PV,el} = Y_{ref} \cdot PR \quad \text{(Eq. 10.2e)}$$

$$FLh_{RPM,el} = FLh_{total,el} - \left(FLh_{PV,el} + FLh_{W,el}\right) \quad \text{(Eq. 10.2f)}$$

$$fuel_i = fuel_{crude\ oil} \cdot cf_i \quad \text{(Eq. 10.2g)}$$

Equation 10.2 Levelized cost of electricity (*LCOE*) for hybrid PV-Wind-RPM-CCGT power plants

The scenario assumptions for calculating LCOE are summarized in Table 10.2 and are based on experience curve assumptions for PV and wind power plants [3, 25]. A major LCOE component of PV and wind power plants is the capital cost, whereas conventional fossil power plants are more dependent on the fuel cost in contrast to fossil CCS power plants showing a higher dependence on capital cost. The fossil fuel cost is indirectly coupled to the crude oil price due to the specific thermal energy [25], which leads to fluctuations and long-term escalation.

The total FLh of the hybrid PV-Wind-RPM-CCGT power plant are superposed by the renewable source, i.e. PV or wind or hybrid PV-Wind, and the balancing CCGT which receives the methane by the RPM component of the hybrid plant. The RPM is assumed to be generated by excess electricity and CO_2 either by extraction of CO_2 from the ambient air (dialysis) or provided by the carbon capture and storage (CCS) component of the CCGT plant. The dialysis and electrolysis components show lower FLh due to their adaption to the availability of fluctuating excess electricity. However, the methanation component can be run in baseload operation mode due to internal hydrogen storage. Some minimum FLh are assumed for practical reasons, i.e. minimum 500 FLh of CCGT component, minimum 500 FLh of NG-CCS component, minimum 500 FLh of critical PV and wind overlap being very conservative (see Section 10.2) and minimum total 5 000 FLh of the hybrid PV-Wind-RPM-CCGT power plant. The latter assumption is rather conservative, since the global power plant capacity is operated for about 4 300 FLh on average [3]. The hybrid power plant in this configuration can use the RPM as both a daily and also a seasonal storage.

PV and wind power plants are on an excellent cost trend and are expected to achieve LCOE of about 40–60 EUR/MWh in regions of good resource quality in the year 2020, as shown in Figure 10.5. Assumed hybrid PV-Wind sub-plants are 1-axis horizontal north-south continuous tracking PV power plants and wind power plants with a hub height of 150 m for the conditions of Table 10.2. The underlying resource data are provided by NASA SSE 6.0 [27] but reprocessed and discussed elsewhere [3, 28]. A major advantage of the hybrid PV-Wind power plant component is the higher FLh compared with only one renewable power source leading to beneficial total LCOE of the hybrid PV-Wind-RPM-CCGT power plant.

Production costs of RPM by PV and wind power supply are visualized in Figure 10.6 for extracting CO_2 from ambient air and for CO_2 accessible by CCS. The calculations are based on scenario assumptions defined in Table 10.2 and visualized for respective hybrid PV-Wind plants in Figure 10.5.

Results for RPM production costs based on PV and wind FLh and LCOE show challenging cost levels even in regions of excellent resource availability. In case of extracting CO_2 from ambient air, a cost level of about 300–400 USD/barrel$_{eq}$ can be reached. For CO_2 being available due to CCS the cost level is equivalent to about 200–300 USD/barrel$_{eq}$. However the cost level might be not too high compared with typical fuel prices in industrialized countries, since 200, 300 and 400 USD/barrel$_{eq}$ are equivalent to diesel prices of 0,90, 1,35 and 1,80 EUR/l, excluding any kind of taxation and subsidies.

The technological route of RPM production offers access to highly valuable fuels in many regions in the world, as shown in Figure 10.6, in contrast to today's fossil fuel resource availability. Based on a more homogeneous solar resource and wind resource distribution in the world, the perspective is given for an additional more homogeneously distributed renewables sourced hydrocarbon fuel availability.

10.4 HYBRID PV-WIND-RPM POWER PLANT ECONOMICS

RPM storage offers three key features: Firstly, a direct renewable energy supply can be used to the maximum possible extent, e.g. PV and wind power. Secondly, storage can be charged by surplus energy in periods when renewable power in the grid exceeds load, in particular for seasonal balancing. Thirdly, a stable power supply can be guaranteed for the entire year. The remaining question will be whether it might be economically feasible to run hybrid

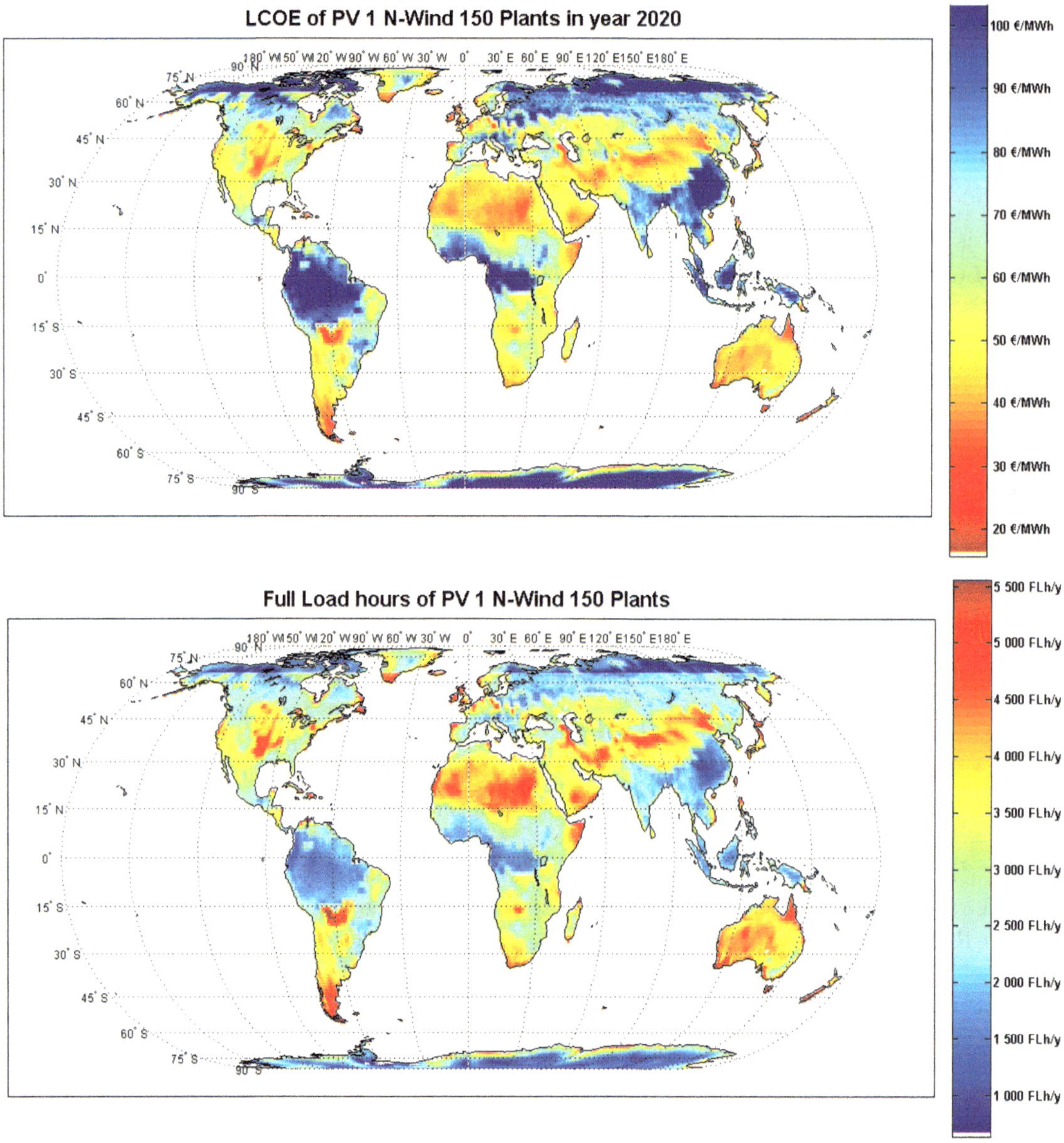

Figure 10.5 Hybrid PV-Wind power plants characterized by their global LCOE (top) and FLh (bottom) projected for the year 2020

PV-Wind-RPM-CCGT power plants as the centrepiece of such a potential fully stable and sustainable renewable electricity future.

The total LCOE of hybrid PV-Wind-RPM-CCGT power plants for 5 000 FLh is calculated on the basis of Equation 10.2 and scenario assumptions in Table 10.2 for the renewable components 1-axis horizontal north-south continuous tracking PV, wind turbines with 150 meter hub height and the hybrid composition of these two sub-plants for the two CO_2 source routes, i.e. from ambient air and accessible by CCS.

As shown in Figure 10.7, the LCOE outcome of both CO_2 sourcing routes is nearly identical, whereas the CO_2 extraction from ambient air seems to be slightly lower in cost for the specific scenario assumptions. From the PV point of view, the hybrid PV-Wind sub-plant leads to lower LCOE for the total hybrid PV-Wind-RPM-CCGT power plant compared with the PV-only or wind-only variation. This is caused by the beneficial effect of extended FLh on a renewable low cost basis thereby reducing costly RPM production to a minimum level [3]. As a result, a hybrid PV-Wind sub-plant reduces the total LCOE to a level of 60–100 EUR/MWh in regions of good and excellent solar and wind resource availability, e.g. the USA, Chile, Argentina, Bolivia, some locations in Europe, nearly the entire MENA region, parts of Central Asia and Australia.

Nearly identical LCOE results for hybrid PV-Wind-RPM-CCGT power plants based on the two different CO_2 sourcing routes (Figure 10.7) are in contrast to the results of RPM production costs (Figure 10.6), where the CO_2 extraction from ambient air route is found to be about 40% higher in cost than the CO_2 accessible from CCS option. This contradiction dissolves as the CCS system approach is significantly higher in cost on the power plant level due to

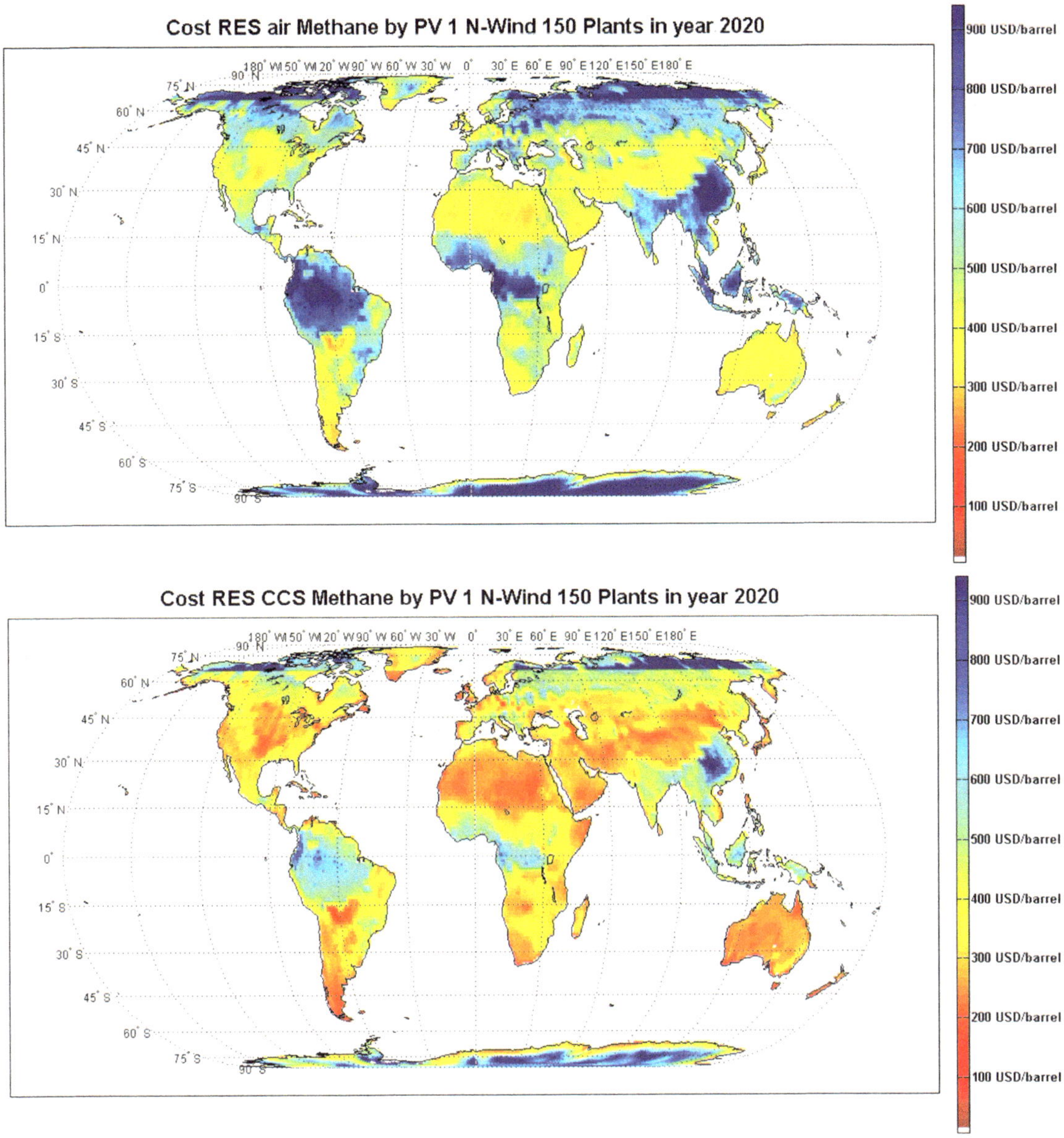

Figure 10.6 Cost of RPM production for CO_2 from the air (top) and supplied by CCS facilities (bottom) projected for the year 2020 as part of the hybrid PV-Wind-RPM plant

higher Capex of the CCGT-CCS versus the CCGT power plant component and lower respective primary energy conversion efficiency of CCGT-CCS versus CCGT (Table 10.2). In short, extracting CO_2 from ambient air is less costly than CCS technology. As a consequence, the higher RPM production cost of the ambient air route is levelled out by beneficial power plant characteristics of the CCGT component.

This quite relevant result needs to be analyzed in a much deeper and broader scope than possible in this chapter. It might be possible that the enormous investments in CCS technology could end up as stranded cost on a macro-economic level, in particular in case of lower total power generation system LCOE of hybrid PV-Wind-RPM-CCGT power plants using CO_2 extraction from ambient air versus fossil fuel powered CCS power plants. Enormous public financial means are needed for establishing the CCS technology and infrastructure, which might be superfluous not only in the end but also right from the beginning of CCS diffusion.

10.5 GLOBAL POWER SUPPLY POTENTIAL

Upgrading NG-CCGT power plants by hybrid PV-Wind power plants typically leads to lower LCOE for fossil fuel prices of about 50–90 USD/barrel$_{eq}$. In case of good availability of solar and wind resources, the hybrid PV-Wind-NG-CCGT power plant is very competitive from fossil fuel prices of 70–90 USD/barrel$_{eq}$ onwards. The remaining natural gas fired in the NG-CCGT sub-plant still leads to CO_2 emissions but the natural gas can be replaced by RPM, however it will be higher in cost than the natural gas option in nearly all regions in the world and

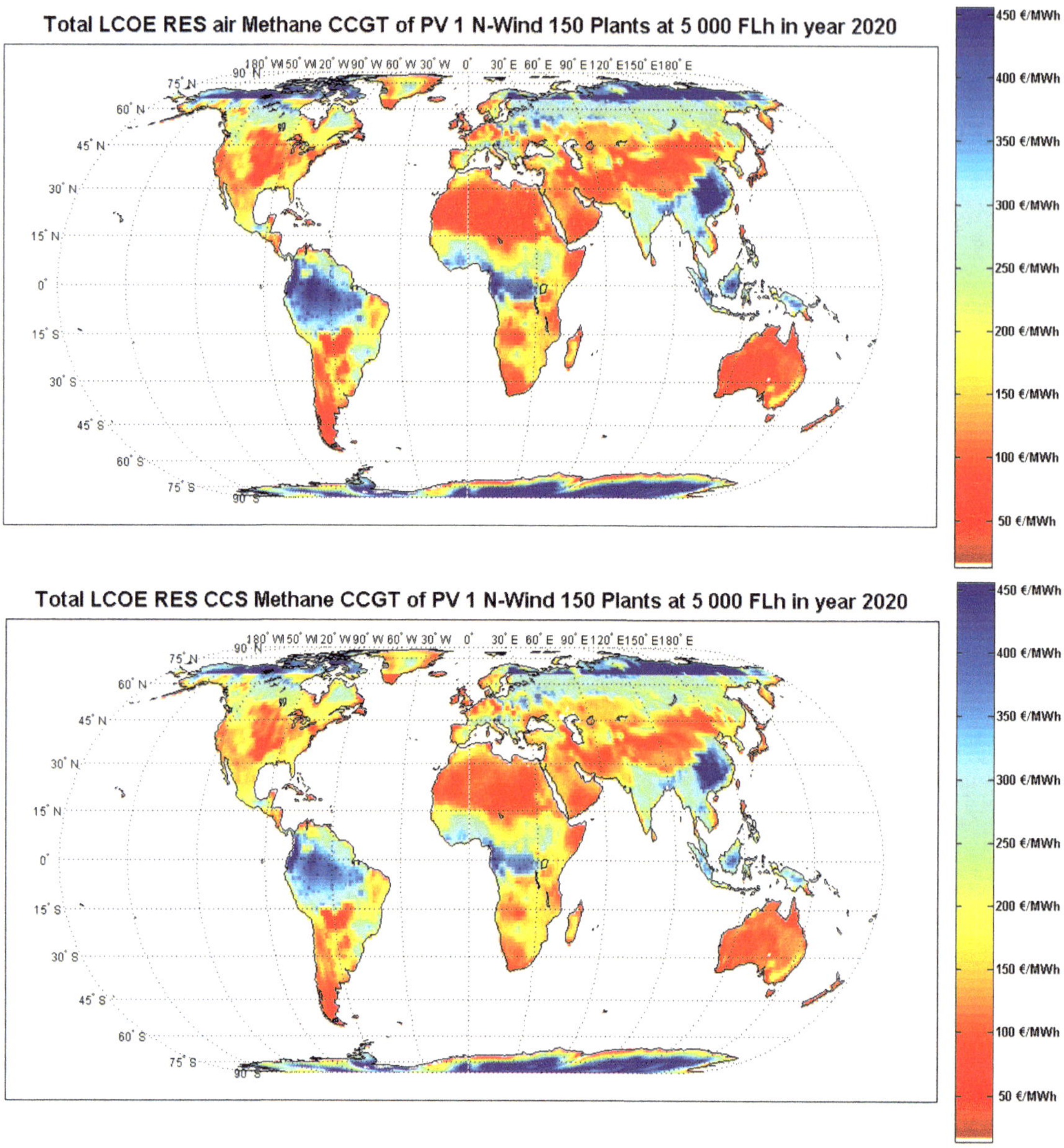

Figure 10.7 Global LCOE of hybrid PV-Wind-RPM power plants for CO_2 from air (top) and for CO_2 from CCS route (bottom) projected for the year 2020

for fossil fuel prices up to 250 USD/barrel$_{eq}$ assuming no carbon emission cost.

The hybrid PV-Wind-RPM-CCGT power plant extracting CO_2 from ambient air is an excellent centrepiece of a 100% renewable power supply, which might be established on a LCOE level of about 80–90 EUR/MWh in regions of very good availability of solar and wind resources. In regions of good solar and wind resources the LCOE ranges from about 100–120 EUR/MWh and for at least one good and one moderate resource the LCOE could be about 140–170 EUR/MWh. These LCOE levels are fully decoupled from fossil fuel impacts and can be considered to represent the full social cost, i.e. no further external costs create an additional financial burden. Moreover, no CCS route is needed. CO_2 extraction from air leads in most cases to lower total LCOE and in cases of cheaper CCS CO_2 sourcing this is only slightly lower in cost. It is not clear whether CCS technology will be really available by 2020 or even in 2030, hence waiting for this route might waste a lot of valuable time. In conclusion, hybrid PV-Wind-RPM-CCGT power plants extracting CO_2 from ambient air enable a 100% renewable power supply in many regions in the world and show favorable economic performance.

The enormous solar and wind resource potential lays the basis for analyses of the global energy supply potential of solar PV and wind power. The previous sections clearly emphasize that 100% renewable energy supply is technically feasible on the basis of PV and wind power using RPM for storage purposes, and in particular for seasonal storage. Economic considerations result in a total power generation cost of below 100 EUR/MWh in many large regions spread over the world.

In reality, more renewable energy sources can be used for power supply. An excellent example how such

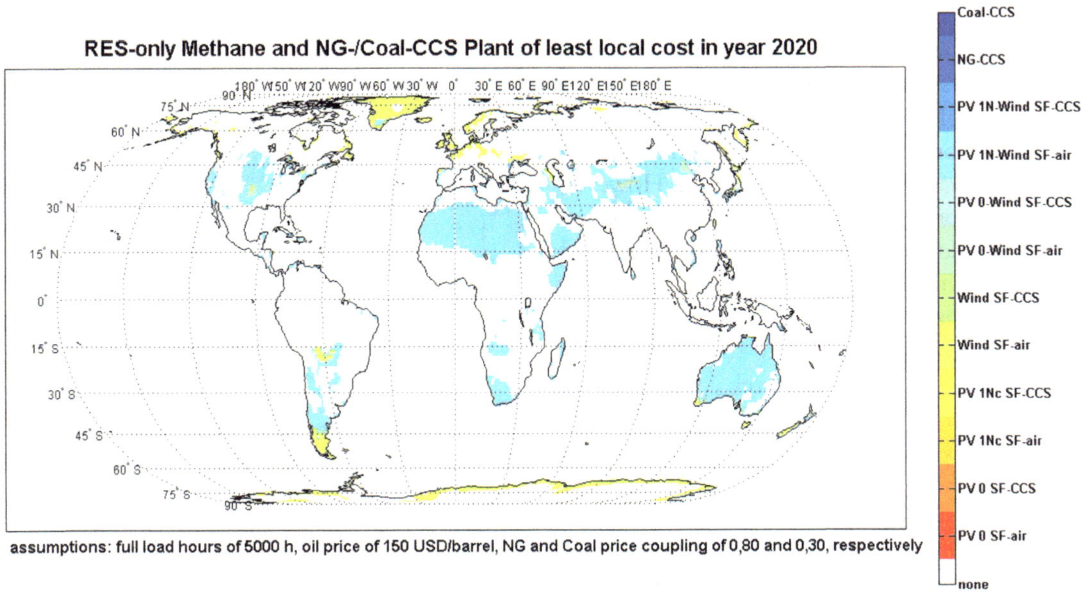

Figure 10.8 Hybrid PV-Wind-RPM-CCGT power plants of least local LCOE in competition to NG-CCGT-CCS and coal-CCS power plants for a fossil fuel price of 150 USD/barrel$_{eq}$ operated 5 000 FLh in the year 2020

a fully renewable powered energy system could work has been analyzed for the Desertec project (see Section 2.2.1). The global energy supply potential for solar thermal electricity generation (STEG) has already been analyzed [29]. This analysis of the global energy supply potential of solar electricity generated only in regions of excellent solar resources, i.e. at least 2 000 kWh/m^2/y direct normal irradiation, clearly shows the true potential of solar power: 90% of world population could be supplied by solar power (solar PV and solar thermal) via high voltage direct current (HVDC) power lines no longer than 3 000 km.

The indication is high that based on STEGs a nearly 100% renewable power supply could be established. However, the STEG economics are not as favorable as the comparable ones for PV and wind power. Therefore it might be of very high relevance to perform a global energy supply potential calculation for hybrid PV-Wind-RPM power plants on the basis of economic competitiveness.

The previous chapters point out that a 100% renewable power supply will be economically feasible at latest at the end of the 2010s in the regions of the world where at least good and very good solar and wind resources are available. The key question in this chapter is the global energy supply potential of hybrid RPM systems. The three major steps for answering this are: Firstly, it needs to be identified where hybrid PV-Wind-RPM-CCGT power plants are lower in cost than natural gas and coal-fired CCS power plants, which would be the major competing power plant technologies in a CO_2 constrained world. Secondly, it needs to be estimated how much energy can be provided by those regions. Thirdly, it needs to be evaluated how many people live in those favorable regions and, more relevant depending on the distance to those regions, how many further people could be supplied.

The regions of least LCOE for 100% renewable power plants on the basis of PV, wind, RPM and CCGT components are shown in Figure 10.8, derived on the basis of cost competition against NG-CCGT-CCS and coal-CCS power plants and shown for a fossil fuel price of 150 USD/barrel$_{eq}$. The two CO_2 source options are extracting CO_2 from ambient air (air) and CO_2 accessible by CCS (CCS), whereas RPM is denoted as SF. The hybrid PV-Wind components are characterized in Figure 10.5. Further assumptions for cost calculation are taken from Table 10.2.

The regions of most competitive hybrid PV-Wind-RPM-CCGT power plants are distributed all around the world and comprise the regions of very good solar and wind resource availability. The global energy supply potential of hybrid PV-Wind-RPM-CCGT power plants can be roughly estimated. The details of the calculation are discussed elsewhere [3].

Notably, good PV sites of about 2 000 FLh generate an annual electricity amount of about 107 GWh/km^2, whereas good wind power sites of about 3 000 FLh generate an annual electricity amount of about 56 GWh/km^2, hence the practical specific energy generation density of PV is by a factor of two higher than that of wind power. However, the entire site needs to be reserved more or less fully for a PV power plant, but the site beneath the wind turbines can be used for similar purposes as before, e.g. crop land, forests, etc.

For a crude oil price of about 150 USD/barrel$_{eq}$, about 37 million km^2 fulfil the criteria of lower hybrid PV-Wind-RPM-CCGT LCOE than comparable fossil fuel fired CCS power plants. Only 1,3% and 6,9% of that area would be needed to fully cover the current power and additional thermal energy demand by the hybrid PV-Wind-RPM-CCGT approach. The number for the thermal energy is a worst case assumption due to enormous efficiency potentials. The required thermal energy might be too high by a factor of four, according to fundamental efficiency reasons. For the worst case it is assumed that there is a conversion of valuable electricity to methane and then subsequent conversions for the various thermal energy services, like transportation, heating, cooking, etc. However, it would be much more efficient to use the electricity in a direct way like electric transportation, electric heating and electric cooking which would be more efficient by roughly a factor of four, or even more.

The numbers for some decades into the future, based on a crude oil price of 200 USD/barrel$_{eq}$, would be 12 billion humans, about 57 million km^2 of least local hybrid PV-Wind-RPM-CCGT LCOE, 3.9% and 16.2% of that area needed for covering the electric and thermal energy demand and similar efficiency considerations for the hybrid PV-Wind-RPM-CCGT but also for the thermal energy demand. The area requirement might not be as high as it appears, since in the future 200 USD/barrel$_{eq}$ case only 11% of the Earth's surface is classified for excellent hybrid PV-Wind-RPM economics, enormous efficiency potentials of about a factor of four in the thermal energy demand are to be realized, the energetic wealth level of the EU today might be too high for 12 billion people and all technologies for the hybrid PV-Wind-RPM-CCGT approach are still significantly improvable. In total, enormous amounts of energy under least local LCOE conditions are available for powering the energy needs of mankind without relevant sustainability criteria restrictions, in particular due to the fact that several other renewable power technologies are able to complement PV and wind power, the two core power technologies in the years and decades to come.

Besides the enormous energy supply potential of the most competitive regions on LCOE basis in the world, it is of highest interest how many people live in these regions and within what distance the other part of mankind lives. Many people live in the regions of least cost 100% renewable power supply based on PV and wind power plants even for low fossil fuel prices. For fossil fuel prices of up to 50 USD/barrel$_{eq}$ about 500 million, and for up to 100 USD/barrel$_{eq}$ about 800 million, live within 100 km of the regions of least cost power supply. These numbers sharply rise for higher fuel prices to about 1 200 million (150 USD/barrel$_{eq}$), 1 800 million (200 USD/barrel$_{eq}$) and 2 200 million (250 USD/barrel$_{eq}$). The aggregated population depending on the distance to the least cost regions is depicted in Figure 10.9 for various fossil fuel price levels. The evaluation is based on least cost hybrid PV-Wind-RPM-CCGT power plants (Figure 10.8) and the distribution of global population density. Data for population density are provided by the Center for International Earth Science Information Network (CIESIN) [31].

Power lines can transmit electricity over several thousand kilometers very efficiently. Distances over 800–1 000 km are economically best bridged by high voltage direct current (HVDC) power lines, being applied for decades for distances of 2 000 km and more. HVDC power

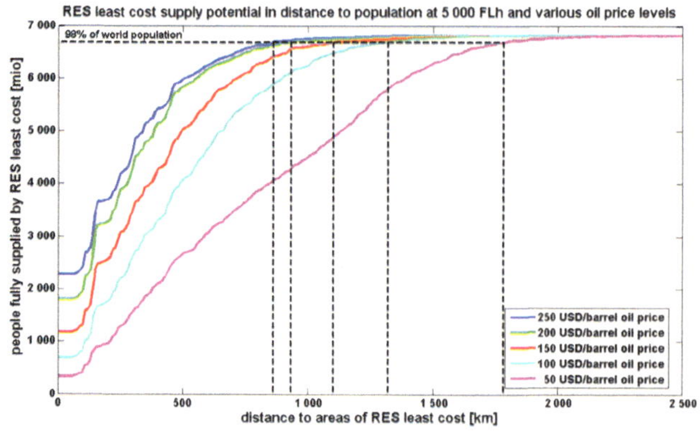

Figure 10.9 Aggregated numbers of people living in regions of least cost 100% renewable power supply and respective distance to these regions in dependence of fossil fuel prices in the range of 50–250 USD/barrel$_{eq}$ in the year 2020

lines show a power transmission efficiency of about 97% per 1 000 km. Below 800–1 000 km conventional high voltage alternating current (HVAC) power lines represent the most cost efficient power transport solution [30].

Within about 800 km to the regions of least cost 100% renewable power supply most people in the world could be supplied depending on the fossil fuel prices, i.e. about 85% of world population based on the least cost situation for 100 USD/barrel$_{eq}$, about 90% for 150 USD/barrel$_{eq}$, about 95% for 200 USD/barrel$_{eq}$ and about 98% for 250 USD/barrel$_{eq}$.

Comparing the results of the global energy supply potential of STEG versus hybrid PV-Wind-RPM power plants shows the beneficial consequences of good solar and wind resource potential accessible in many regions in the world. The supply potential for more than 90% of world population is lowered from 3 000 km [29] to about 500–1 000 km. This reduction in distance is very important for lowering the political obstacles to the issues of transmitting large amounts of power through various countries and the time-consuming construction of HVDC power lines. Moreover, hybrid PV-Wind-RPM power plants enable 100% renewable power supply and guarantee the least LCOE option.

REFERENCES

1 EPIA (European Photovoltaic Industry Association) (2011) *Global Market Outlook for Photovoltaics until 2015*, EPIA, Brussels, www.epia.org/publications/photovoltaic-publications-global-market-outlook/global-market-outlook-for-photovoltaics-until-2015.html

2 GWEC (Global Wind Energy Council) (2010) *Global Wind 2009 Report*, GWEC, Brussels, www.gwec.net/fileadmin/documents/Publications/Global_Wind_2007_report/GWEC_Global_Wind_2009_Report_LOWRES_15th.%20Apr.pdf

3 Breyer, Ch. (2011) Economics of Hybrid Photovoltaic Power Plants, Dissertation, University of Kassel.

4 Gerlach, A.-K., Stetter, D., Schmid, J., and Breyer, Ch. (2011) PV and Wind Power – Complementary Technologies, *26th EU PVSEC*, Hamburg, September 5–9.

5 IEA (International Energy Agency) (2010) *World Energy Outlook 2010*, IEA, Paris.

6 Breyer, Ch., Rieke, S., Sterner, M., and Schmid, J. (2011) Hybrid PV-Wind-Renewable Methane Power Plants – A Potential Cornerstone of Global Energy Supply, *26th EU PVSEC*, Hamburg, September 5–9.

7 Sterner, M. (2009) Bioenergy and renewable power methane in integrated 100% renewable energy systems, Dissertation, University of Kassel.

8 Specht, M., Sterner, M., Stürmer, B., Frick, V., and Hahn, B. (2009) Renewable Power Methane – Stromspeicherung durch Kopplung von Strom- und Gasnetz – Wind/PV-to-SNG, Patent No: 10 2009 018 126.1, patent filed in March 9.

9 Specht, M., Brellochs, J., Frick, V., Stürmer, B., Zuberbühler, U., Sterner, M., and Waldstein, G. (2010) Speicherung von Bioenergie und erneuerbarem Strom im Erdgasnetz, *Erdöl Erdgas Kohle*, 2010(10), 342–6.

10 Sterner, M., Saint-Drenan, Y.-M., Gerhardt, N., Specht, M., Stürmer, B., and Zuberbühler, U. (2010) Erneuerbares Methan – Eine Lösung zur Integration und Speicherung Erneuerbarer Energien und ein Weg zur regenerativen Vollversorgung, *Solarzeitalter*, 2010(1), 51–58.

11 Pengg-Bührlen, H. (2011) Business model and market strategy for PtG-facilities in the German market, 6th International Renewable Energy Storage Conference, Berlin, November 28–30.

12 Otten, R. (2011) The Audi e-gas project – using PtG for sustainable mobility, 6th International Renewable Energy Storage Conference, Berlin, November 28–30.

13 Bandi, A., Specht, M., Weimer, T., and Schaber, K. (1995) CO_2 Recycling for Hydrogen Storage and Transportation – Electrochemical CO_2 Removal and Fixation, *Energy Conversion and Management*, 36, 899–902.

14 Arnhold, O., Hlusiak, M., Möhrke, F., and Breyer, Ch. (2011) Mobility Concepts for the Use of Excess Power from the Renewable Energy System on the Island of Graciosa (Azores Archipelago), *6th International Renewable Energy Storage Conference*, Berlin, November 28–30.

15 Jacobson, M.Z., and Delucchi, M.A. (2009) A Path to Sustainable Energy by 2030, *Scientific American*, 2009(11), 58–65.

16 Weingart, J.M. (1978) The Helios Strategy: An Heretical View of the Potential Role of Solar Energy in the Future of a Small Planet, *Technological Forecasting and Social Change*, 12, 273–315.

17 WBGU – German Advisory Council on Global Change (2003) *World in Transition: Towards Sustainable Energy Systems*, WBGU, Berlin, Earthscan, London, www.wbgu.de/fileadmin/templates/dateien/veroeffentlichungen/hauptgutachten/jg2003/wbgu_jg2003_engl.pdf

18 WBGU – German Advisory Council on Global Change (2011) *Welt im Wandel: Gesellschaftsvertrag für eine Große Transformation*, WBGU, Berlin, www.wbgu.de/fileadmin/templates/dateien/veroeffentlichungen/hauptgutachten/jg2011/wbgu_jg2011.pdf

19 Teske, S., Schäfer, O., Zervos, A., Béranek, J., Tunmore, S., Krewitt, W., Simon, S., Pregger, T., Schmid, S., Graus, W., and Blomen, W. (2008) *Energy revolution: A Sustainable World Energy Outlook*, Greenpeace International and EREC, Amsterdam and Brussels, October, www.greenpeace.org/denmark/PageFiles/205220/new-global-energy-r-evolution.pdf

20 Sawin, J.L., and Moomaw, W.R. (2008) An enduring energy future, in: Sarke L. (ed.): *State of the World: Into a Warming World*, The Worldwatch Institute, Washington, www.worldwavgjnorg/fimesotdg/SOW39_chaq4.pdg

21 Lu, X., McElroy, M.B., and Kiviluoma. J. (2009) Global potential for wind-generated electricity, *Proceedings of the National Academy of Science*, PNAS early edition, www.pnas.org_cgi_doi_10.1073_pnas.0904101106

22 IPCC – Intergovernmental Panel on Climate Change (2011) *Special Report on Renewable Energy Sources and Climate Change Mitigation (SRREN),* IPCC WG III – Mitigation of Climate Change, Geneva, http://srren.ipcc-wg3.de/report

23 Kohn. W. (2010) A world powered predominantly by solar and wind energy, in: Schellnhuber, H.J., Molina, M., Stern, N., Huber, V., Kadner, S., (eds), *Global Sustainability – A Nobel Cause*, Cambridge University Press, Cambridge.

24 Hubbert. M.K. (1949) Energy from Fossil Fuels, *Science*, 109, 103–9.

25 Breyer. Ch., Görig. M., Gerlach. A.-K., and Schmid. J. (2011) Economics of Hybrid PV-Fossil Power Plants, *26th EU PVSEC*, Hamburg, September 5–9.

26 Short, W., Packey, D.J., and Holt, T. (1995) *A Manual for the Economic Evaluation of Energy Efficiency and Renewable Energy Technologies*, NREL, NREL/TP-462-5173, Golden.

27 Stackhouse, P.W., and Whitlock, C.H., (eds) (2008) Surface meteorology and Solar Energy (SSE) release 6.0, NASA SSE 6.0, Earth Science Enterprise Program, National Aeronautic and Space Administration (NASA), Langley, http://eosweb.larc.nasa.gov/sse/

28 Breyer, Ch., and Schmid. J. (2010) Population density and area weighted solar irradiation: global overview on solar resource conditions for fixed tilted, 1-axis and 2-axes pv systems, *25th EU PVSEC/ WCPEC-5,* Valencia, September 6–10, DOI: 10.4229/25thEUPVSEC2010-4BV.1.91

30 Trieb. F., Schillings. C., Kronshage. S., Viebahn. P., May. N., and Paul. C., (eds) (2006) *Trans-Mediterranean Interconnection for Concentrating Solar Power*, Research of German Aerospace Center (DLR) by order of Federal Ministry for the Environment, Nature Conservation and Nuclear Safety, Berlin, www.dlr.de/tt/trans-csp

29 Breyer. Ch., and Knies, G. (2009) Global Energy Supply Potential of Concentrating Solar Power, *Proceedings SolarPACES 2009*, Berlin, September 15–18.

PART IV
VLS-PV case studies

CHAPTER ELEVEN

VLS-PV case studies

11.1 VLS-PV IN WEST AFRICA: A POSSIBLE ROADMAP

11.1.1 Introduction

The implementation of VLS-PV systems in West Africa is quite a challenge given the current situation of the power sector in the region. Indeed, the recent development of the interconnection systems, the difficult financial situation of the national utilities and the absence of an efficient information management system make it difficult to implement any power transaction in West Africa, let alone new types of transactions such as large scale photovoltaic power plants.

The goal of the present proposal is to identify a possible roadmap for West Africa to achieve a high penetration rate of photovoltaic in the regional energy mix and to determine the key factors to increase the value of this technology for the national economies.

11.1.2 The energy situation in West Africa

With most electricity generation facilities having been commissioned in the 1950s and 60s (and having undergone little service since then); with transport and distribution infrastructure development traditionally undertaken on a national basis; with national electrification rates ranging from 4 to 40%; and with recurrent power shortfalls averaging 56 days per year [1], the situation of the West African power sector is quite similar to that of the rest of sub-Saharan Africa.

With regards to the electricity grid, the energy consumption in the 15 ECOWAS (Economic Community of West African States) Member States in 2005 was about 42 TWh [2], and peak power demand reached 6,83 GW while the market's installed capacity was estimated to be 6,4 GW [3]. Additionally, small privately owned diesel generators represent an estimated auto generation capacity of 1 GW.

Furthermore, most of the West African countries have small power utilities. Only Ghana and Nigeria utilities have installed generating capacity superior to 1 GW. Meanwhile, landlocked countries such as Mali, Burkina Faso and Niger display a fairly small electricity production and other small economies characterized by past and current political strife display limited energy development (Liberia, Sierra Leone, Guinea and Guinea-Bissau).

As a consequence, the regional average annual per capita electricity consumption rises only to 155 kWh, which is slightly higher than Central Africa (151 kWh) and twice as much as Eastern Africa (65 kWh), but still close to the world's smaller electricity consumption levels, mainly due to deficient and most often nonexistent access to electricity in rural areas where electrification rates are lower than 3%.

In parallel, the distribution of energy resources follows the pattern of power sector development in the region. The energy resources are oil, natural gas (mainly in Nigeria, but also some in Côte d'Ivoire and Mauritania) and hydro. The main hydroelectricity potential is located in Nigeria, Ghana and Guinea and, to a lesser extent, in Sierra Leone, Liberia and Mali; hydroelectricity development is undertaken on the Niger (Nigeria), Volta (Ghana), Bafing (Mali) and Bandama (Cote d'Ivoire) rivers. The remaining countries rely on expensive imported heavy fuel oil and gas oil, or on electricity imports from bordering states. The latter applies to countries such as Togo, Benin, Burkina Faso and Niger.

Forecasts for West Africa electricity demand growth rates are over 7% annually. This will result in expected regional peak power demand at around 22,5 GW by 2020. Meanwhile, the estimates for suppressed demand range from 20 to 30% of combined grid and auto generation supply. Based on current grid connected capacity and planned decommissioning, by 2023 West Africa will need

Table 11.1 Actual and forecasted peak demand by country (MW) [4]

	2003 (actual)	2007	2011	2015	2020
Benin & Togo	191	275	352	448	607
Burkina Faso	105	135	178	223	282
Cote d'Ivoire	606	737	930	1 175	1 572
Ghana	1 303	1 616	1 945	2 354	2 971
Niger	60	103	131	168	230
Nigeria	3 500	5 768	7 560	9 910	13 900
Gambia	48	88	132	156	186
Guinea	147	204	397	959	1 098
Guinea-Bissau	21	28	41	53	61
Liberia	44	58	105	122	122
Mali	98	127	170	223	284
Senegal	319	414	572	785	1 057
Sierra-Leone	35	45	57	72	97
	6 477	9 698	12 670	16 638	22 467

to upgrade its generating capacity by at least 27 GW in order to meet estimated demand.

Finally, the major breakthrough in power sectors across West African nations (and also across all Africa from a broader perspective) is power pooling, which is the development of a cross-border energy network through the gathering of 14 ECOWASS nations (all of the ECOWASS countries except Cap-Vert). Created in October 2000, the West African Power Pool (WAPP) sets the target of completing the interconnection between all the countries and upgrading capacity of the existing interconnections. The advantages of this mechanism are a better allocation of energy resources across nations and an overall generation costs optimization.

Compared with other African regional power pools (SAPP, South African Power Pool; PEAC, Pool Énergétique d'Afrique Centrale; and EAPP, Eastern African Power Pool), WAPP has developed more efficiently because of similarities in the legal frameworks of its member states which set a better basis for harmonization and foster local cooperation. In the long run, ECOWAS forecasts that 5 600 km [5] of power lines interconnecting segments of national grids will be put in place. And, whilst projects are at different stages of development, by 2016 most interconnections should have been finalized, ensuring greater security of supply for the region.

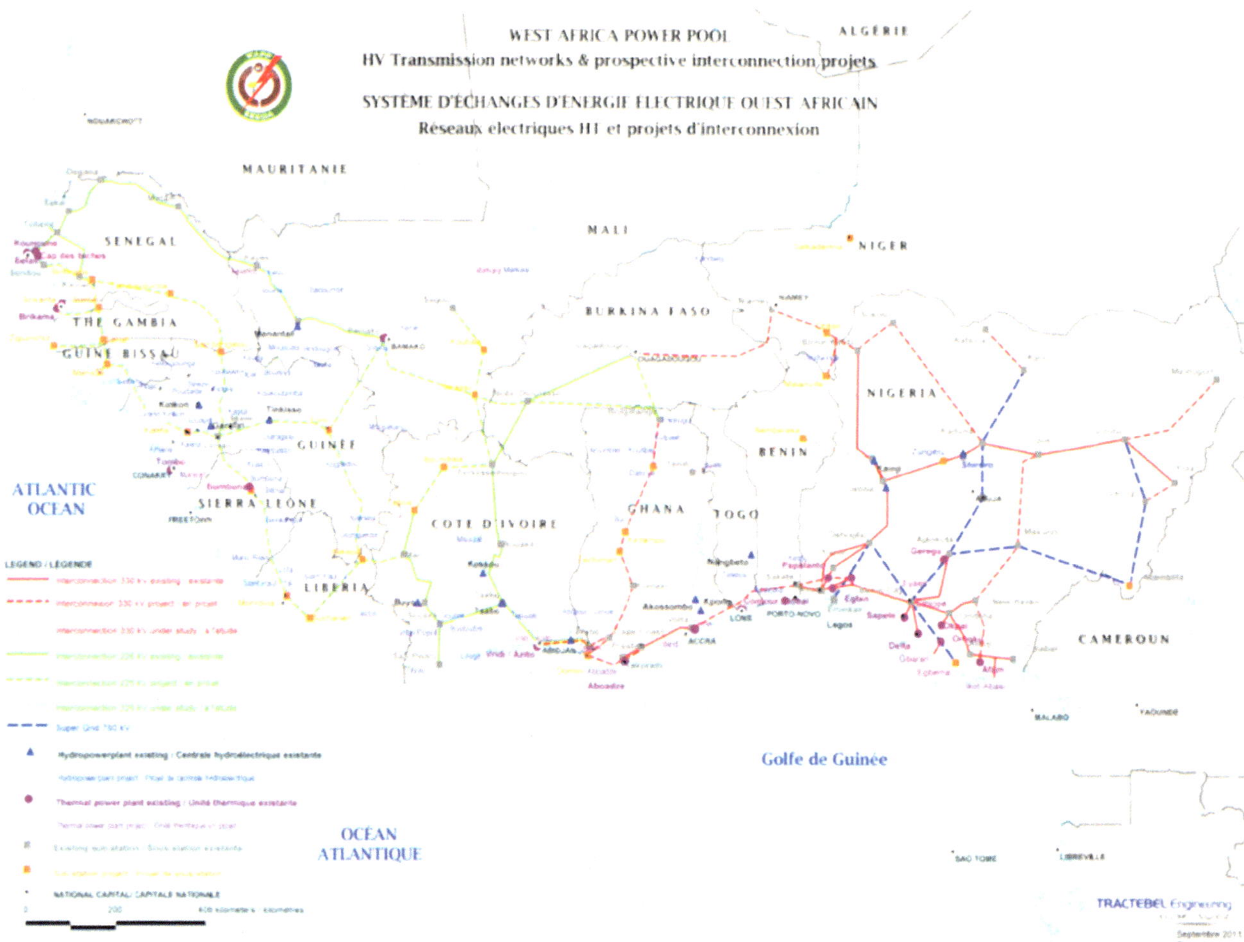

Figure 11.1 West Africa high voltage transmission network and interconnection projects [4]

11.1.3 The use of photovoltaic in West Africa

(1) DEFINING THE BEST STRATEGY FOR WEST AFRICA

West Africa countries cannot afford to subsidize photovoltaic. The plants have to produce electricity which is useful for the grid operators and which represent a reliable, adapted and cost competitive solution. Our proposal to define an efficient policy consists of the following points:

DEFINING THE MARKET: SEGMENT, CAPACITY AND PRICE

The first step for West African countries would be to define the role that PV can play in the energy mix of each country as well as on the regional level: large scale systems connected to interconnected grids, distributive plants installed near consumers and mini grid or off grid systems. For each of these segments, it is possible to define targets in terms of kWh price and capacity installed.

- The definition of capacity and prices should be based on the current and future energy mix and grids. It is key to define these targets from single State's and/or grid operators' points of view, considering the technical and economic situation of the regional and national energy sectors;
- For example, if the PV penetration technical limit of a country is only a few tens of MW, and considering the continual improvement of the PV technology, then this State should not reach this technical limit too fast if they want to benefit from the decrease in costs of the sector (see the next paragraph to see what happened in French overseas territories where there was no limit in terms of yearly capacity installed);
- This first step is the most important and should be undertaken independently from the commercial targets of the PV sectors.

Then a "reality check" on these figures should be done. In particular, the States can run scenarios in order to identify the main conditions which lead to the achievement of the pre-defined capacity and target prices: financial conditions, size of the plants, development costs, EPC and O&M prices, development of the grid and interconnections, etc. If these conditions are to be met today or in the short term, then the States can use these figures as a reference scenario for their politics. If those figures are not realistic, but will become so in the mid-term, then the States should ask themselves if they want to subsidize the sector during this first period in order to prepare the local industry for the time when they will become competitive in a non-subsidized environment. In this case, the efficient adjustment of price and capacity is very important so that the States limit their financial burden. It is also possible, to help the definition of these markets and see what the PV industry is capable of, to identify the best sites and a first sequence of projects to "test" the market.

SELECTING THE PROJECT: TENDERS AND PROPOSALS

West African countries should completely avoid the feed-in-tariff system based on a cost-approach that would be given to all projects, even with an adjustment formula that would take into account the technology improvement or the radiation. We see two major mechanisms that could be implemented:

- Reverse unction, such as in India – approach based on costs of PV;
- PPA proposals, such as in the USA – approach based on the value of PV.

The development phase could be undertaken by the States when the projects are identified and when the expertise exists – and in this case the development will lead to a tender process, such as in Peru – or by the private sector, and in this case the proposal should be validated carefully by the States and also the regional institutions – since the markets are small, projects should not disturb the regional interconnection strategy. Section 11.1.4 gives more details about the implementation strategy.

(2) THE EXAMPLE OF EUROPE: LESSONS TO LEARN FROM EXPERIENCE

The implementation of photovoltaic plants in a country, when it is not properly designed, can lead to great financial losses for the national economy. The case of Europe is a good example. Almost all the countries that have implemented a "feed-in-tariff'" mechanism have stopped their support. In most cases, the PV plants installed on the grid only represent a source of problems for the European utilities: they are now stuck with capacities installed in total disorder in their grid, potentially disturbing the management of the grid, and connected at far too high a price.

Analyzing the case of the French overseas territories could be extremely useful. Indeed, the situation of their grid is close to that of many West African countries: high electricity production costs, isolated grid, small demand. The only difference is that the information management system is much more developed there, providing therefore all the required information to set up an efficient energy strategy.

France launched a "feed-in-tariff" mechanism in 2006 with the following specificities:

- No entry barrier;
- No selection criteria applicable to projects;
- No capacity cap or timeframe;

- No specific permitting or authorization process;
- For the French overseas territories, the feed-in-tariff was higher than in France.

If this support can be seen as quite generous for the industry, unexpected outcomes were observed:

- There is a bonus for the first projects connected. When there is an obligation for the utility to buy the electricity at a tariff that is too high, without any selection criteria applicable to projects, all actors have a huge incentive to be the first ones to close a deal. Indeed, they see, rightly, that this situation will not last for a long time and they want to benefit from the bigger margins. In small grids, this competition is even bigger since there is a technical limitation on the capacity installed of intermittent systems relative to grid capacity. This leads in many cases to a shortening of the development and construction period, which is dangerous in a new-born industry at the beginning of its learning curve;
- A high feed-in-tariff without any barrier (which can be seen as a generous policy) creates a financial bubble and only a piece of the value is kept by the solar industry. When the value of the projects is too high, the state is no longer capable of directing its financial support towards the targeted group. In some cases, a huge portion of the value benefits a group which is not going to reinvest its profit in the solar business – which may be good for the economy in itself but is not the goal of the support mechanism;
- When there are no selection criteria for the projects, the plants are concentrated where it is easier to get the permits. Concentration of plants in Europe is not really an issue given the size of the grid and the interconnection system; although, even in Europe, at a local level, it may create grid instability. On islands, disseminating the plants along the grid is essential to stabilize the PV production and partially reduce the disturbances generated by clouds.

The application of the French PV law in the French overseas territories therefore had strong negative impacts on the energy sector of these islands, but also on the social climate. On the one hand, the local utility is now facing an energy situation where PV plants are concentrated in only a few sites, regardless of the capacity of the grid to absorb this electricity, and provide electricity at a tremendous price. Since the maximum possible installed capacity of intermittent energy peak power (such as wind and solar) has been reached in less than 2 years, these islands will now have to wait several years before installing new capacities of PV at the right places and at the right price. The sector will therefore face a break for several years, while the utility will lose a lot of money for the next 20 years.

On the other hand, the social climate has worsened. Some people, outside the energy sector, made fortunes thanks to the state support for PV. Also most of the developers that have been successful in these territories have taken back their profits to re-invest in the development of other markets. In the end, nothing is left locally.

Although this situation may be the "worst case" in Europe, it is interesting to note that some of these outcomes are reflected at a local level in many European countries. West Africa should therefore carefully study these examples to make the most benefit of PV, since this technology could represent a great asset for the utilities when it is well developed, even though it remains difficult to manage in small market environments.

11.1.4 Implementation strategy

Based on the above remarks, we believe that the strategy should be defined at a regional level, and then spread and managed at a national level, while still being audited and monitored by the regional institutions.

Concretely, these principles could be embodied in the following strategy (see Figure 11.2):

STEP 1: SETTING UP A SOLAR TASK FORCE WITHIN ECREEE, WAPP AND UEMOA

These three institutions include some of the best experts and energy professionals of the energy sector in West Africa. Nonetheless, they do not have a "solar task force" and their approach to PV is mostly based on external consultancy work. We suggest that one of these institutions should take the lead in the definition of the strategy, without the help of third party experts, and then propose it to the others so that they may validate it. In this respect, a "solar task force" should be set up within these institutions and receive the highest level of training: visits to international laboratories, private companies, PV plants, international experts, public institutions, financial institutions, etc. The training may last several months. International agencies can help in the design and financing of the training program.

STEP 2: DEFINITION AND VALIDATION OF THE PRELIMINARY REGIONAL STRATEGY

We believe that the Task Force should be directly responsible for the definition of the strategy. Based on their knowledge of PV and the regional energy context, the Task Force will be able to define a preliminary strategy (as mentioned, a strategy is mostly based on segments, capacity and price). The Task Force will then submit the strategy to a consortium of international independent experts that could comment and validate the approach. The international agencies could help in the financing of the audit, but should not be part of the initial definition work led by the Task Force, who are solely responsible for it.

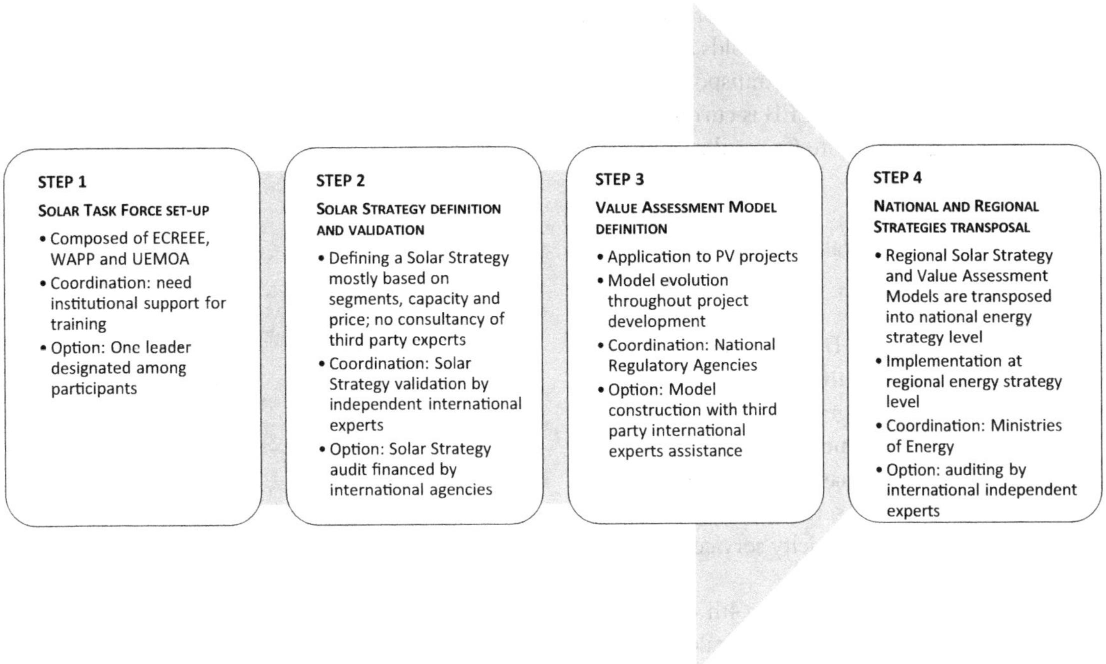

Figure 11.2 Steps for implementation strategy

STEP 3: DEFINITION OF THE VALUE ASSESSMENT MODEL

The Task Force could set up and manage a value assessment methodology of PV projects. The Task Force could work with international experts to build the model. International agencies could help in the financing of this process. Alongside the development of the PV sector, the assessment model will evolve; the Task Force will monitor the outcomes of the model and suggest improvements every year. The Task Force will disseminate the model at the national level, for example in coordination with the National Regulatory Authorities. The NRA will be responsible for running the model and assessing the local proposals.

STEP 4: DEFINITION AND VALIDATION OF THE NATIONAL STRATEGIES

Based on the "information package" set up at a regional level, comprising the regional strategy and the assessment methodology, all States should be able to design a national strategy. Again, Ministries of Energy would be in charge of the design. A national solar task force could even be set up to do so. Given the small size of the markets, we believe that the national strategies should be validated at the regional level. These national strategies could also be audited – but not designed – by international independent experts, while the financing of this auditing process could be supported by international agencies.

At the national level, the strategy is more detailed than at the regional one. Indeed, if segments, prices and capacities are still key, each State should also propose an implementation strategy: tenders, bilateral agreements, financing of the development, legal procedures for permits and authorizations.

11.1.5 Example of Benin

Benin is an energy dependent country that relies mainly on expensive imported hydrocarbons for its electricity production. Furthermore, electricity contributes to a very small portion of Beninese primary energy: only 2,2%, which represents the third most important source of energy after oil products (38,4%) and biomass (59,4%).

For coping with this underpowered status, Benin's national energy policy has the following objectives:

- For oil: increase research and secure the supply;
- For electricity: increased the electrification rate (especially in rural areas), improve the quality of service and promote regional integration (e.g. WAPP);
- For renewable energy: promote the development of technical solutions adapted to the local context.

(1) ELECTRICITY SECTOR

REGULATORY CONTEXT

The national power company in Benin is the SBEE (Société Béninoise d'Energie Électrique). SBEE holds a monopoly on thermal power generation as well as on the electricity distribution Beninese sector. With regards to hydropower resources, its exploitation was transferred in 1968 to a bi-national organization (Benin and Togo), the CEB

(Communauté Électrique du Benin). Governed by the Daho-Togolese Code of electricity, CEB holds a monopoly on hydroelectric power generation, transportation and electricity imports in each country. CEB is currently under the supervision of a High Interstate Council composed of ministers in both countries.

More recently, the Beninese electricity sector legal framework has undergone several in-depth modifications that have led to a market opening. This is marked by:

- The 2004 revision of the Benin-Togolese Code of Electricity, which ended the historical operator's monopoly, thus opening segments of the production and distribution to private operators; and
- The development of the national electricity supply through the realization of a series of projects that aim to improve the quality of electricity services.

In particular, Benin's Code of Electricity 4th chapter sets a comprehensive framework for electricity sale by private producers, which requires drawing up a contract between retailer and supplier according to terms of reference that must be approved by the Regulation Authority.

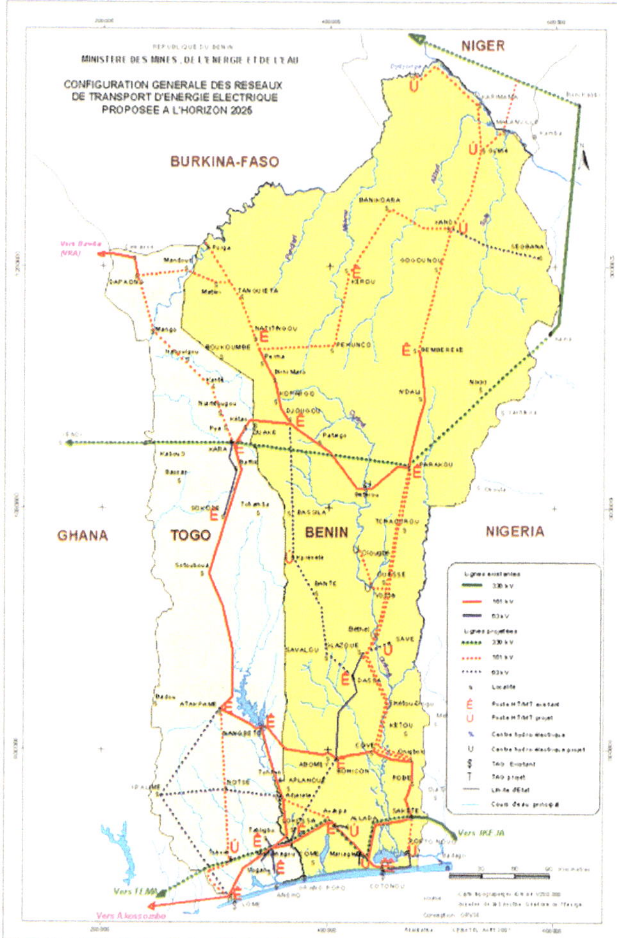

Figure 11.3 Benin transmission network, current (continuous lines) and future infrastructure (dotted lines) [6]

ELECTRICITY GENERATION

In 2005, 85% of Benin's production was imported by the CEB. The remaining 15% of production was locally generated by the SBEE, mainly by thermal power plants. For the same year, peak power demand was 105 MW.

ELECTRICITY TRANSMISSION NETWORK

Located at the heart of West Africa, Benin benefits from several high-voltage interconnections under the WAPP framework. In this context, by 2016 Benin will feature, given African standards, a reasonably dense transmission network (see Figure 11.3). This situation favors the regulation of transits from industrial scale variable sources of production such as solar.

ELECTRICITY DEMAND AND GROWTH TRENDS

Over the period considered in Figure 11.4, from 1996 to 2006, annual electricity consumption growth figures reached an average of 10,8%.

In the "Benin 2025, Agenda for an Emerging Economy" strategic master plan, the baseline scenario with regard to the development of the electricity sector implies a rapid growth of the country in order to meet the following objectives:

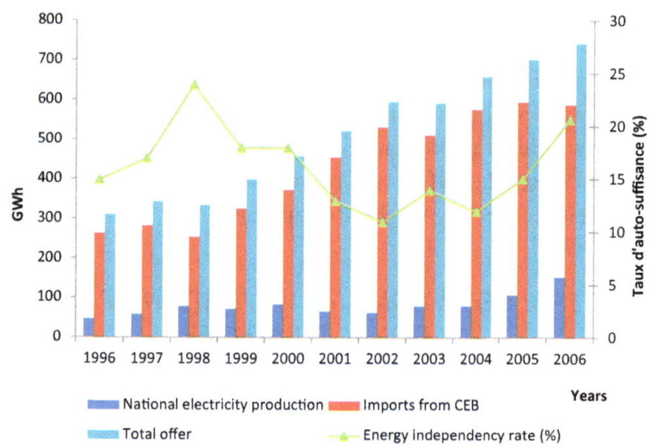

Figure 11.4 Evolution of national electricity production, imports and energy independence rate from 1996 to 2006

1. Increase access to electricity in rural areas from the current 2% to 65% in 2025, with a growth rate of 4% per year: this means electrifying 150 communities and villages every year;
2. Reducing losses in electricity transmission and distribution networks from 18% to 14%;
3. Gradually increase wealth in order to reach 11% GDP growth rate in 2025 (5% in 2010).

This scenario will lead to more than tripling Benin's consumption and peak demand by 2025 (see Table 11.2). Such an increase is likely to cause economic tensions with respect to imported hydrocarbons and international relations.

Table 11.2 VLS-PV 2025 reference scenario for Benin

Year	2005	2010	2012	2015	2020	2025
Total electricity consumption (GWh)	560	1 002	1 260	1 647	2 484	3 781
Transmission and distribution losses (%)	18	18	17	15	14	14
Total generation (GWh)	661	1 182	1 467	1 894	2 832	4 310
Average annual growth (%)		9	9	8	7	7
Peak power demand (MW)	105	188	233	300	449	683
Average annual growth (%)		9	8	7	7	7
Per capita GDP (×1 000 FCFA)	133	149	168	196	279	408
Average annual growth (%)		2	3.2	5	6	6
VLS-PV installed capacity (MWp)	–	–	6	20	44	68

(2) BENIN'S PROPOSED STRATEGY

Given economies of scale in the PV industry, hydrocarbon price volatility and its upward trend, Benin electricity transport and distribution infrastructure development as well as significant demand growth forecasts and, finally, considerable local solar resource, Benin possesses all the assets required for the development of electricity production based on solar technologies.

In this favorable context, we have established a reference scenario (see Table 11.2) that supposes very large scale PV (VLS-PV) plants installed capacity representing 10% of peak power demand by 2025. According to our experience, the development of such PV capacity in Benin (68 MW) is reasonably achievable from a technical, economic and "project development" perspective.

(3) CASE STUDY: 6 MWP PROJECT IN KANDI, BENIN, DEVELOPED BY HELIOS ENERGIE

Helios Energie, a French PV project developing company operating in international markets such as sub-Saharan Africa, has recently conducted a technical and financial feasibility study for the implementation of a 6 MWp photovoltaic plant on the site of Kandi, located in the north of Benin. This study was finalized in April 2011.

The assets of the Kandi project for Benin are: supplying electricity during day time for the northern region of the country, contributing thereafter to the grid stability in these periods and off-setting part of the thermal generation that currently supplies the region. Furthermore, this 6 MWp project displays an intrinsic environmental, economic and social value for the region as the first very large scale PV project in the region.

Today, the Kandi project is undergoing its financial closing process at an advanced stage, following the favorable opinion of the French Development Agency (AFD). The selected builder-operator is Siemens Energy France. Construction itself should begin subsequent to the loan agreement by AFD in mid-2012. The electricity produced by this 6 MWp installation, currently expected to be the largest grid-connected PV power plant in sub-Saharan Africa, will be amortized over a 25-year period through production sale to SBEE.

Figure 11.5 Computer image of the 6 MWp PV power plant project in Kandi, north of Benin [8]

Once commissioned, this project will represent a significant step forward for the development of VLS-PV power plants in sub-Saharan Africa, illustrating the feasibility of such a business model in this region of the world. Figure 11.5 shows a computer-generated image of the plant.

11.2 VLS-PV FOR ISRAEL: A MICROCOSM OF A SOLUTION FOR A GLOBAL PROBLEM

11.2.1 Introduction

World electricity production is currently undergoing a quadratic growth with time. This may be inferred from Figure 11.6, which plots the past 20 years of data [9].

This relentless upward trend in electricity production is presumably driven by a combination of population growth and rising industrialization. It is consequently a fact that must be taken into consideration for any long-term planning. For example, by extrapolation through only the coming ten years, Figure 11.6 implies that by 2020, electricity generation will probably be nearly triple what it was a decade ago in 2000.

If we examine this curve in greater detail we see that its slope for the year 2010 is 736 TWh per year. Taking a typical capacity factor for a modern fossil-burning power

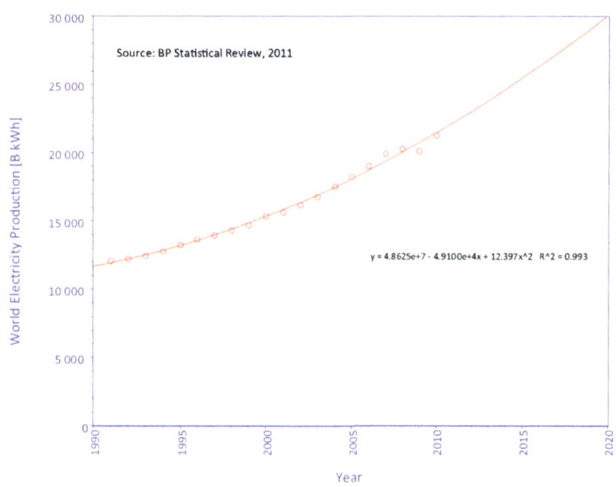

Figure 11.6 World electricity production during the past 20 years [9]

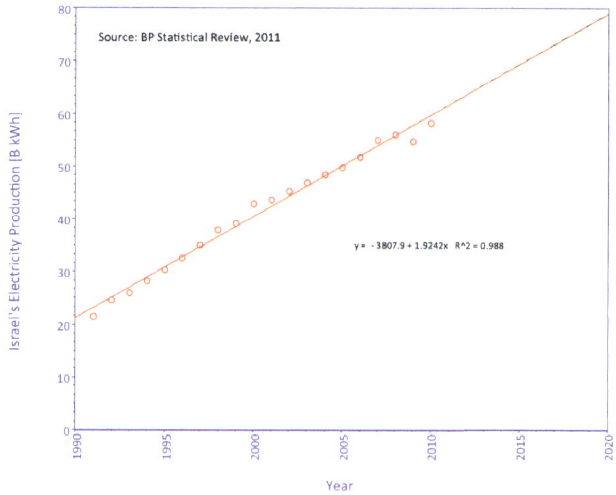

Figure 11.7 Israel's electricity production during the past 20 years [9]

plant as being approximately CF = 0,6, then this translates into 140 GW of new generating capacity for the year 2010. However, since most of these additional 736 TWh were in all probability fossil-fueled, and much of it involving power plants with capacity factors considerably less than 0,6, the 140 GW is probably an underestimate of the actual generating capacity that was installed that year. The consequences for reducing greenhouse gas emissions on a global scale are thus not promising unless renewables can be introduced on a massive scale.

As an example of the kind of scale required, suppose that the world had built VLS-PV plants in 2010 instead of fossil-burning plants. Taking a nominal annual output of 2 TWh/GWp for a typical desert location [10], this would have entailed the construction of 368 GWp during that year. Had such a strategy been adopted by the world, the consumption of fossil fuel for electricity generation during 2010 could have remained at approximately its 2009 level.

Although VLS-PV technology is essentially mature, unfortunately the largest PV plant that has yet been built is 100 MWp. We thus see that there is an enormous gap between what is being done to reduce greenhouse gas emissions and what needs to be done. Basically the world needs to be installing the equivalent of *at least* 368 000 MWp *per year* of PV or other renewables.

On the world energy scale, the state of Israel is of course minuscule, 60 TWh being its estimated 2010 production [9] – i.e. less than 0,3% of the world total. Figure 11.7 illustrates this point.

However, Israel is a "sunny" country, with the relatively empty Negev desert, and it has the industrial capability of implementing VLS-PV on a relatively massive scale. Moreover, its *per capita* use of electricity is comparable to that of several of the developed European countries. The purpose of this study is accordingly to treat Israel as a microcosm of the world's emissions problem, to study a novel strategy for how the country could implement VLS-PV on the required scale, and to consider some possible extensions of this strategy to other countries. The wide-spread adoption of such a strategy could help solve the fuel supply problems of individual states on a scale that would have massively positive global consequences.

11.2.2 An economic barrier

In the case of Israel, and its linear electricity growth curve, Figure 11.7 indicates that VLS-PV would need to provide 1,92 TWh per year in order to replace the country's present average increase of 366 GW each year of new fossil-fueled generating capacity. In terms of VLS-PV, this means installing some 960 MWp each year. This is a tall order, when set against the world's first 100 MWp PV plant. It is therefore important to examine why there is so much talk – in the guise of newspaper reports – about very large photovoltaic plants, but much less action.

The answer would appear to be an economic barrier – possibly, to a certain extent psychological but probably merely cautionary until there is more experience with the economics of large systems. Until recently the installed cost of large grid-connected PV systems was in the vicinity of 5 EUR/Wp. This means that the turn-key cost of a 100 MWp system would have been approximately 0,5 billion euros. This figure may have constituted some kind of invisible barrier above which banks were not prepared to put up loans. If this is indeed the case, then since system costs have subsequently fallen to around 2 EUR/Wp, we may expect to see some 100–250 MWp PV plants built in the near future, but probably not larger ones.

Naturally, when a bank offers a loan, it expects to be paid back, and there are two classic instruments for paying back the loan on a PV system. One is the so-called feed-in-tariff (FIT) that many governments have adopted. This instrument requires electrical utilities, by law, to purchase PV-generated electricity at a higher rate than it costs them to generate conventional power. This enables a return on the loan within a time frame that is deemed acceptable to banks. The FIT was originally introduced in Germany to help promote the spread of PV, and has subsequently been adopted by other European countries and beyond.

The second method of financing PV is for the government to offer tax incentives to help the purchaser of a PV system cover part of the cost of the initial investment.

In both of the above cases the PV system is treated as an up-front investment that has to return its capital outlay over a period of years. In the case of relatively small PV systems – say domestic or commercial scale – the amounts of money involved are sufficiently small that many people are prepared to take the risk (i.e. that the system may not continue to function in a trouble-free manner over its expected lifetime). But evidently the big lenders – i.e. the banks – have their own ideas about how large the risk may be allowed to become. This may be wisdom on the part of the banks, but if the world is urgently in need of constructing hundreds of GWp of VLS-PV plants *each year*, financing of the required investment is unlikely to become available in time to avert catastrophe.

11.2.3 The public good vs. investment

Although there is still some debate in scientific circles as to whether the presently perceived era of global warming is entirely due to anthropogenic causes, there can be little doubt that the increasing use of fossil fuel is definitely a non-negligible contributory factor to this warming. As such, it is in the public good for the world to try and cut down on the burning of fossil fuel. Now many people wrongly assume that this automatically means impeding the development of developing countries. "After all", they argue, "the developed countries are the ones responsible for global warming, so why should the developing countries be made to pay the price of the formers' mistakes?" They clearly have a strong case. However, our claim is that VLS-PV can allow full development of developing countries with a low carbon footprint, while at the same time enabling the developed countries to clean up their act. The only problem is how to implement VLS-PV on a large enough scale and at a fast enough rate.

Traditionally, the public good is paid for by taxation. In developed societies, such goods as health, education, safer roads, etc. are normally paid for, at least in part, by taxation. Society does not inquire whether a hospital or a school is a good "investment". Neither does it consider how long it will take to return the initial capital cost. It is taken for granted that such capital outlay is for the long-term good of all. And so it should be in the case of so-called "green energy" – particularly in a world that is urgently in need of reducing its use of fossil fuel.

11.2.4 An electricity consumption tax (ECT)

In light of the above discussion, we examine the consequences that could emerge if, for example, Israel were to impose an ECT on all use of electricity in the country. As will be seen, such a tax could not only fund the introduction of VLS-PV on a massive scale, but it would almost certainly lead to a number of additional benefits that could also contribute to lowering the global carbon footprint.

At the present time, the cost of electricity to Israeli consumers is approximately 0,08 EUR/kWh (slightly rounded down for simplicity). Suppose a nominal ECT of 20% were to be imposed upon electricity *consumption* (treated here, again for simplicity, as being identical to *production*). From Figure 11.7, such a tax during 2010 would have realized 0,96 billion EUR: enough funding – in a single year – for the construction of a 480 MWp VLS-PV plant.

Under desert conditions, such a plant would generate, on average, 0,96 TWh of electricity per year. Hence, if no additional fossil-fueled plant were added to the country's existing approximately 10 GW of generating capacity, the production/consumption for 2011 would reach 60,96 TWh. The ECT on this would in turn yield a revenue of 0,975 billion EUR, which could fund a second VLS-PV plant, this one of capacity 488 MWp.

This process could clearly continue indefinitely, but there are a number of questions that must first be addressed.

11.2.5 Discussion

(1) IS THIS GROWTH RATE FAST ENOUGH?

Figure 11.7 indicates that the historic annual growth rate for Israel is slightly less than 2 TWh per year. By contrast, an ECT of 20% would only cover approximately half of this rate. So our first question is how to overcome this problem. There are two possible directions.

FOSSIL BACKUP

As was previously indicated, Israel's historic annual increase of around 2 TWh has been brought about by the construction of, on average, 366 MW of new generating capacity each year. This was paid for by tax money that was available in the past and, in a business-as-usual scenario, would presumably continue to be available. Therefore, if the 0,96 TWh of VLS-PV, paid or by tax money, does not turn out to be sufficient, the shortfall could be made up by a supplement of additional fossil power. Such a step, although disappointing, would at least cut the addition of new fossil-burning generators by 50% compared with what would have been the case without VLS-PV.

PUBLIC EDUCATION AND INCENTIVES

A more interesting possibility would be to take advantage of the massive enthusiasm there is for "green energy" among large sectors of the public – particularly the youth. In particular, a public education campaign could be undertaken in the media, stressing the importance of not wasting electricity and offering examples of how easy it is to cut down on unnecessary usage. Simple steps such as extinguishing unnecessary lighting, turning off computers at night-time or air-conditioners when one goes out for lunch, etc. could be emphasized – backed up by an ongoing competition among consumers, with cash prizes for those who save the most energy. It is not unlikely that the combination of human nature and cash prizes for the largest energy savers would quickly render the desire to use electricity sparingly an almost automatic reflex for many people. If such were to be the case, then an annual increase of around 1 TWh might suffice, and VLS-PV plants, constructed annually, would be able to provide 100% of this requirement.

(2) IS THIS GROWTH TOO FAST?

The thought of any nation constructing something like 500 MWp per year may be somewhat daunting, particularly for a country like Israel where barely a handful of MWp represents the largest plant that has yet been constructed. There would obviously need to be a learning period, for decision makers, for the electric utility and for the public. To this end, it might be deemed expedient for the ECT to be imposed for a one-year trial period, and its proceeds used to construct a single 480 MWp VLS-PV plant. Continuation of the program would then proceed only after all of the consequences of the first VLS-PV plant had been digested.

(3) MIGHT THERE BE GRID STABILITY PROBLEMS?

There have been reports [11] that the fluctuations in the PV output of grid-connected systems, caused by passing clouds, can lead to instabilities in the power lines. In the case of Israel, the potential matching of VLS-PV to the grid system has been studied extensively ([12–16] and Chapter 8). According to those studies, one or two VLS-PV plants of the size considered here could be accommodated by the existing grid without the need for major changes in operating strategy. However, for more such plants, it would be necessary to incorporate appropriate storage and to place greater reliance on fast-reacting gas turbines for short-time matching purposes. To this end, some of the avoided cost from the no-longer-needed annual construction of 366 MW of conventional generating capacity could be used for funding storage and increasing the number of gas turbines that are currently available to the power company.

(4) WHAT ABOUT LAND AVAILABILITY PROBLEMS?

The Springerville experience [11] indicates that a ground allowance of 273,3 kW(AC) per ha provides "generous access space for construction and maintenance activities". In our case, 480 MWp(DC) may be translated into 379 MWp(AC), assuming a system efficiency similar to that observed at Springerville, which in turn would require the availability of 1 387 ha of land per year. If this area were to be divided among the 40 *kibbutzim* (agricultural settlements) that are currently scattered around the Negev desert, it would amount to approximately 35 ha of land per kibbutz per year, each accommodating a modest 12 MWp conventional flat-panel PV plant. At the present time, kibbutz Qetura has a grid-connected 5 MWp PV plant – the largest in the country. Furthermore, the Negev desert occupies approximately one million ha. Therefore, the scope of the project that is here proposed, although large by current world standards, should not provide any land availability problems, even for a small country like Israel.

11.2.6 Applicability elsewhere

In principle the idea of treating VLS-PV as a public good rather than as an investment, has implications far beyond the example of our case study. Naturally the prescription for worldwide adoption would vary from country to country according to local requirements and political structure. For example, in the case of Australia, where electricity generation has stabilized at approximately 250 TWh during the past five years (Figure 11.8), future needs may be expected to rise with population growth, currently estimated at 1,15% per year [17]. This will require the average construction of approximately one 1,68 GWp VLS-PV plant per year. Such plants could be paid for by an almost unnoticeable 5% rise in the Australian electricity tariff, which at present is approximately 0,2 AUD/kWh. Naturally, with a country the size of Australia, the availability of land would not be a serious consideration.

Israel and Australia are, of course minuscule examples (on a global scale) of CO_2 emitters; respectively 0,25% and 1%. At the other extreme would be China. That country's electricity production is plotted in Figure 11.9. The rate at which China's electricity production was rising in 2010 is seen to be 387 TWh per year, which under our previous assumptions translates into 74 GW of new fossil-burning capacity per year. This was 50% of the 2010 world growth in electricity production. For China to employ PV plants to maintain fossil consumption for electricity generation at the 2010 level would have required the construction of a VLS-PV complex of 193 GWp rating, or a combination of PV and other renewable technologies as discussed in Section 4.2. However, like Australia, land issues would not be a consideration for China, but unlike the Israel and Australia examples, the raising of existing electricity tariffs might involve different considerations. However, it is encouraging to note that land area (i.e. the Gobi Desert), combined with the world's largest PV industry and a political structure that can make relatively rapid decisions, could enable China to emerge as the world leader in the employment of renewables on the scale of VLS-PV.

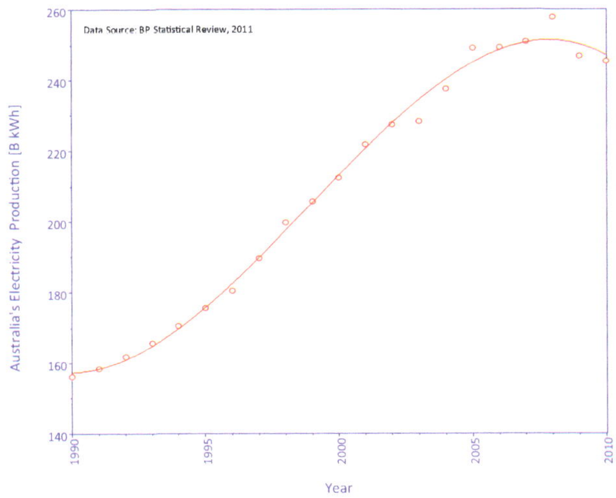

Figure 11.8 Australia's electricity production during the past 20 years [9]

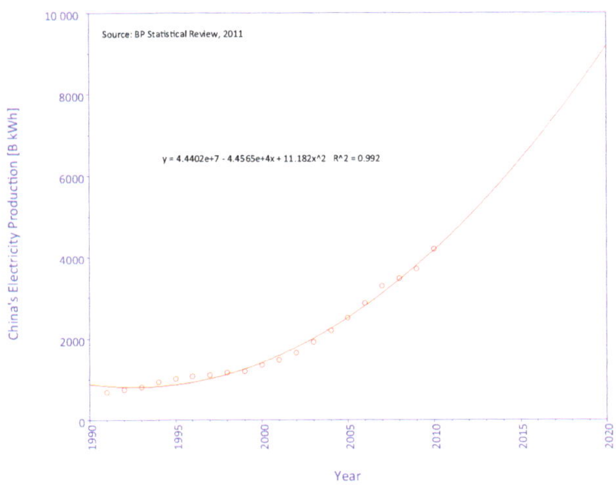

Figure 11.9 China's electricity production during the past 20 years [9]

11.2.7 Conclusions

In this chapter we have argued that the world needs to install VLS-PV plants at a rate of approximately 368 GWp per year in order for fossil-fuel consumption for electricity generation to cease its annual increase. We have further argued that feed-in-tariffs and other instruments that treat PV as an "investment" actually impede this growth instead of encouraging it. We have proposed, instead, that VLS-PV be regarded as a public good rather than as an investment and, as such, that it should be funded by public money. Taking Israel as a case study we have shown that: (a) the construction of 480 MWp per year of conventional PV plant could be paid for by the equivalent of raising the present electricity tariff by 20%; (b) with appropriate public education, such a program could actually promote a more sparing use of electricity than has hitherto been the case, leading to a situation in which no further fossil-fueled plants need be constructed; (c) the amount of land area required is relatively modest and, if divided among existing desert settlements, would result in PV systems of a size comparable to those that have already been built in the country. We have also given two examples of how, with suitable modifications in each specific situation, the principle of public funding for VLS-PV could contribute to a serious decrease in the rate at which world fossil fuel consumption is rising.

REFERENCES

1 Eberhard, A. *et al.* (2008). *Underpowered: The State of the Power Sector in Sub-Saharan Africa*. World Bank. Public documentation.

2 Infrastructure Consortium for Africa (2008). *Power Supply Situation in Africa*. Background paper prepared for ICA Annual Meeting, 13–14 March 2008. Public documentation.

3 KPMG, Global Power & Utility Centre of Excellence in Budapest (2011). *Sub-Saharan Africa Power Outlook*. KPMG. Public documentation.

4 West African Power Tool (WAP)

5 Madamombe, I. (2005). *Energy key to Africa's prosperity: Challenges in West Africa's quest for electricity*. Africa Renewal, United Nations. Vol.18 #4. page 6. Public documentation.

6 Agence Béninoise d'Electricfication Rurale et de la Maîtrise de l'Energies (ABERME), 2009

7 Société Béninoise d'Energie Electrique (SBEE)

8 Helios Energie

9 BP Statistical Review, 2011: www.bp.com/sectionbodycopy.do?categoryId=7500&contentId=7068481

10 Faiman, D., D. Raviv and R. Rosenstreich (2007) "Using solar energy to arrest the increasing rate of fossil-fuel consumption: The southwestern states of the USA as case studies" *Energy Policy* 35: 567–76.

11 Moore, L.M. and H.N. Post (2008) "Five years of operating experience at a large, utility-scale photovoltaic generating plant. *Prog. Photovolt: Res. Appl.* 16: 249–259.

12 Solomon, A.A., D. Faiman and G. Meron (2010) "An energy-based evaluation of the matching possibilities of very large photovoltaic plants to the electricity grid: Israel as a case study". *Energy Policy* 38: 5457–68.

13 Solomon, A.A., D. Faiman and G. Meron (2010) "The effects on grid matching and ramping requirements, of single and distributed PV systems employing various fixed and sun-tracking technologies". *Energy Policy* 38: 5469–81.

14 Solomon, A.A., D. Faiman and G. Meron (2010) "Grid matching of large-scale wind energy conversion systems, alone and in tandem with large-scale photovoltaic systems: An Israeli case study". *Energy Policy* 38: 7070–7081.

15 Solomon, A.A., D. Faiman and G. Meron (2001) "Properties and uses of storage for enhancing the grid penetration of very large photovoltaic systems". *Energy Policy* 38: 5208–22.

16 Solomon, A.A., D. Faiman and G. Meron (2011) "Appropriate storage for high-penetration grid-connected photovoltaic plants". *Energy Policy* (2011), doi:10.1016/j.enpol.2011.10.019.

17 CIA World Factbook, 2011: <www.cia.gov/library/publications/the-world-factbook/geos/as.html>

PART V

Possible scenario and strategy for VLS-PV

CHAPTER TWELVE

Expected role of international cooperation

12.1 INTRODUCTION

Electricity remains the key driver of social economic development at national and international level. The energy demand is growing fast. Experts predict that by 2035, the energy demand will increase by 65% compared with 2008 [1]. To meet such fast growing energy demand without damaging our environment is not possible without significant increases in the share of renewable energy. It is very fortunate that many of developing countries in Asia and Africa have abundant resources of solar energy, low-cost desert and arid land areas which lead to the favourable economic conditions of large scale PV deployment. Wider utilization of solar energy should not only provide the foundation for reliable energy supply, but it may also bring a renaissance in the development of developing countries and can significantly contribute to enhancing their social economic development.

On the other hand, in order to accelerate the current transition towards renewable energy, principal changes are needed in the ways that energy is supplied and used globally. Development of VLS-PV power generation systems that can be deployed on a massive scale is one of the ways that this transformation can be achieved.

Significant advances and dramatic cost reductions have been demonstrated in PV technologies over the last decade. Further cost reductions of PV are expected due to further technological innovations, economies of scale and increased market competition among PV industries worldwide. It is clear that supporting high level research and facilitating exchange of information and experiences are crucial if technologies are to be continually expanded and costs lowered. It can be realized only through the active collaboration of all stakeholders at local, regional and global levels. IEA PVPS Implementing Agreements are multilateral technology initiatives through which IEA member and non-member countries, industries, international organizations and non-government organizations share research on breakthrough technologies, fill existing research gaps, build pilot projects and carry out deployment or demonstration programmes [2].

12.2 WAY FORWARD TO LARGE SCALE PV DEPLOYMENT WORLDWIDE

The PV market developments detailed in EPIA studies forecast that the global total installed PV capacity in 2015 could reach between 131 GW and 196 GW depending on the considered scenario. The 100 GW mark could be reached in 2013 according to an EPIA scenario [6]. Different sources have confirmed that there are over 200 large PV power plants in the world with a capacity of 10 MW prior to the end of 2011 [7].

When the IEA PVPS Task8 study into VLS-PV was initiated in 1999, the idea of VLS-PV was considered a long-term, futuristic concept [3–5]. However, industrialized countries have made significant efforts to promote development of MW-scale PV installations, and related infrastructures at local, regional and international level. During the last few years, hundreds of MW-scale PV systems have been successfully constructed in many countries of the world [7–9]. It has been demonstrated that large MW-scale PV can be a feasible option of the current and future energy mix and especially VLS-PV potentially could provide a substantial portion of global energy needs. However, although the development of large scale PV worldwide has experienced unprecedented growth during the last years and the cost of PV is declining significantly, PV currently provides a tiny fraction of global energy sources. The future expansion of VLS-PV depends on successful international cooperation to advance knowledge on development of VLS-PV systems. In addition, a clear political commitment of governments and a stable regulatory framework towards creating a long-term stable policy is essential for further deployment of VLS-PV systems worldwide.

The successful and faster deployment of large scale PV market development in recent years has been driven exclusively by policies to promote PV by setting effective feed-in tariffs (FIT) in Germany and other European countries, and other incentives that have helped develop markets and improve the economics of PV investments. Germany, Japan, Spain, Italy and the United States are pioneers in this field. These countries account for more than 2/3 of the total global large scale PV capacity. Other countries that have developed or are developing substantial PV capacity due to new policy and economic support schemes include Australia, the People's Republic of China, France, Greece, India, the Republic of Korea, Portugal and Canada [6, 7]. Germany and Spain have been the two most active promoters of large scale PV market development through their highly attractive FIT laws. In Germany, the Renewable Energy Sources Act (EEG) is recognized as the best example of an effective legal instrument to promote deployment of PV technologies and it was the first pioneering legal act of its kind among the European Union countries, inspiring and triggering the creation of similar regulatory frameworks across the world [7]. Roof-top programmes and FIT laws introduced in many countries have stimulated high growth in deployment of the on-grid PV market in recent years. As a result, the installation of the large scale PV systems has been rising rapidly every year and extensive research has now accumulated to advance knowledge on the concept of VLS-PV and enhance further expansion of large scale PV power generation systems around the world [3–7].

According to recent studies, the development of PV has experienced the third PV market diffusion phase – grid-parity in residential PV markets throughout the world which has been enabled by the political invention – roof-top programmes and feed-in tariff (FIT) laws. And now PV is entering the fourth large scale PV market diffusion phase: commercial utility-scale PV power plants all over the world driven by technological innovation and cost reduction of PV systems as a whole [8]. It is true that cost reduction is the key driver for the transition of PV from a subsidy-driven, niche energy technology to the natural cost reduction-driven development of PV as an energy technology capable of competing with conventional power sources. This forward transition of PV grid-parity is characterized by further advances and technological innovations, high rate of economies of scale for PV components along the value chain from solar cell production to large scale PV plant construction and increased market competition among PV industries worldwide. Global grid-parity for PV technology should occur throughout 2012–2020 in the majority of all market segments in the world, starting in remote island countries and regions with good solar insolation and high electricity prices. The cost of PV electricity generation in regions of high solar irradiance is expected to decrease from 0,16 EUR/kWh to 0,06 EUR/kWh in the years from 2010 to 2020 [9].

An excellent overview of the PV cost reduction pattern is given by Photon Consulting [10] and Solarbuzz [11], global market research companies focusing on solar energy. According to these reports and other key sources, PV technology is the fastest growing power generation technology globally, spread across over 100 countries [10]. Over 40 years of development of the PV industry has demonstrated its capability to achieve fast price reduction and to tend to become the least cost option in both centralized and decentralized, and on-grid and off-grid markets. Table 12.1 summarizes the PV technology cost reduction potential in the next two decades. One of the key parameters which is driving PV system cost reduction is the power conversion efficiency of solar cells. Its improvement automatically results in savings on all components along the value chain of PV systems from processing of initial semiconductor materials and structures to PV systems or PV plants.

PV systems in general are widely deployed, with global capacity now exceeding 40 GW [6]. Currently, more than 4 GW capacity of PV plants are in operation, the number of MW-scale PV plants are increasing rapidly and several VLS-PV plants with capacity over 100 MW are under construction. As a consequence of this growth PV module prices are continuing to fall, and as a result PV electricity generation costs are continuously decreasing (see Table 12.1).

The development of large scale PV has experienced unprecedented growth worldwide and hundreds of MW scale PV plants have been established during the last few years. Worldwide, the average price of 92 large scale PV plants installed in 2010 was estimated as 4,7 USD/W (Figure 12.1) [11]. The global weighted average price had decreased to 3,93 USD/W for crystalline silicon PV systems by the end of 2011, with a market segment specific average price range for the year spanning from 2,81 USD/W for utility scale (>20 MW fixed ground mount, integrated crystalline silicon PV plants in Germany) to 6,20 USD/W (Japan residential rooftop PV system) [10]. But the global weighted average system cost had decreased to 3,41 USD/W by the end of 2011 for crystalline silicon PV systems with a market segment specific average cost range for the year spanning from 2,47 USD/W for utility scale (>20 MW large commercial rooftop PV system in Germany) to 5,27 USD/W (Greece small ground mounted PV system) [10].

The global weighted average levelized cost of electricity from PV systems was estimated as 0,26–0,35 USD/kWh in 2010 and 0,22–0,34 USD/kWh for crystalline silicon PV systems. For the First Solar CdTe thin film system, the global weighted average price was 2,64 USD/W in Germany and 3,07 USD/W in southwestern USA in 2011 [10]. The global weighted average levelized cost of electricity of thin film PV system was estimated as 0,16 USD/kWh in 2010, and it has decreased 0,14 USD/kWh in 2011 for utility scale (>20 MW) fixed ground mount, integrated CdTe thin film

Table 12.1 PV technology cost reduction potential in the next two decades

Cost reduction potential	2011	2012	2015	2020	2030
World PV market, GW	26,5		54,7		
Crystalline Si PV module efficiency, %	>15%	>18%	>20%	>20%	18–25%
Thin film PV module efficiency, %	5–11%	6–12%	8–14%	12–16%	14–19%
Concentrated PV module efficiency, %	>25%	>25%	>30%	>35%	>40%
Large scale c-Si PV system price, USD/W	3,65		2,55		
Large scale thin film PV system price, USD/W	3,50		1,86		
PV electricity cost, USD/kWh	<0,35	<0,28	<0,15	<0,07	<0,05
PV module lifetime, years	>25	>25	>30	>35	>40
Balance of systems lifetime, years	>10	>15	>20	>25	>35
Energy payback period, years	<3	<2	<1	<0,5	<0,5

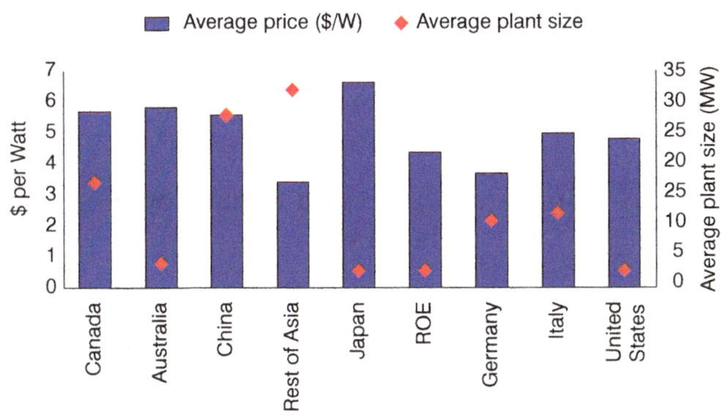

Figure 12.1 Average prices of large-scale PV plants by countries [11]

PV plants [10]. An important driver of the cost of utility-scale PV plants is whether the system is roof mounted or ground mounted, and whether the system is fixed or has a tracking mechanism. Fixed PV systems mounted on the ground are the cheapest (an average of 2,81 USD/W in 2011), while adding tracking increases the cost significantly (an average of 6,39 USD/W in 2011) [10]. According to a recent analysis, large utility scale PV systems can produce electricity at 0,12–0,14 USD/kWh or less, while the middle range for residential systems comes in at 0,13–0,19 USD/kWh with respect to the following parameters: 18% capacity factor, 25-year period of cost amortization, 6% borrowing rate, 30% Federal and 35% state tax [12].

Faster development of large scale PV during the last few years has been possible due to the large demand from the European market, and to some extent US and Japanese markets. However, these markets offer only a modest growth opportunity compared with the exponential growth possibility in the new and emerging markets of Asia and Africa. Market development in the region is thus critical to the next wave of global investments in solar energy applications that will accelerate the global progress of solar energy toward grid-parity. There are some ambitious plans to develop several hundreds of MW large scale solar power plants in deserts in North Africa, Southern Europe, the USA and Asia [6].

Despite the cost of PV declining continuously, a VLS-PV system of several hundred MW or GW scale capacity will require increasing amounts of investment. In order to promote VLS-PV systems worldwide, especially in developing countries located in deserts, innovative and favourable financing approaches and active international collaboration are required. Major international collaboration is needed to expand R&D in all areas of VLS-PV systems, especially in policy, regulation, standards and business model development. International cooperation in these areas stimulates the strengthening of international collaboration of existing institutions and activities, as well as the creation of new joint initiatives at regional and global levels.

12.3 POSSIBLE ACTIONS IN ASIAN COUNTRIES – THE CALL FOR INTERNATIONAL COLLABORATION

Regional and national programmes promoting PV as a mainstream energy source are crucial for further development of large scale PV worldwide. There is an emergence of some of the largest solar promotion programmes globally. If such programmes are able to achieve the scale envisaged, they have the potential to create a huge wave of demand, which, in turn, will spur increased innovation, efficiency in production, and further cost reduction in PV. The competitiveness of PV against traditional and other alternative energy sources around the world will consequently be increased.

12.3.1 Asian Development Bank launches the Asia Solar Energy Initiative

The energy demand for Asia and the Pacific will grow at an annual average rate of 2,4% until 2030, more than double the world's average over the same period according to the ADB study [13]. As shown in Chapter 2, in May 2010, the Asian Development Bank (ADB) announced the launch of the Asia Solar Energy Initiative (ASEI), targeted to assist in identifying, developing, and implementing 3 000 MW of solar power in Asia and the Pacific over the next 3 years.

In order to assist in providing adequate, affordable access to clean energy for the large, growing population of Asia and Pacific countries, ADB plans to finance up to 2,25 billion USD directly under the ASEI and leverage an additional 6,75 billion USD in solar investments over the same period using soft loans, donor contributions, grant funds, innovative risk mitigation mechanisms, carbon market support measures and direct support. The overall goal of ASEI is to demonstrate the feasibility and sustainability of various forms of public and private solar power generation projects.

The ADB has launched the Asia Solar Energy Forum to facilitate solar technology transfer to Asia and the Pacific, deliver quality inputs to policy and decision makers in solar power development, exploit economies of scale that the large regional market offers, and facilitate innovation in financing mechanisms. The Asia Solar Energy Forum is an exclusive regional knowledge management platform linked with knowledge hubs globally.

In order to mitigate risks associated with solar energy projects and buy down the up-front cost of solar energy development to catalyze solar energy growth, which will accelerate progress toward achieving grid-parity, ADB aims to establish an Asia Accelerated Solar Energy Development Fund and to raise 500 million USD from financing partners, which will then be used to design innovative mechanisms so that commercial banks and the private sectors are encouraged to invest in solar technologies and projects.

12.3.2 Solar mission in India

India's Solar Mission, an ambitious plan for scaling-up solar power generation, aims to install 20 000 MW of solar power over the next 12 years – a 6 666-fold leap from the current 3 MW of installed capacity, and the equivalent of 13% of India's total installed power generation capacity [14]. The objective of the Solar Mission is to create an enabling policy environment for rapid diffusion of solar technology across the country through rapid scale-up of capacity and technological innovation to drive down costs towards grid-parity. The Solar Mission anticipates achieving grid-parity by 2022 and parity with coal-based thermal power by 2030 – enabling a rapid scale-up of both PV and CSP technologies [14].

The Solar Mission's call for greater international collaboration on technology, research, and finance could be met through enhanced international cooperation between India and other countries that could attract the necessary level of private finance as well as through joint research and development leading to enhanced cooperation [14].

The Mission states that although India aims to install 2 000 MW of utility scale solar power, including roof-top by 2017, this capacity could reach 10 000 MW or more by that date with the support of enhanced and enabled international financing and technology cooperation. Strategic international collaborations and partnerships aimed at meeting the priorities set out under the Mission would be developed, along with effective technology transfer mechanisms and strong IPR protection. Cooperation will be encouraged at the level of research organizations along with industry partners and also at individual level to generate new ideas. Wherever feasible, cooperation through bilateral and multilateral arrangements would be facilitated. In addition to the multilateral cooperation, India has been increasingly strengthening its bilateral relationships with China, the EU, Australia, Japan and the USA. India's Ministry of New and Renewable Energy is also implementing bilateral solar projects with Japan and Australia under the Asia-Pacific Partnership Programme. Meanwhile, the establishment of the US–India green partnership confirmed that collaboration on solar energy would be a significant dimension of that effort [14].

12.3.3 China moving forward to develop PV at a large scale

China is known for its large share of coal-based generation (around 67% of total primary energy supply in 2010). However, the Chinese government has made significant changes to their energy policies and set up ambitious targets for scale-up solar power generation. The Chinese government has increased their target for PV. The new target for cumulative installed PV power is 10 GW by 2015, instead of the earlier target of 5 GW, and 50 GW by 2020 instead of the previous target of 20 GW [16]. China has become a world leader in the manufacture of PV cells and modules, but it was mostly exported. Recently, China has been moving to expand its domestic market for solar power. Following the new incentive scheme of Golden Sun announced by the government in 2009, numerous plans have been announced by the Chinese government and industry players that have become part of the milestones for solar industry and technology development in China. Development of large scale PV projects will play a decisive part in this. The two largest projects planned are a 5 GW solar park in Wulan (Qinghai province) and a 2 GW project in Ordos (Inner Mongolia) with First Solar [16]. Table 12.2 gives a list of the largest PV projects under construction and announced in China

The State Grid Corporation of China aims to integrate 10 MW PV power plants before 2015, which gives a huge development opportunity to the PV industry. The integration of high capacity PV plants to the electricity network in short period may cause a significant problem. However, China is overcoming these difficulties by making the necessary investments in grid infrastructures, using leading edge technologies, such as constructing long distance High-Voltage Direct Current transmission lines

Table 12.2 Largest photovoltaic projects under construction and announced in China

1st round national concession bidding	
10 MW Dunhuang Gansu	China Guangdong Nuclear Power
2nd round national concession bidding	
20 MW Yulin Shangxi	Shenhua Group Guohua Energy
20 MW Yulin Shangxi	China Power Investment Corporation
20 MW Yulin Shangxi	China Power Investment Corporation
30 MW Gonghe Qinghai	China Power New Energy Development
20 MW Henan Qinghai	China Huaneng New Energy
20 MW Baiyin Gansu	China Power New Energy
20 MW Jinchang Gansu	China Guodian Neimeng
20 MW Wuwei Gansu	Shenhua Group
20 MW Alashan Neimeng	China Guodian Neimeng
20 MW Baotou Neimeng	Shenhua Group
20 MW Bayanzhuoer Neimen	China Guodian Neimeng
30 MW Qingtongxia Ningxia	China Huaneng New Energy
20 MW Hami Xinjiang	China Power Investment Corporation
20 MW Tulufan Xinjiang	China Power Investment Corporation
20 MW Hetian Xinjiang	China Power Investment Corporation

and introducing smart grids and storage capacity. China will basically have a strong background and experience for the deployment of long distance High-Voltage Direct Current transmission lines and meanwhile, in order to support the nation's wind and solar power development plan, the State Grid Corporation of China issued a goal to integrate 100 million kW wind power and 50 million kW solar power before 2015. By the end of 2010, there has been invested 418 million CNY to construct 23 200 km of power lines for wind and solar power integration.

12.3.4 Call for a 1 GW photovoltaic project in Gobi desert of Mongolia

The idea of the establishment of VLS-PV in the Gobi desert region and connecting it via high-voltage transmission lines to populated areas in northeast Asian countries was first described by a Task8 study [3]. Within the framework of the IEA PVPS Task8 activity, the project proposal to build a large scale PV plant with capacity reaching 1 GW in Gobi desert, Mongolia was developed [4]. A VLS-PV plant with capacity of 1 GW will require around 8–10 km² of land area depending on the efficiency of the PV module and the tilt angle of the PV area. The expected cost of a 1 GW PV plant is estimated between 2 billion USD and 2,5 billion USD to build, including grid connection and infrastructure. It is proposed to build the plant in four phases, including a 1 MW pilot demonstration plant in the first year, expanding to hundreds of MW (40 MW, 500 MW and 1 GW) over the following few years.

For the selection of a suitable site for construction of a VLS-PV plant, the following technical, environmental and economical factors have to be considered:

- Solar resource, level of global and direct solar radiation (DNI)
- Land requirement, and site topography, land cover
- Land surface flatness, land slope
- Limited agriculture value
- Availability of transmission network, railway transportation and other infrastructures.

The potential of solar resource in the Gobi desert region is abundant. Large scale PV plants in deserts can provide a large amount of electricity to meet growing energy demands in an economically feasible and sustainable manner. A number of sites in the Gobi desert have been identified which can satisfy the above-mentioned requirements for construction of VLS-PV. About 40% of land in the Gobi desert area has been identified as suitable for construction of the VLS-PV power plants according to the Task8 study (Figure 12.2).

The solar resource is the most important siting factor for a cost effective VLS-PV power plant, and it is related directly to the energy generation price. The solar energy resource distribution map estimates the total daily radiation using inputs derived from satellite and surface observations of weather parameters. The daily solar irradiation in the Gobi desert area is in the range 5,5–6,0 kWh/day/m² [17]. It can be seen that, regardless of its northern position, the Gobi desert in Mongolia has solar conditions complying with the above-mentioned requirement for construction of VLS-PV power generation systems. Compared with countries at the same latitude, the solar resource in Mongolia is similar to those in the USA and better than in Europe and Japan. In addition to the potential of solar resources, the cost of the VLS-PV project depends on the land area requirements and land slope. The land slope of selected sites for VLS-PV is less than 1%, which is the most cost effective and most efficient. Land cover is also an important characteristic for siting of a VLS-PV project. Mongolia has an unpopulated waste desert land area of 0,66 million km² (41,3% of Mongolia's total land area), among which 25,96% is dry steppe zone, 21,9% is semi-desert, 15,34% is desert uncovered rock [17]. Most of this wasteland is located in the south and south eastern part of the country, which has the best solar resources. These areas would be suitable locations for large scale VLS-PV projects.

A PV plant is intended to be located near Sainshand city, which is located along the high voltage transmission line and main railway connecting Ulaanbaatar city to China and Russia. The Sainshand city area of the Gobi desert in which the PV plant is intended to be established has an annual global solar radiation at tilted surface ranging from 1 700 to 2 200 kWh/m²/year, and is suitable for VLS-PV according to existing feasibility criteria for construction. The generation cost of the VLS-PV plant with 1 GW capacity was estimated

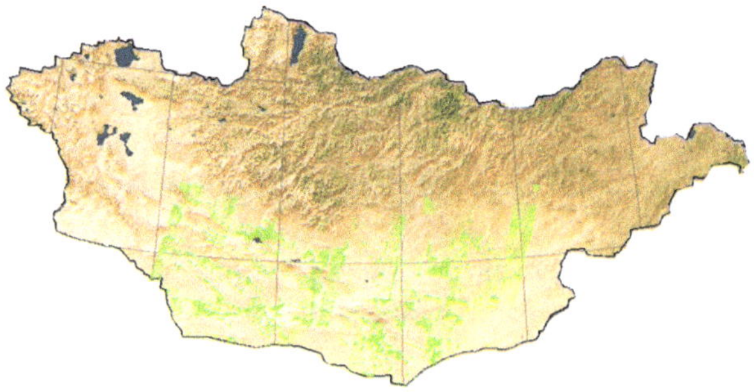

Figure 12.2 The identified land area in Gobi desert suitable for the construction of VLS-PV

at 0,10–0,11 USD/kWh according to studies by the IEA PVPS Task8 activity (see Figure 12.3) [3, 4]. The results of the study show that a VLS-PV power plant to be established in the Gobi desert can be economically feasible today.

High-voltage DC transmission grids can carry the high volume of electricity over thousands of kilometres from the Gobi desert region to the cities and industry in northeast Asia. Thus, it is assumed that VLS-PV plants installed in the Gobi desert can be connected to the existing power lines of Mongolia, as well as to the utility network of north and west China to feed parts of the energy demand in China and northeast Asia. The project is in line with the ambitions of the Mongolian Government, which have been adopted by the National Renewable Energy programme (2006–2020). It aims to generate electricity and comply with sustainable development and intends to demonstrate that, under certain conditions, such a VLS-PV project can be profitable and thus be an illustration that coal-based power generation is not the only option for the future of the Mongolian energy industry. By relying on renewable energy power sources, Mongolia could become more independent of expensive energy imports.

The introduction of VLS-PV systems in Gobi desert regions may have quite a large impact on the future development of Mongolia. As studied by the IEA PVPS Task8 activity, the introduction of VLS-PV systems in Gobi desert regions will stimulate wider use of PV power generation systems to increase the efficiency of using water

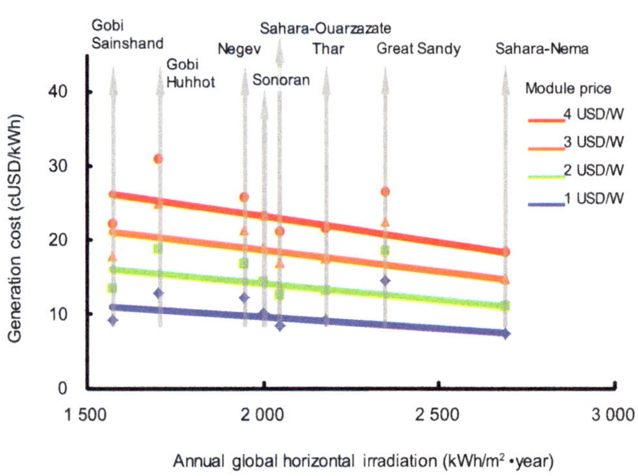

Figure 12.3 The estimated generation cost of VLS-PV systems for different world deserts

resources for agricultural purposes. As agriculture is a key economic sector of Mongolia, and agriculture is the largest user of water, developing VLS-PV systems in desert areas in combination with modern irrigation and desalination techniques would help greatly in greening desert land and agricultural development of the country. VLS-PV systems will also prevent further desertification of the Gobi desert, which could have disastrous effects for its inhabitants.

Mongolia is experiencing an economic boom due to the recently discovered minerals in the south of the country. This economic growth will inevitably imply higher energy needs. On the other hand, revenues from the mining industry should support the further successful introduction of VLS-PV deployment in the country.

12.4 EXPECTED ROLE OF IRENA IN PROMOTING INTERNATIONAL COOPERATION

A transition to a renewable energy based on low-carbon economy with higher shares of PV would require increasing efforts of international cooperation to promote R&D by increasing investments in PV technologies, to promote technology and knowledge transfer via relocating PV manufacturing to lower cost developing countries, to overcome various economic, technological, institutional and policy barriers related to existing industry, infrastructure and regulations of energy systems for introducing large scale PV power generation systems.

The International Renewable Energy Agency (IRENA) is an intergovernmental organization with global coverage (as end of December 2011, in total 148 states and the EU have signed the Statute of the Agency and 84 states and the EU has ratifying members of the Agency), created to act as global focal point for international cooperation and to underpin the effort to fundamentally increase accelerated deployment of renewable energy technologies on global scale [18].

As the world's single international intergovernmental agency dedicated to promote and accelerate current transformation of world energy systems towards renewable energy, IRENA endeavours to become a global catalyst in leveraging international cooperation and policy action related to innovative and cost-effective renewable energy policies, technologies and services. IRENA has a diverse membership. This diversity provides a wealth of experience, resources and specialist knowledge in all areas of renewables. There are great opportunities for IRENA to increase the cooperation with all countries and different stakeholders where lessons may be learned and replicated from one place to other parts of the world. Wider utilization of PV is one priority that the agency considers as a potential solution to the existential challenges that the global community faces today. Advances in PV technologies and dramatic cost reductions during the last few years now make PV an economically viable energy supply option, particularly in view of the large scale of its dissemination.

Large scale PV application is considered to be a key solution to overcome both climate change and energy issues. Additionally, a number of beneficial applications are possible for VLS-PV, including greening of desert regions and generating power for water supply (desalination) or other productive uses (agriculture, hydrogen production and so on) [4]. In addition, large scale PV dissemination will be a key to building local knowledge and human capacity as well as developing technical and industrial bases through successful technology transfer and as a result to trigger massive investments for large scale deployment of PV worldwide, particularly in developing countries.

IRENA has started developing scenarios and strategies, as well as technology roadmaps for different types of renewables. IRENA should have a strong role in supporting its member states in accelerating the large scale PV applications in different regions and countries and implementation of optimal policies able to trigger massive investments for deployment of VLS-PV power generation in different parts of the world, particularly in developing countries. IRENA is currently setting up partnerships and cooperation mechanisms with different stakeholders, and in particular with the industry and the finance communities. With the support of its member countries, and in partnership with the leading experts in the field, a regional scenario for Africa has been developed and concrete forward-looking strategies are proposed [19]. These are based on regional assessments of the readiness of countries to deploy renewables, and an in-depth knowledge of their technical potentials and current energy mix.

The results of the readiness assessments, as well as proposed scenarios and strategies will allow countries to formulate most effective policies needed. Once the policy framework is in place, the relevant financial instruments will be mobilized, through a dedicated interface with the finance community. The next stage, project development, requires mobilizing the industry, providing solutions to scale-up the local capacities, as well as a detailed understanding of the economic and financial conditions for developing sound projects. In regard to large scale dissemination of PV power generation systems, advanced technology provides potential solutions to the existential challenges that the global community faces today. With the right policies in place, the current transformation towards renewable energy can be equitably spread to all regions of the world where PV should increasingly contribute.

The key areas of international cooperation where IRENA should participate to promote VLS-PV worldwide are identified as follows:

- *Investment in research and development.* In order to advance the transition to a global energy system for sustainable development, it is very important to orient substantial and increasing investments towards the introduction of renewable energy. Research and Development is a major driving force for constant cost reduction in PV and consequently an enabler for the fast diffusion of PV.
- *Knowledge sharing and technology transfer.* Since industrialized developed countries with the exception of some of the newly emerging countries such as China, India, and Brazil are currently the leaders in PV technologies, it is important for developed countries to work with their counterparts in developing countries to advance the technologies in PV markets. There is an urgent need to accumulate and analyse operational experiences of existing MW-scale commercial PV plants and demonstration projects and share the results among countries and stakeholders. IEA PVPS Task8 activity and other IEA implementing agreements on related tasks are examples of global or regional initiatives that need to strengthen their collaboration as they monitor the development and implementation of the standardization in different countries and regions in different areas of the PV supply chain. IRENA and other organizations focusing on developing renewable energy could be used to help capacity-building efforts and to share lessons learned and experiences of VLS-PV development.
- *Government commitment and policies.* VLS-PV systems of several hundred MW or GW scale capacity will require increasing amount of investment. True government commitment and political will to promote wider development of PV with targeted policies at the governmental level can provide investors and the PV industry with confidence in their own investments and undertakings. That will then enhance market certainty, lower the investment risks, and support and ease access to capital. A key role of IRENA as the world's single international intergovernmental agency dedicated to promote and accelerate the transition to renewable energy is to facilitate implementation of an effective international policy framework to continuously increase the share of PV in worldwide electricity generation. There is an urgent need for individual governments to formulate appropriate and effective policies that promote large scale PV development under local conditions, by overcoming specific local barriers and capitalizing upon unique local opportunities. By linking national efforts to the broader international community, a country can reap the benefits of PV technology improvements and cost reductions at home, while contributing to the collective scale-up of large scale PV power generation worldwide.
- *Global atlas of solar resources.* The comprehensive data of available solar resources (atlas) as a global decision-making instrument is essential for the development of large scale PV power generation systems. This information combined with data on available land area and other meteorological parameters can direct and enhance cooperation on global scenarios and strategies on large scale PV development worldwide. Within the scope of the activity on developing of the Global Wind and Solar Atlas program, IRENA

has started development of the Global Atlas of Solar and Wind resources – a collaborative internet-based Geographic Information System (GIS) in cooperation with UNEP, WMO and other leading research centres [20]. The solar resource data provided by the Atlas can be used as a trusted data source for large scale PV market prospection, by cross-checking the PV potential and the available socio-economic conditions, laws and regulations, and policy data including up-to-date financial support and tax incentives. Data can also be used to identify the potential of different PV technologies and their most suitable locations as well as for policy making or defining support schemes to provide confidence and stability to the private sector for investing in large scale PV development.

- *Development of international standards.* International collaboration on standards is crucial to ensure that the needs of various countries and regions are included in the development of large scale PV power generation systems and to reduce repetition and overlap in the development of standards. Development of improved universal design standards for large scale PV system components, installation procedures and streamlined installation and O&M services can increase competition, accelerate innovation and reduce costs. Several countries, organizations, and research groups are involved in developing related standards and already working to harmonize existing standards at a national or regional level. But these activities and other aspects related to developing standards need to be shared globally. One of the key tasks of IRENA will be to support governments in sharing experiences on developing internationally accepted technology standards, and standardization is in varying stages of development to ensure uptake of PV technologies across the world.
- *Development of smart grids.* Smart grids and large scale electricity storage are key enablers of integrating VLS-PV into the power generation system. In order to enhance further deployment of large scale PV power generation systems, smart grids need to be rapidly developed and widely deployed. Many countries have made significant efforts to develop smart grids, but the lessons learned are not being shared in a coordinated fashion. Major international collaboration is needed to enhance further deployment of large scale PV power generation systems, especially in utility standards, policy, regulation and business model development.

These are the major themes for international cooperation identified as key activities that can further advance international collaboration to stimulate the global deployment of large scale PV power generation systems.

12.5 CONCLUSION

In more than 12 years operation of the IEA PVPS Task8 study, a strong and efficient network has been established within research, industry and utility contributing countries. Currently, there is no other research network in the field of VLS-PV technologies. IEA PVPS Task8 activity study team work has been effective in achieving its purposes: maintaining the PV community throughout the world by facilitating exchange of information and experience, and promoting an awareness of VLS-PV technology and potential worldwide. Active international collaboration during the last decade within the framework of the IEA PVPS Task8 activities has enabled significant progress in deepening of knowledge concerning technological, economic and environmental issues related to potential deployment of the VLS-PV concept worldwide.

In the near future, VLS-PV systems should become an economically viable option to meet the electricity needs of communities in remote desert regions around the world where conventional power plants cannot be built. The rate of deployment of VLS-PV is greatly influenced by the perception of the general public and utilities, local, national and international policies, as well as the availability of suitable standards and codes to govern it. Among a variety of support mechanisms, feed-in-tariffs (FITs) could be the main reason behind the strong growth in large scale PV in developed countries.

In the long term, with a solid strategy for implementing VLS-PV systems for the northeast Asian region, further expansion of large scale PV generation can satisfy not only the increasing energy demand of the northeast Asian region in a sustainable way but also will stimulate the local PV industry and create enough need to justify local manufacturing of PV system components. In January 2012, the International Energy Agency (IEA) and the International Renewable Energy Agency (IRENA) signed a partnership agreement to strengthen cooperation between the two organizations. Although the agencies had collaborated since IRENA's inception in 2009 – including developing a joint approach on renewable energy statistics – this agreement launched a number of new initiatives, such as a joint database of renewable energy policies. There will also be increased collaboration between the two agencies at the Secretariat level, and in energy technology networks, including the IEA's Multilateral Technology Initiatives, e.g. Implementing Agreement such as PVPS.

Some themes for international cooperation are identified in this chapter. The agreement would further advance international collaboration to stimulate the global deployment of large scale PV power generation systems.

REFERENCES

1 IEA (2011), *World Energy Outlook 2011*, Special Report, IEA/OECD, Paris, France.

2 IEA (2003) Implementing Agreements: Background and Framework as of 2003, IEA, Paris, France.

3 Kurokawa, K. *et al.,* (2003) *Energy from the Desert: Feasibility of Very Large Scale Power Generation (VLS-PV) Systems*, James and James.

4 Kurokawa, K. *et al.*, (2006) *Energy from the Desert: Practical Proposals for Very Large Scale Photovoltaic Systems*, Earthscan.

5 Komoto, K. *et al.*, (2009) *Energy from the Desert: Very Large Scale Photovoltaic Systems: Socio-economic, Financial, Technical and Environmental Aspects*, Earthscan.

6 EPIA (2011) *Global Market Outlook for Photovoltaics until 2015*, EPIA.

7 PV Power Plants (2011), *Industry Guide*, Solarpraxis, Berlin, Germany.

8 EEG Government of Germany (2000) Erneuerbare-Energien-Gesetz, Berlin.

9 Breyer, Ch. and A. Gerlach (2011) 'Global overview on grid-parity event dynamics' *Progress in Photovoltaics: Research and Applications,* 19 (7) Special Issue 25th EU PVSEC, WCPEC Valencia Spain.

10 Photon Consulting (2011) *True Cost of Solar Power: The Pressure's On*. Boston, USA.

11 Solarbuzz (2011) *Annual World PV Market Review*.

12 Wiser, Ryan, *et al.*, (2011)*Tracking the Sun II: The Installed Cost of Photovoltaics in the U.S. from 1998–2010*, Lawrence Berkeley National Laboratory, Environmental Energy Technology Division.

13 Asia Solar Energy Initiative (2011) *A Primer*, April 2011, Asian Development Bank.

14 Australian Government (2011) Large Scale Solar Deployment in Australia, Discussion Paper, Enhancing Australia's Economic Prosperity, .

15 Jawaharlal Nehru National Solar Mission (2010) Towards Building SOLAR INDIA.

16 Grau, Thilo, Molin Huo, and Karsten Neuhoff (2011) *Survey of Photovoltaic Industry and Policy in Germany and China*, CPI Report, Climate Policy Initiative, DIW Berlin and Tsinghua University.

17 Dagvadorj, D., L. Natsagdorj, J. Dorjpurev, and B. Namkhainyam (2009) *Mongolia Assessment Report on Climate Change*, MARCC 2009, Mongolia.

18 International Renewable Energy Agency (IRENA) (2011) *Bulletin*.

19 International Renewable Energy Agency (IRENA) (2011) *Prospects for the African Power Sector: Scenarios and Strategies for Africa Project*.

20 International Renewable Energy Agency (IRENA) (2012) *Global Atlas for Solar and Wind Energy Proposal for the Implementation*, Summary Paper.

CHAPTER THIRTEEN

Possible contribution of VLS-PV to sustainability

13.1 OBJECTIVE

According to the United Nations, the global population will reach 9,3 billion (approximately 35% increase from that today) in 2050 and continues to increase until 2100 [1]. The Asian Development Bank (ADB) estimates that global GDP will expand to 292 trillion USD in 2050 from 63 trillion today [2].

Those changes in demography and economic activity are anticipated to have negative impacts on the global environment. The energy consumption will increase considerably and, naturally, so will associated greenhouse gas emissions. Although the impacts of climate change cannot be predicted precisely, various sectors, such as water, ecosystem, food and health are projected to have negative impacts as the temperature increases [3].

Even if the impacts of climate change on food or water are not as serious as anticipated, population increase and economic growth will demand more food and water consumption. In order to fulfill the food demand, input of resources such as water and land for production may also increase.

This study is trying to answer the following key question: *To what extent can PV/VLS-PV contribute to these sustainability issues?*

In order to answer the question, a simplified model was developed. The model looked carefully at the relations between various sustainability factors to keep the consistency among those factors.

It should be noted that the aim of this study is not to predict the future precisely but to roughly understand the effect of PV/VLS-PV. Therefore, the model neither developed new methodologies for the analysis in academic terms nor employed optimization frameworks. Instead, relevant coefficients or econometric equations which have been used in existing studies are applied where necessary.

13.2 METHODOLOGY AND BASE SCENARIO ANALYSIS

13.2.1 Model structure

The conceptual structure of the model is presented in Figure 13.1. The main drivers of the future changes are GDP and population. The model is composed of five sub-modules, namely energy supply module, water consumption module, food supply and demand module, land-use module, and greenhouse gas emission module. The details of each module are described below.

13.2.2 Driving forces

IPCC SRES B2 was used as datasets for the drivers [4]. In the SRES B2 scenario, the world population is expected to grow continuously until 2100 and reach 10 billion. The increasing rate is very close to the latest "World Population Prospects" from the United Nations. In this scenario, the current level of GDP is projected to increase 3-fold by 2050 and 8-fold by 2100. The downscaled data presented from the Center for International Earth Science Information Network (CIESIN), Columbia University was used for the analysis [5] (see Figure 13.2).

13.2.3 Energy supply module

Projection of energy supply is relatively complicated in general. However, the approach used in this study is simple. It is well known that energy intensity (energy supply per GDP) decreases as economic status (per capita GDP) improves. Figure 13.3 shows the historical change in the energy intensity as a function of per capita GDP in various regions. As shown in Figure 13.3, energy intensity

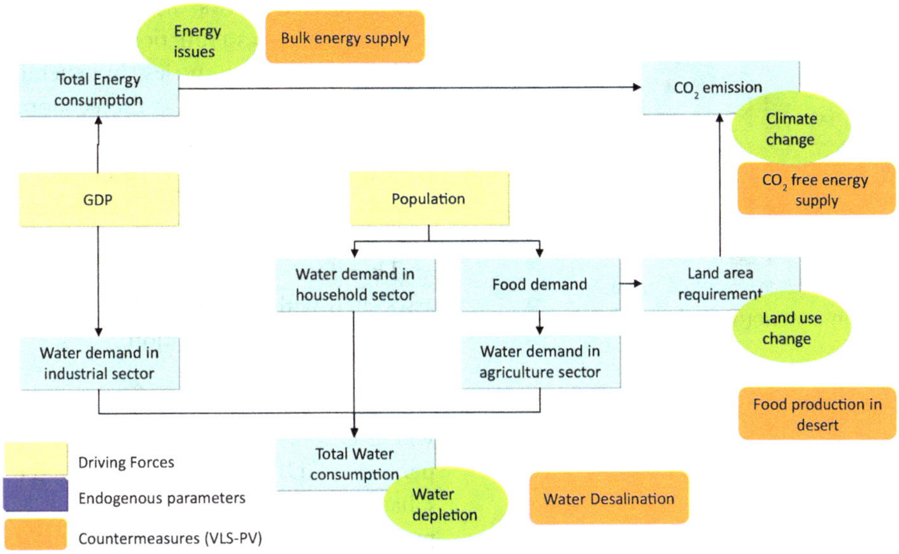

Figure 13.1 Model structure

decreases drastically when per capita GDP is under 5 000 USD, but the improvement rate becomes moderate after it passes the threshold. Taking this factor into consideration, change in energy intensity is calculated as follows,

$$EI = \alpha GDPC^{\beta}$$

where EI is energy intensity, GDPC is per capita GDP, and α and β are constant coefficients.

Figure 13.4 shows a shift in energy intensity by region during the projected period for the base scenario. As shown in the figure, the regional gap in the energy intensity should become smaller in the future.

Another important factor which affects the future sustainability issue is change in energy mix. CO_2 emission per unit of energy consumed varies greatly by energy sources. Coal has the highest emission factor, followed by oil and natural gas. In the study, the share of each energy source was kept identical to the one in the SRES B2 scenario.

Using the model above, the future changes in regional and global primary energy supply were estimated. As presented in Figure 13.5, the total primary energy supply is projected to increase continuously until 2100 and reach 45 800 Mtoe, which is approximately four times the current primary energy supply. The increasing rate is moderate compared with that of GDP, in other words, decoupling of energy and economy is incorporated in the scenario through the intensity improvement.

High growth rates are observed in Africa, India, and China. The share of energy supply from those regions becomes 36% and 45% in 2050 and 2100 respectively, from 25% today.

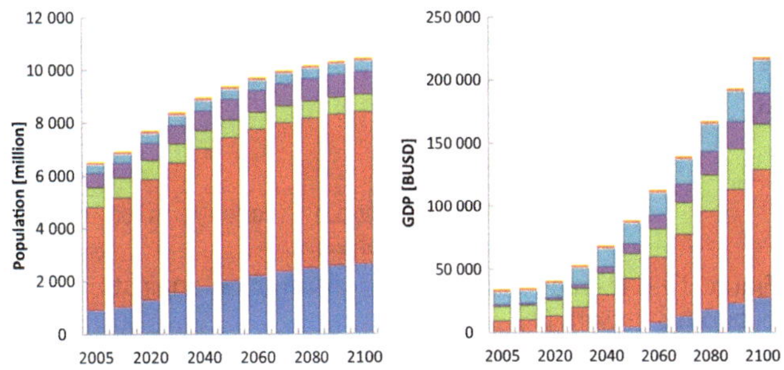

Figure 13.2 Population (left) and GDP (right) in the IPCC SRES B2 scenario [5]

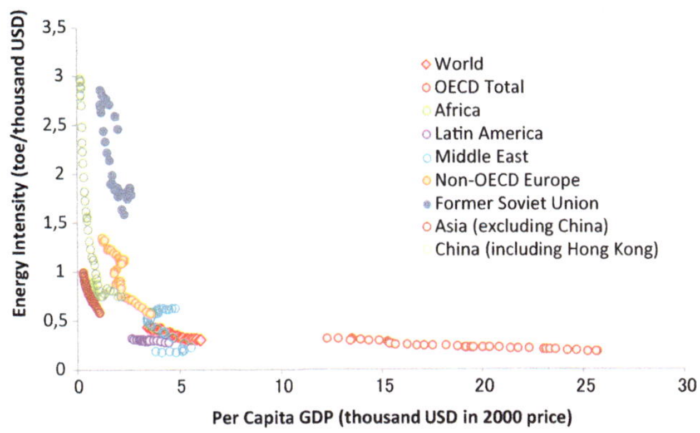

Figure 13.3 Relationship between energy intensity and per capita GDP [6

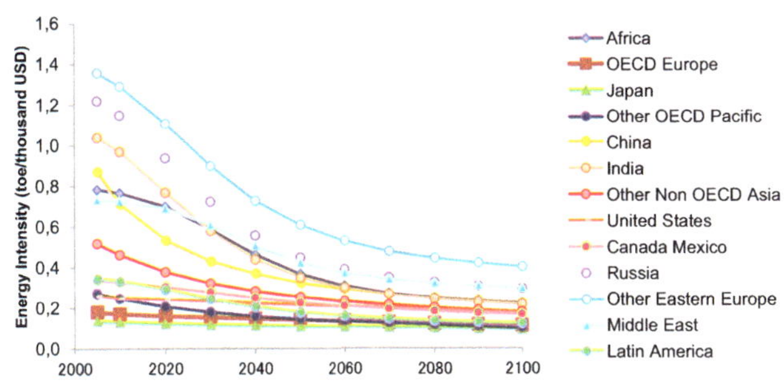

Figure 13.4 Changes in energy intensity by region

13.2.4 Food production and consumption

Production of food requires many resources as inputs, such as water, land, labor, machinery, fertilizers, pesticides,

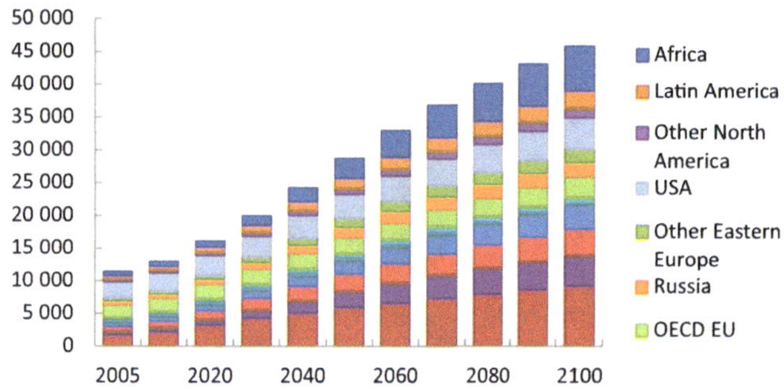

Figure 13.5 Projected primary energy consumption by region (Mtoe)

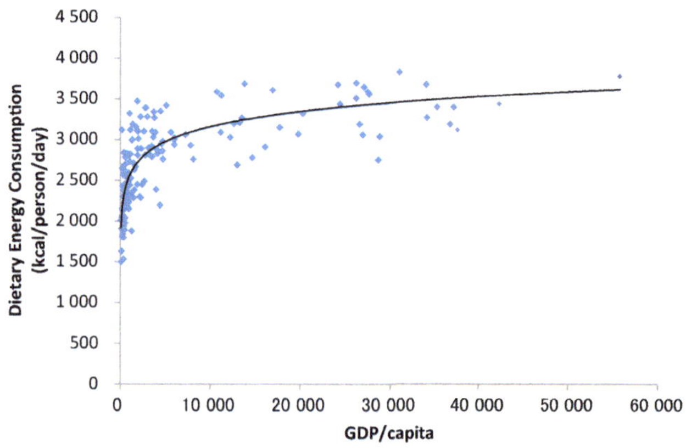

Figure 13.6 Estimation of dietary consumption against economic growth [7]

energy and so on. Clearly, as the food demand increases, the resources extracted from nature to fulfill the demand also increase.

The projection of food production in the future can be influenced by various parameters; however, the main driver would be food demand. Changes in economic status and food preference of the people as well as population increase are expected to have great impacts on food demand and production.

Figure 13.6 shows the changes in per capita calorific value consumed as a function of economic status (per capita GDP) in various countries. In general, the dietary energy consumption increases as the economic status improves. The phenomenon is more evident when per capita GDP is relatively lower. When the dietary energy consumption reaches approximately 3 000 kcal/person/day, the increasing rate becomes moderate.

Therefore changes in dietary energy consumption associated with economic development can be calculated using the following equation,

$$DEC_i = \gamma \ln(GDPC_i) + \delta$$

where DEC is dietary energy consumption and GDPC is per capita GDP in each country. γ and δ are constants. Changes in food preference also influence the food production. The more people eat meat, the more calories need to be produced since animal fat production requires much more input as feed.

In order to include this factor, the model incorporates regional food balance sheets presented by FAO [7]. The food balance sheets are a very powerful tool to calculate impacts of changes in dietary patterns on food production demand. The results are shown in Figure 13.7.

13.2.5 Water consumption module

AQUASTAT, statistics on water resources and consumption provided by FAO [8], was used as an original dataset for the water consumption module. The water consumption analysis can be split into three sectors, namely agriculture, municipal and industry sector.

Since historical data on water consumption is quite limited, it is difficult to simulate future water consumption from historical data. However, a few studies have developed

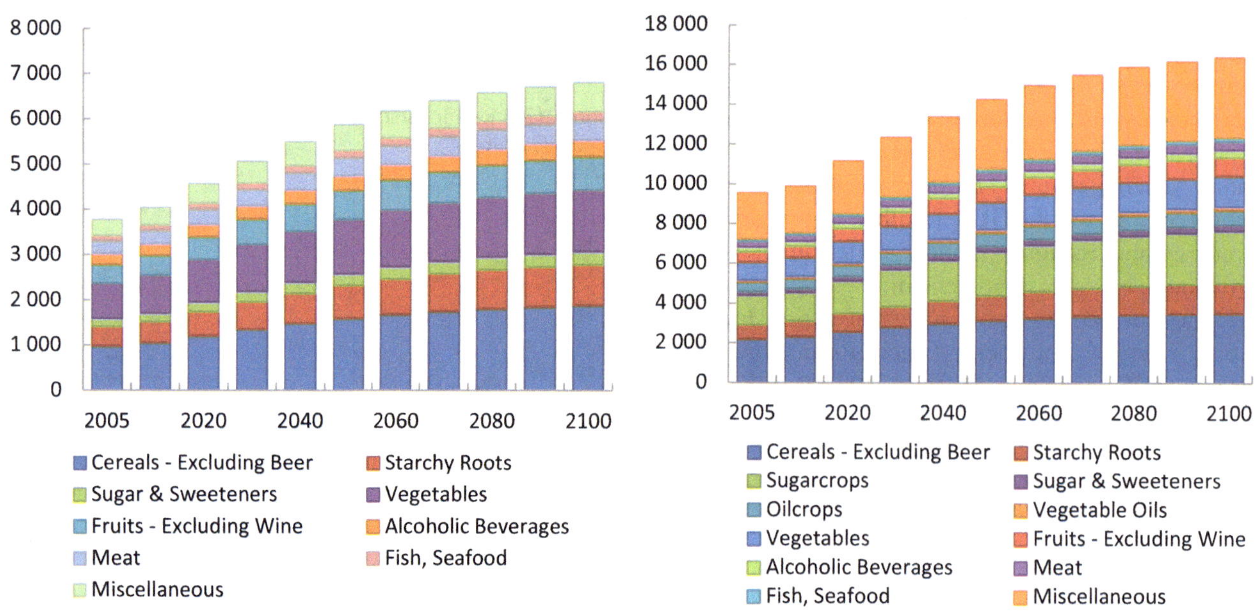

Figure 13.7 Food demand (left) and production (right) projections in the base scenario (million tonnes)

methodologies to estimate future water use in each sector. This study follows some of those methodologies.

INDUSTRIAL WATER CONSUMPTION

For the industrial sector projection, equations proposed in a existing study [9] were applied in this study.

According to the paper, industrial water consumption (Q_{ind}) can be estimated by

$$Q_{ind,t} = Q_{ind,t0} \times \frac{ELC_t}{ELC_{t0}} \times \frac{EI_t}{EI_{t0}}$$

where Q_{ind} is industrial water consumption, ELC is electricity consumption, EI is energy intensity (energy consumption per unit GDP), and t_0 and t are suffixes for the base year and the estimated year respectively. Electricity consumption and energy intensity values are calculated based on the energy projection presented above; hence the sub-modules are consistent with each other.

MUNICIPAL WATER CONSUMPTION

For municipal water consumption, it is assumed that water consumption in developing countries will follow the consumption pattern of developed countries as their economy grows. Additionally, the consumption will be heavily dependent on country-specific climatic conditions such as water resource availability. Taking these factors into consideration, municipal water consumption was estimated by

$$\ln(QC_{mun,t,i}) = \varphi \ln(GDPC_{t,i}) + \omega_i$$

where QC_{mun} is per capita municipal water consumption and GDPC is per capita GDP in country (i). φ is a constant whilst ω is country specific parameter. The parameter ω is set by taking the difference between the simulated value and statistical data in the base year.

AGRICULTURAL WATER CONSUMPTION

The agricultural sector consumes approximately 70% of the current global water consumption. The consumption is heavily dependent on factors such as water productivity and food production volume. The water productivity is water demand to produce a unit of the crop in question, and is influenced by applied irrigation technologies and soil conditions.

Regarding the volume of crop production, data calculated from food production and consumption data was used as an input for the model in order to keep consistency among the sub-modules. Agricultural water consumption can be calculated as

$$Q_{agr,t} = WP_{i,j,t} \times P_{i,j,t}$$

where Q_{agr} is water consumption from the agricultural sector, WP is water productivity and P is production volume. Suffixes *i, j, t* indicate country, crop, and time.

SUMMARY OF THE OUTPUT

Using the methodologies above, future water consumptions were simulated as shown in Figure 13.8. The total water consumption is estimated to reach approximately 7 350 billion m^3 in 2050 and 9 760 billion m^3 in 2100. Industry and municipal sectors consumption increase drastically; however, the agricultural sector still consumes more than half of the total consumption. From a regional perspective, Asia dominates the consumption demand, while Africa also increases its consumption.

13.2.6 Land use change

Increase in food demand and agricultural production may demand more land area for production. If forest lands are exploited in order to expand agricultural land, it may affect the local environment such as biodiversity

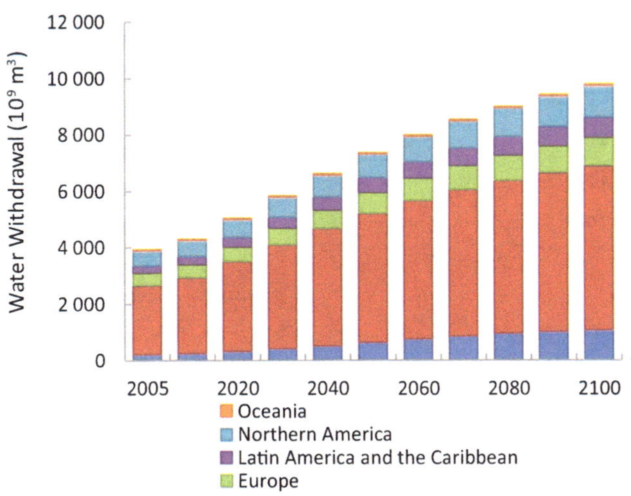

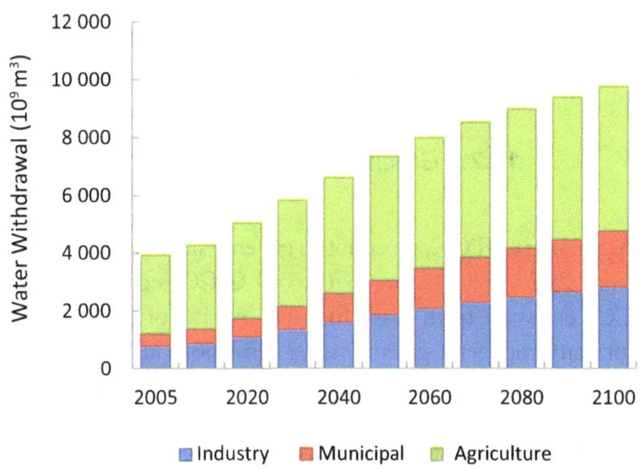

Figure 13.8 Water consumption by region (left) and by sector (right)

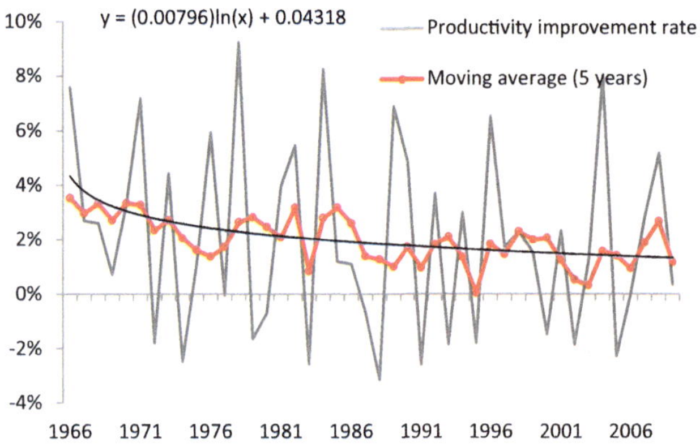

Figure 13.9 Historical changes in land productivity for cereals (global average) [10]

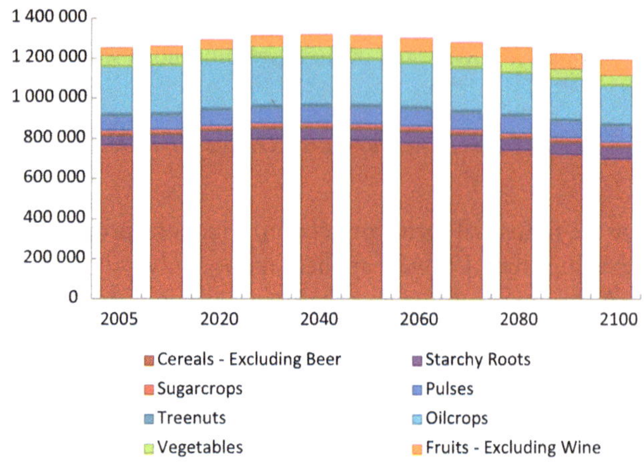

Figure 13.10 Arable land use by products (1 000 ha)

and water resources. In some cases, the land use change may lead to more severe impacts such as desertification through erosion. In addition, land use change from forest land to agricultural land will emit carbon dioxide to the atmosphere, and hence accelerate climate change.

On the other hand, as production technologies in agriculture improve, land productivity would also improve, that is, land input for unit of agricultural production decreases. Figure 13.9 shows the historical change in land productivity.

Taking into account the land productivity improvement rate, the land requirement for food production is calculated by

$$A_{i,j,t} = \frac{P_{i,j,t}}{LP_{i,j,t}}$$

$$LP_{i,j,t} = LP_{i,j,t0} \times \prod_{k=0}^{t} (1 + PI_k)$$

$$PI_t = \varepsilon \times \ln(t) + \theta$$

where A is agricultural land area used for productions, LP is land productivity, PI is productivity improvement rate. ε and θ are constants.

Figure 13.10 illustrates the agricultural land use change associated with this scenario. In spite of the increase in food demand, the arable land use decreases. The results imply that the change in land use will be a carbon sink rather than carbon emission source after 2050.

13.2.7 Greenhouse gas emissions

According to IPCC, greenhouse gas emissions have increased from 28,7 $GtCO_2$-eq in 1970 to 49 $GtCO_2$-eq in 2004 [3]. CO_2 emission from fossil fuel use consists of 56,6% of the total anthropogenic greenhouse gas emissions in 2004, and the remainder are CH_4 (14,3%), N_2O (7,9%), F-gases (1,1%), and CO_2 released from non-energy sources (18,3%).

The future CO_2 emission from energy use can be calculated simply by multiplying emission factors by consumption of each fuel.

CO_2 emissions from land use and land use change and forestry (LULUCF) was simulated from the additional demand for arable land, which was calculated from the food and demand projection module. The assumption underlying the calculation is that forest lands were used for the additional arable land demand. The average carbon stock in the forestry was assumed using the data on “global forest resources assessment”.

The simulated results of CO_2 emissions from land use change are given in Figure 13.11. The red line in the figure represents the result while the gray line is the emission scenarios from other studies presented in the fourth assessment report of IPCC [3] and other sources [11].

In order to analyze the impact of emissions on climate change, the climate model SEEPLUS (Central Research Institute of Electric Power Industry), was applied [12]. The SEEPLUS model is a web application tool developed by the Central Research Institute of Electric Power Industry in Japan. The tool allows users to simulate the CO_2 concentration in the atmosphere for defined scenarios. The CO_2 emissions after 2100 are assumed to decrease at the rate of 1%/year.

Figure 13.12 shows the simulated results from SEEPLUS based on base scenario emissions. As indicated in the figure, the CO_2 concentration in the atmosphere becomes higher than 900 ppm at the stabilization stage. The simulated temperature change associated with the emissions scenarios indicates that the temperature rise will

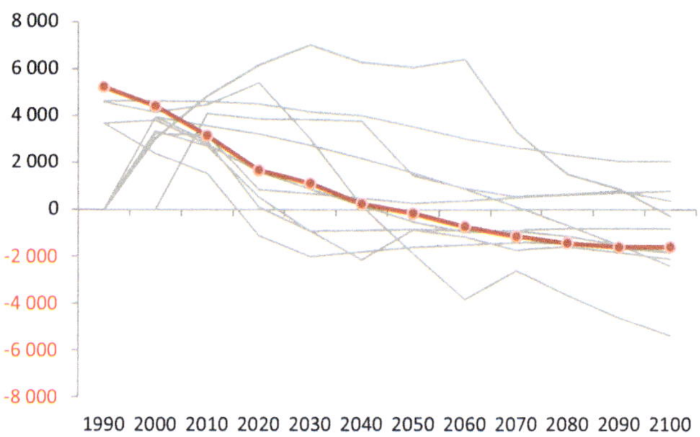

Figure 13.11 Greenhouse gas emissions from LULUCF sector (Gt-CO_2 eq)

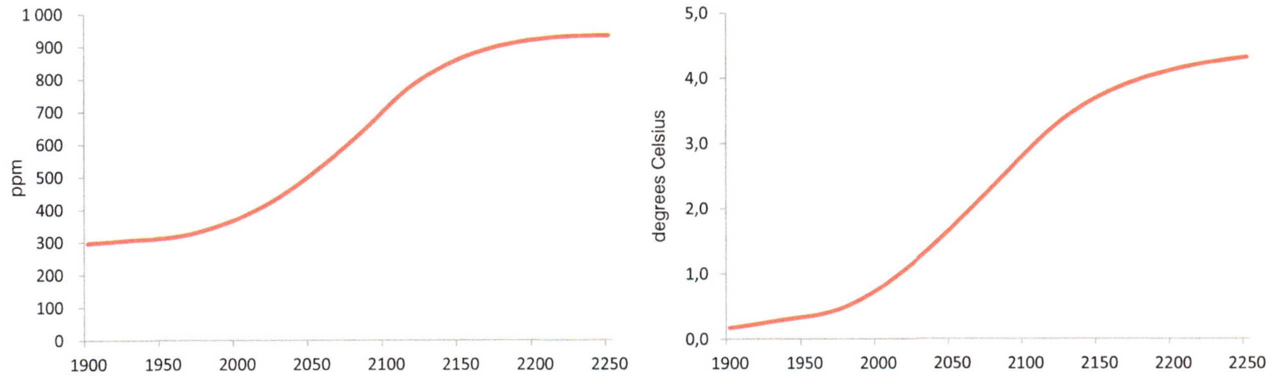

Figure 13.12 Changes in CO_2 concentration in the atmosphere (left) and global mean temperature (right) [12]

reach 2 degrees Celsius in 2050 and continues to increase further to 5 degrees Celsius in the long-term future.

13.3 SCENARIO ANALYSIS

The future projection of the model presented above is the "base scenario" in this study. Using the model above, three additional scenarios (countermeasures scenarios), namely SC-1, SC-2, and SC-3, were developed to evaluate the impacts of VLS-PV. For all three scenarios, a boost in VLS-PV was assumed. The installation scenario of VLS-PV was identical to the VLS-PV roadmap presented in the previous *Energy from the Desert* book [13] (Figure 13.13).

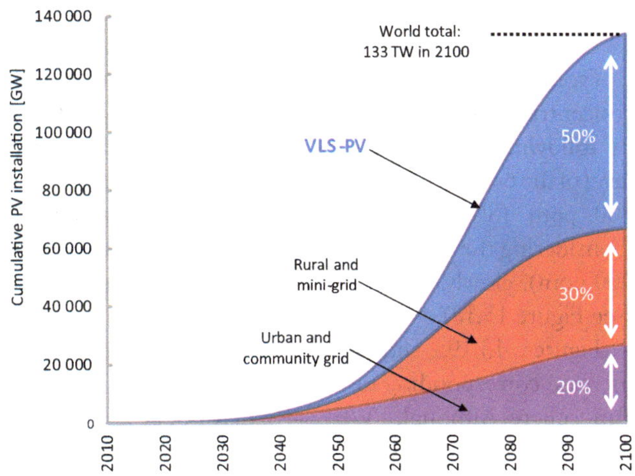

Figure 13.13 VLS-PV roadmap [13

13.3.1 Countermeasures scenarios

The countermeasures scenarios are summarized as Table 13.1.

In the SC-1 scenario, power generated from PV/VLS-PV is used directly in the energy mix. The remaining energy demand is supplied by other energy sources but the share of other energies kept the same as the base scenario.

In SC-2, on the other hand, electricity from PV/VLS-PV replaces only that from fossil fuels and other renewable energy and nuclear do not change. The aim of this scenario is to explore the enhanced CO_2 reduction potential from PV/VLS-PV.

SC-3 assumes a completely different application of VLS-PV. Power generated from PV is used for water desalination or purification. If the power generated from PV/VLS-PV is larger than the demand for water desalination, the excess electricity is fed into the grid.

Table 13.1 Summary of the scenario descriptions

VLS-PV scenario	Description
BaU	Business as usual. It is assumed that the current trend continues until 2100
SC-1	Power generated from PV/VLS-PV is used directly in the energy mix. All kinds of energy are replaced by VLS-PV
SC-2	Power generated from PV is used directly in the energy mix. Only fossil fuels are replaced by PV/VLS-PV power generation
SC-3	Power generated from PV is used for water production (reverse osmosis). The remaining energy is used for power supply

13.3.2 SC-1 and SC-2 scenarios in comparison with base scenario

With the increase in PV/VLS-PV, consumption of other energy resources decreases considerably. As is shown in Figure 13.14, PV/VLS-PV can provide 25% of primary energy in 2100. The contributions of PV/VLS-PV in energy saving for coal, oil and natural gas are approximately 2 600 Mtoe, 450 Mtoe, and 2 900 Mtoe in 2100 respectively.

Cumulative CO_2 emission reductions in the projected period reach 531 Gt CO_2-eq in 2100, which is equivalent to 19,5 times today's annual emissions (see Figures 13.15 and 13.16).

If electricity from VLS-PV is substituted only by fossil fuel, the effect of GHG emission savings would be even higher. SC-2 assumes only fossil fuels are replaced by PV/VLS-PV and other energy sources such as hydro, geothermal, biomass and waste, and nuclear remain the same as the base scenario. In this case, the cumulative GHG

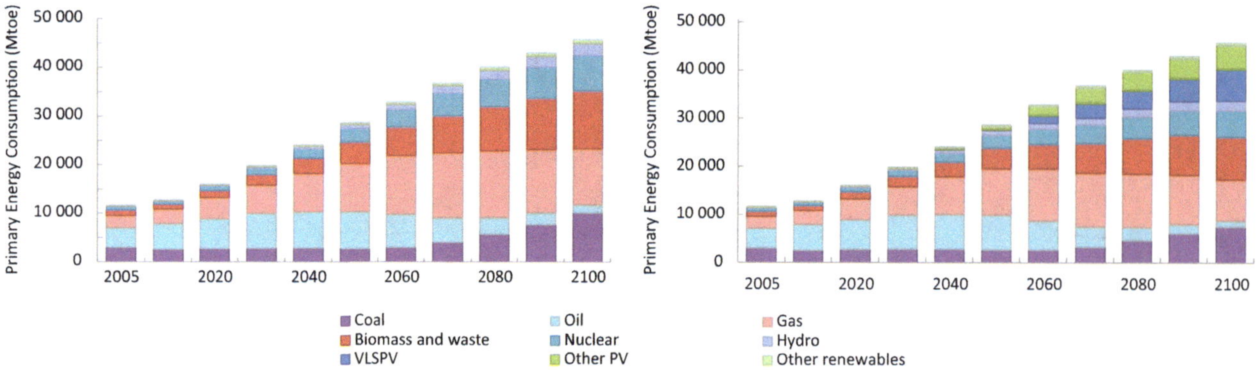

Figure 13.14 Primary energy consumption of base scenario (left) and SC-1 scenario (right) [Mtoe]

emission saving is increased to 929 Gt-CO_2, approximately 34 times current annual emissions (see Figure 13.17).

From the results above, it is clear that the CO_2 concentration in the atmosphere would be greatly reduced by following the PV/VLS-PV roadmap. The stabilization level of the two countermeasure scenarios are approximately 800 ppm (SC-1), and 650 ppm (SC-2) respectively. Considering the change from the base scenario (more than 900 ppm), clearly, the impact of PV/VLS-PV is enormous (see Figure 13.18).

Figure 13.19 shows the projected temperature increase corresponding to the GHG concentration on the atmosphere. Although PV/VLS-PV may not be a single and perfect solution for the climate change issues, it is clearly shown that the potential contribution is considerable. Combining with other renewable energy technologies or energy efficient measures would be necessary for a sustainable future.

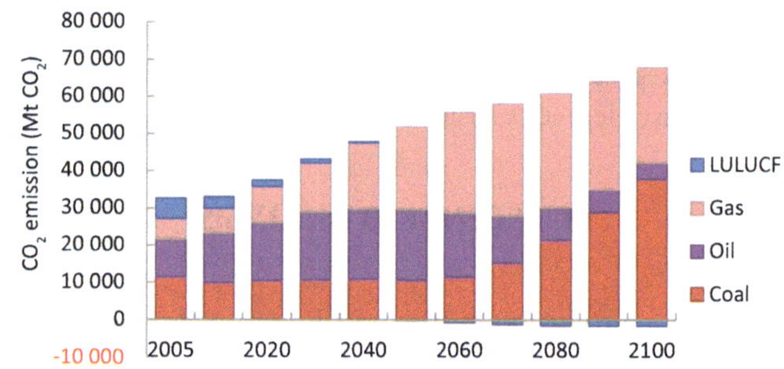

Figure 13.15 GHG emissions of base scenario [Mt-CO_2eq]

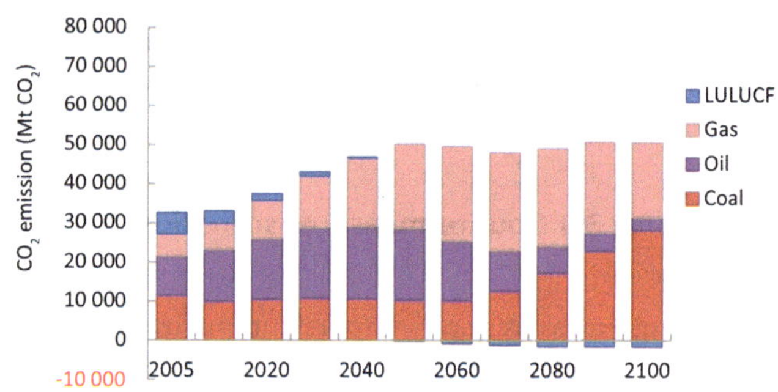

Figure 13.16 GHG emissions of SC-1 scenario [Mt-CO_2eq]

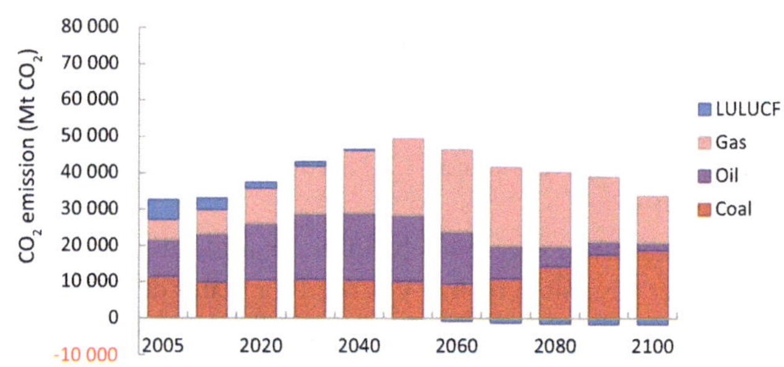

Figure 13.17 GHG emissions of SC-2 scenario [Mt-CO_2eq]

13.3.3 Base scenario versus SC-3 scenario

The SC-3 scenario, on the other hand, is unique. Under this scenario, the power generated from VLS-PV is used for water production. Reverse osmosis (RO) technologies would be used in order to fulfill the increasing global water demand. The overall concept of the scenario is to produce water in the most arid and solar resource rich areas, the deserts. The remaining energy is used for power supply. Only VLS-PV in desert regions is used for desalination applications. Other applications such as rooftop PV in urban areas or PV for rural electrification are for conventional electricity use.

Figure 13.20 illustrates the additional water demand projected in the base scenario and water production potentials from VLS-PV. The water supply potential from VLS-PV will be enough to fulfill the additional global water demand in 2060 and will be much higher afterwards. The potential will be more than 3,4 times in 2100.

Even if the priority of the VLS-PV power use in this scenario is water desalination and only surplus energy is used for electricity supply, the energy consumption and GHG emission reduction is still substantial. Figure 13.21 shows the primary energy mix in the SC-3 scenario. In this scenario, the share of PV/VLS-PV in the primary energy mix would be 22%.

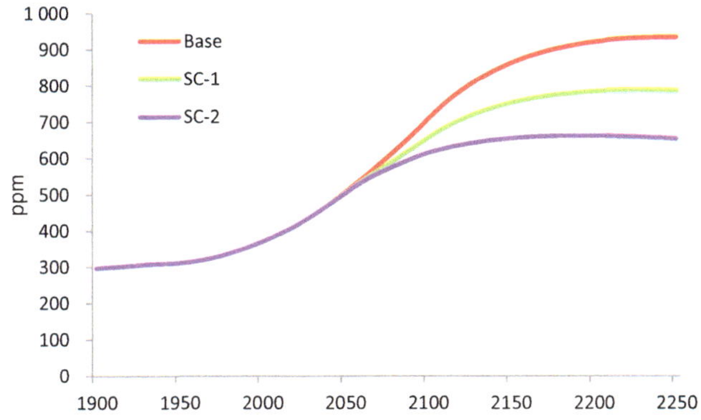

Figure 13.18 CO_2 concentration in the atmosphere of the three scenarios

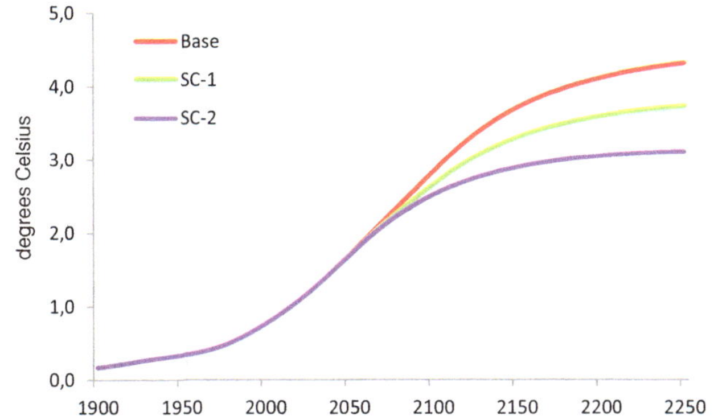

Figure 13.19 Temperature change caused in the three scenarios

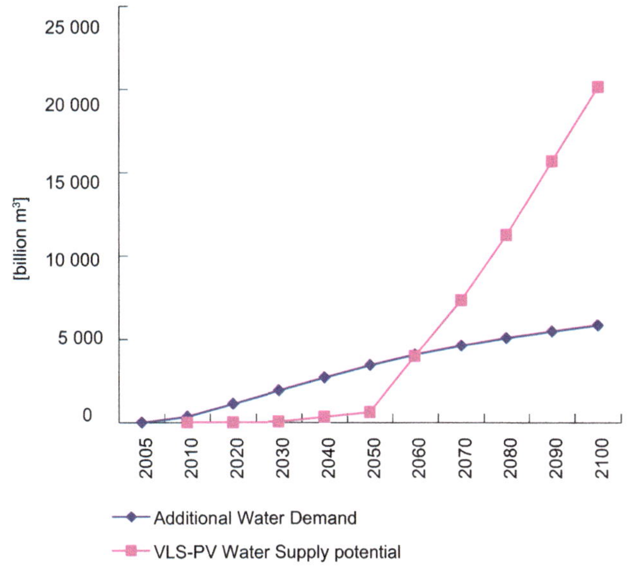

Figure 13.20 Water demand and supply capacity (billion m³)

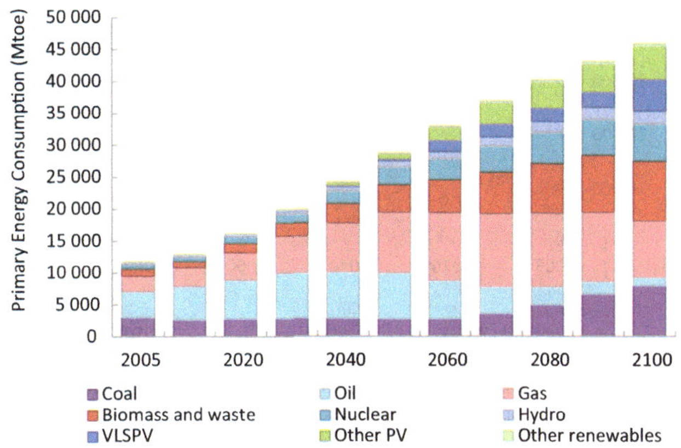

Figure 13.21 Primary energy consumption by energy type (SC-3 scenario)

13.4 IMPACTS OF DRIVING FORCES ON THE ANALYSIS

Our future society is full of uncertainty. The analysis above shows the great potential of VLS-PV; however, the question still remains: what if the driving forces such as population and economy are different in the future? To answer this question, a sensitivity analysis was implemented.

Figure 13.22 shows driving force indices, e.g. population and economic growth rate, of IPCC SRES A1 and B2 scenarios. IPCC SRES A1 scenario was used for an alternative driving force. In the scenario, the population reaches a peak in 2050 and starts decreasing thereafter. On the other hand, the economic growth rate is higher than in B2 scenario. The economic size in 2100 is more than double of that in B2 scenario and approximately 17 times higher than the current economy.

Figure 13.23 shows the primary energy consumption for the SC-1 scenario in which IPCC SRES A1 driving forces are applied. The impacts of lower population and larger economy offset each other, and as a result, the primary energy consumption is approximately 25% higher in A1 than B2.

On the other hand, the CO_2 emissions have a completely different trajectory as shown in Figure 13.24. The emission in 2050 is higher in A1 than B2, however in 2100, the emission in A1 becomes lower than B1.

One of the reasons is the change in share of energy mix. The share of CO_2 free energy such as nuclear power and biomass energy was increasing gradually, whilst the share of coal decreases in the A1 scenario.

Another remarkable change is derived from the LULUCF sector. The arable land area needed for food production is greatly decreased in A1 because of the land productivity improvement and population decrease.

13.5 CONCLUSIONS

In this study, a simplified model for sustainability issues was developed in order to evaluate the impacts of VLS-PV on the issues. The findings and conclusions of the study include:

- VLS-PV can contribute greatly to global sustainability issues, but the main contribution comes after 2050.
- The contribution of VLS-PV is heavily dependent on the BaU scenario assumed. Energy efficiency and other renewable measures are also important.

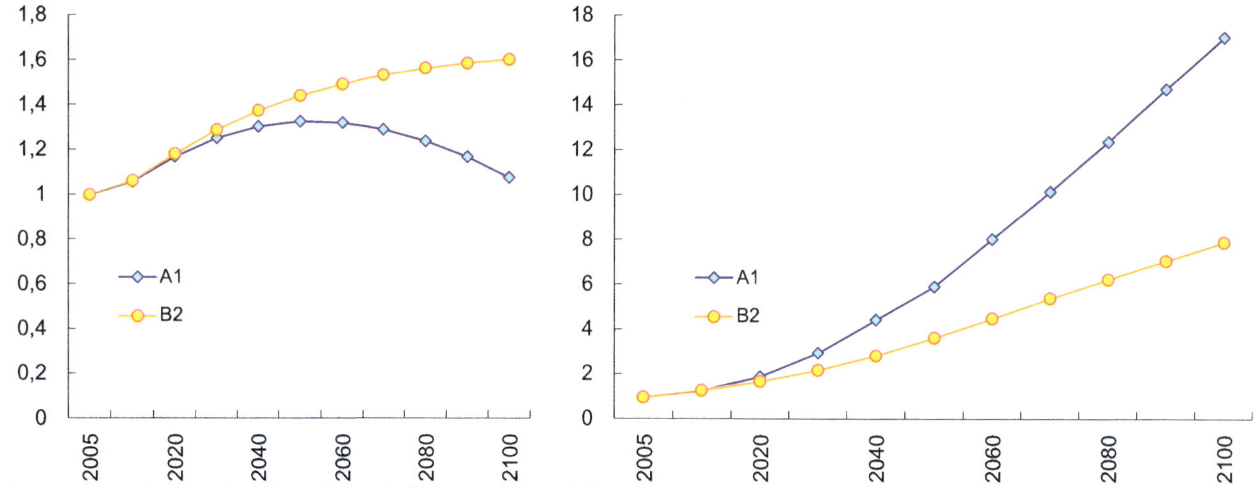

Figure 13.22 Changes in driving forces index: population (left) and economic growth rate (right) (2005 = 1,00) for the IPCC SRES A1 and B2 scenarios [5]

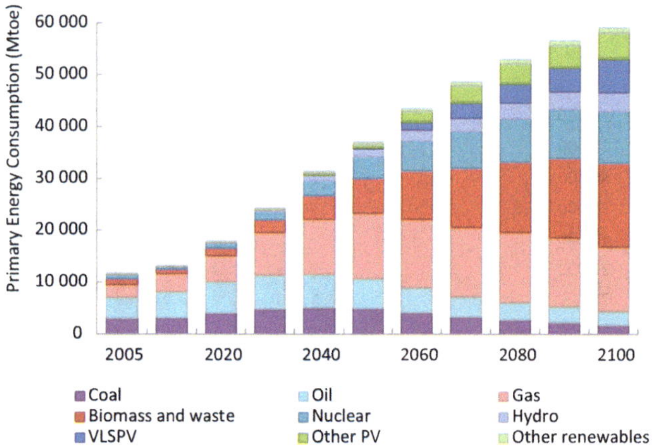

Figure 13.23 Primary energy consumption by energy (scenario: SC-1, driving forces: A1)

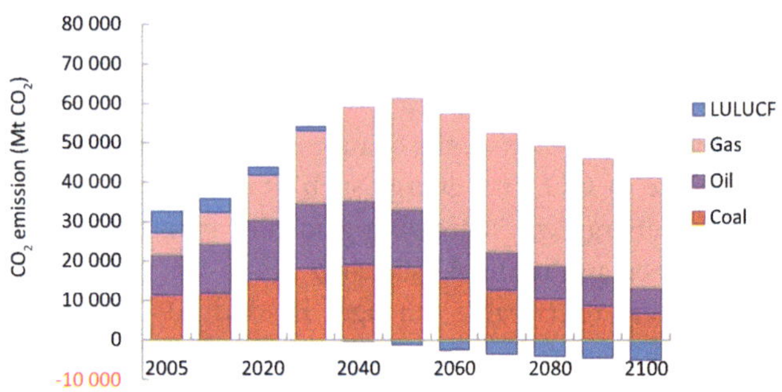

Figure 13.24 CO_2 emissions of SC-1 scenario with A1 driving forces

- Taking into account the energy depletion issues (increase in energy cost) in the timeframe, VLS-PV could be a cost-effective option, although it still needs continuous improvement.
- It should be kept in mind that such innovative changes in the energy supply system require a certain lead time. Also, delay in actions would lead to more serious issues since the driving forces are growing.
- Although the contribution of VLS-PV would be marginal in the next 10 years, actions in the first 10 years would be very important from the sustainability perspective.

REFERENCES

1 United Nations. (2011). *World Population Prospects – The 2010 Revision*. Retrieved August 2011, from http://esa.un.org/unpd/wpp/index.htm

2 Asia Development Bank. (2011). *Asia 2050 – Realizing the Asian Century*.

3 IPCC. (2007). *Climate Change 2007 Impacts, Adaptation and Vulnerability*.

4 IPCC. (2000). *Special Report on Emission Scenarios*.

5 Center for International Earth Science Information Network (CIESIN). (2002). *Country-level GDP and Downscaled Projections based on the A1, A2, B1, and B2 Marker Scenarios, 1990–2100*. Retrieved June 2010, from http://ciesin.columbia.edu/datasets/downscaled/

6 IEA. (2010). *Energy Balances of Non-OECD Countries*.

7 FAO. (2010). *FAOSTAT Food Balance Sheets*. Retrieved June 2010, from http://faostat.fao.org/

8 FAO. (n.d.). *Aquastat Website*. Retrieved October 2011, from http://www.fao.org/nr/water/aquastat/dbase/index.stm

9 Shen, Y., T. Oki and N.K.S. Utsumi (2010). Projection of future world water resources under SRES scenarios: water withdrawal/Projection. *Hydrological Science Journal, 53:1* 1–33.

10 FAO. (n.d.). *FAOSTAT*. Retrieved October 2011, from http://faostat.fao.org/site/291/default.aspx

11 National Institute for Environmental Studies, Japan. (2011, 3). *Emission Scenario Database*. Retrieved from http://www.cger.nies.go.jp/db/scenario/index.html

12 Central Research Institute of Electric Power Industry. (n.d.). *SEEPLUS Model*. Retrieved October 2011, from http://www.jreap.org/seeplus/

13 Komoto, Keiichi, Masakazu Ito, Peter van der Vleuten, David Faiman, Kosuke Kurokawa (2009) *Energy from the Desert: Very Large Scale Photovoltaic Systems: Socio-economic, Financial, Technical and Environmental Aspects*, Earthscan.

CHAPTER FOURTEEN

Implementation strategies for solar power plants in desert regions

Strategic niche management, social acceptance, and business models

14.1 INTRODUCTION

After many years of study and preparatory work [1–3], solar power plants in desert regions increasingly count as a realistic energy option. VLS-PV plants have started to appear around the world. From the perspective of the existing global energy system, however, large scale solar power plants remain a promising yet very small niche. They still require financial and political support, entrepreneurs still search for viable business models, and acclaimed benefits such as relieving energy scarcity, job creation, poverty alleviation, and climate change mitigation are still to materialize. Accordingly practitioners and researchers increasingly focus on the challenges of socio-economic feasibility and market implementation to turn desert solar power plants into the "powerhouses of the world".

So far, most studies of socio-economic VLS-PV implementation have taken an informed practitioner's perspective. This chapter taps into the social sciences, and the innovation sciences in particular, to suggest ways of systematically thinking about VLS-PV implementation and aggregate existing insights. In doing so we need to go beyond conventional socio-economic assessment tools such as cost-benefit analysis and multi-criteria analysis, which will be familiar to most readers. These tools chart and weigh the pros and cons of a given technological option, and can work well for stable technologies in equally stable societal contexts. However, they fit poorly with the case of desert solar power plants. Radical innovations (as opposed to incremental innovations) such as desert solar power plants tend to develop so fast that evaluations at a fixed point in time tend to be instantly outdated and lack serious predictive value. In addition these conventional assessment tools fail to appreciate the intimate relationship between innovation and its societal context. In the case of desert solar power plants this context is in constant flux. For instance, the current financial crises and the political turmoil in a number of North African states (which count as promising solar power plant sites) may greatly affect the odds, forms, implementation trajectory, and benefits of desert solar power plants.

Clearly we need to search the innovation sciences for ways to think about VLS-PV implementation that are more dynamic and take changing contexts into consideration. In the following we take our cue from the booming field of sustainability transition studies. This particular field within the innovation sciences asks questions that solar power plants face today: Why do some radical innovations transform from technical possibilities into viable market niches and eventually become part of (or even displace) the dominant technological system, while others do not? What lessons can we draw for the management and implementation of sustainable technologies in uncertain contexts [4, 5]? We here discuss three such approaches. Strategic niche management is an analysis, policy, and practitioner tool that seeks to study and strengthen the dynamics of emergent sustainable technology niches and their transition to full-blown technological systems or "regimes". ESTEEM is a practitioner tool for social acceptance of novel technologies in local projects [6,7]. New approaches in the business models literature discuss how entrepreneurs may systematically search and design business models for radical innovations in unstable contexts [22–25]]. The former approach focuses on the overall level of the niche, in our case desert solar power plants, where individual projects are aggregated; the latter two target individual projects. All take uncertainties in technology and context head-on and emphasize the role of experimenting and learning. The perfect strategic management, social acceptance, or business model for desert solar power plants needs to be actively developed in an ongoing learning process, profiting systematically from lessons of well-designed pilot projects. These approaches suggest what types of challenges we should take into consideration during this learning process, i.e. what we need to focus on and learn about.

In this chapter we introduce these approaches and mobilize insights from other technology domains for the case of desert solar power. We consider this a first and introductory step towards a more VLS-PV specific inquiry and the development of more fine-tuned implementation tools for practitioners and policy makers.

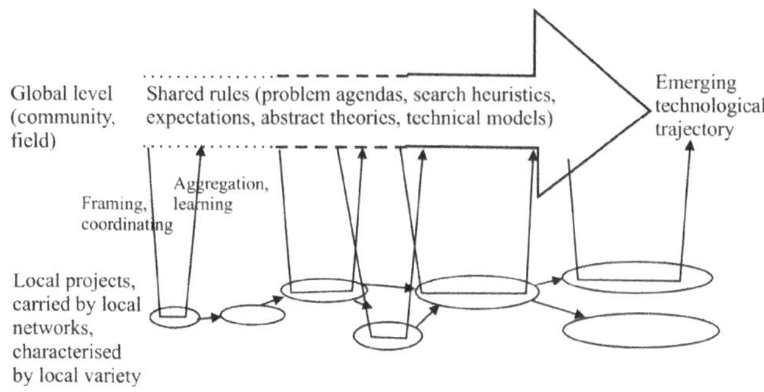

Figure 14.1 A technological niche organized to facilitate interaction between local projects and a global niche level, where the niche community accumulates knowledge and best practices [10]

14.2 BASICS OF STRATEGIC NICHE MANAGEMENT

Strategic niche management (SNM) is an analysis, policy, and practitioner framework to understand and manage radical innovations that promise a transition towards more sustainable ways of fulfilling societal needs. The SNM literature has focused particularly on sustainable energy and mobility issues [8, 9]. The point of departure is that such radical innovations may be desirable in the future from a sustainability perspective, but face grave competitive disadvantages today. These disadvantages partly stem from the technological and economic uncertainties that accompany most new technology. Other, and arguably more fundamental, disadvantages follow from the fact that radical technologies may not fit well with existing dominant technological practices – precisely because they represent a break with conventional ways of thinking and doing. Consider how national employment, industrialization, and energy security policies support conventional energy players and technologies. So do existing technological capabilities in most countries. Even such mundane matters as insurance rules, planning and construction procedures, energy cost calculation methods (how to account for externalities such as climate change or nuclear risk?), energy user behavior, and so on have been tailored over many years to suit conventional energy stakeholders and technologies. Through these mechanisms and others, existing energy regimes tend to lock-in on conventional power plants and lock-out radical innovations that are at odds with its well-entrenched economic, organizational, legal, technological and other practices. This is why many promising energy innovations eventually succumb or gain only marginal importance.

Alternatively radical energy innovations may develop in niches, where they are temporarily shielded from dominant energy regime forces and pressures. SNM seeks to understand and promote sustainable innovation through the deliberate creation, sustenance, and up-scaling of such niche processes. It suggests how proponents can use protected niche spaces to experiment with technical and non-technical components of their novel technology, and thereby find a constellation that fits with opportunities and demand in the outside world. If the learning process works well, lessons learned in individual projects accumulate into a "global niche level". In this perspective, the IEA PVPS Task8 group, working on the introduction of VLS-PV strategies in desert regions, is an example of a global niche player that accumulates insights on the level of desert solar power niche as a whole. This work includes selection, comparison, framing, contestation, and interpretation of local lessons. Conversely, global niche level insights can be used to frame individual local projects so as to optimize the learning process from an overall niche perspective (Figure 14.1).

SNM does not merely consider internal niche processes, however. In order to systematically chart the opportunities and barriers for niche technologies in the outside world, it adds the analytical layers of the *sociotechnical regime* and the *sociotechnical landscape* (Figure 14.2). While niches nurture radical energy innovations, incumbent regimes host dominant energy stakeholders with vested interests and stable social networks; regulations, standards, and cognitive routines that may blind actors to developments outside their habits and focus; and conventional energy technologies with their sunk investments and historically grown technical complementarities, such as fits with existing infrastructure. For the niche of desert solar power plants, one may think of the incumbent regime of centralized electricity generation in fossil fuel or nuclear power plants. This regime has long been entrenched within national borders, though in some parts of the world it is rapidly internationalizing [11]. Regime components such as dominant stakeholders, government legislation, and conventional power plants tend to interlock, keep each other in place, and resist radical change. SNM managers need to learn about the barriers and opportunities for niche technologies in incumbent regimes and feed these back into niche development processes.

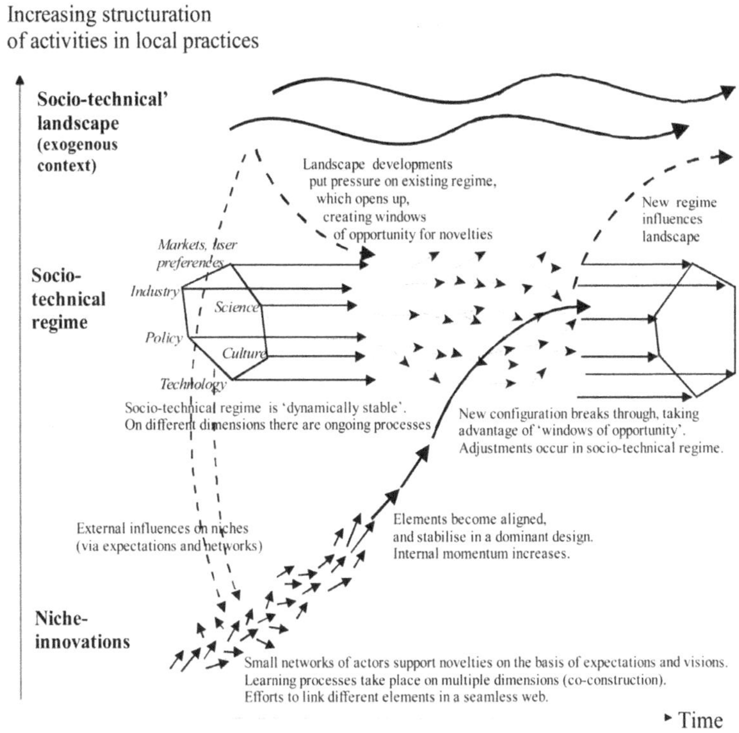

Figure 14.2 A dynamic multi-level perspective on radical innovation journeys [12]

The socio-technical landscape, finally, draws into the analysis exogenous forces on which both niche and regime players have little influence. In some cases such forces seem to constitute additional barriers for change. In the case of desert solar power, think of political turmoil in desert regions, or financial crises that make financiers wary of risky investment. Other landscape forces and events, however, may put pressure on stable energy regimes and open up opportunities for niche technologies to enter. Examples that may benefit desert solar power include global CO_2 emission reduction agreements, national or international (e.g. European Union) renewable energy targets, or disasters; the Tohuku tsunami of 11 March 2011 and subsequent crisis of the Fukushima nuclear power plants in Japan inspired – at least temporarily – a public and political distrust of nuclear power and renewed interest in renewable energy sources worldwide, as the Chernobyl disaster did decades ago.

In this three-layer analytical framework, the challenge is to propel promising energy technologies from deliberately protected technological niches into viable market niches and eventually latch on to (or even displace) incumbent energy regimes. Landscape pressures on existing regimes provide opportunities for such change. Successful innovation in this perspective is about navigating niche, regime, and landscape dynamics, so as to find a workable innovation journey for the novel technology. To give just one historical example, in the 1970s and 80s wind turbine projects failed to take off in Germany and the United States despite heavy investment and impressive technological capabilities. By contrast the Danish global success in this field was prepared on the Danish home market, where an engaged wind turbine community boosted internal niche processes and gained momentum. Simultaneously the wind turbine niche linked up with landscape forces such as government industrialization and employment concerns; politicians of different persuasions supported the emergent domestic wind turbine industry with subsidies and other measures. The niche technology also hooked up to the existing electric power regime and its lock-in on centralized power production, for a trajectory of individual back-garden projects transformed into an innovation pathway of large-scale wind turbine parks exploited by traditional electric power companies. Internal niche dynamics linked up with regime and landscape dynamics, and Denmark became a leader in terms of its share of wind power in its energy mix [13].

Today we can see similar processes emerging in the desert solar power niche. We interpret the recent reframing of desert solar power plants from a utopian very large scale technical possibility to a "utility-scale energy option" as an alignment with the existing electric power regime, and electric power producers in particular. In the EU-MENA region, linking up with forces for a Mediterranean Ring and a European super-grid makes Sahara desert power plants increasingly attractive. Simultaneously the connection of VLS-PV to sea water desalination plants links up the desert solar power niche with quite another regime, that of drinking water supply in desert regions. On the other hand desert solar power plants face particular challenges, especially when they involve international energy systems. Again the EU-MENA region illustrates this: Sahara solar power exports to Europe might prove difficult when European governments provide massive visible and invisible support to domestic conventional power plants that provide domestic employment, turnover, and energy independence. VLS-PV practitioners are already grappling with such opportunities and barriers. SNM is designed to learn about and manage such processes systematically.

One major disclaimer is in place. So far SNM, and the field of sustainability transitions at large, have overwhelmingly focused on country studies. VLS-PV practitioners may certainly learn from these SNM insights. Conversely, SNM researchers can learn from the transnational character of the desert solar power plant niche (and, as in the case of EU-MENA, its embedding in transnational energy systems) to improve their analytic framework. The encounter between VLS-PV and SNM could be mutually beneficial [14, 15].

14.3 NICHE DYNAMICS AND LESSONS

In discussing strategies for successful niche development, we distinguish between the shielding of the niche from the outside world and the nurturing of internal niche processes.

14.3.1 Shielding

As stated above, path-breaking energy innovations may develop in niches that shield those innovations from mainstream selection criteria of the dominant energy regime [16]. Important selection environments that the niche needs to be shielded from include:

- Established *industry structures* form a selection environment through, for example, established network relations, industry platforms, strong user–producer interactions, shared routines and heuristics, existing capabilities and resource allocation procedures. Path-breaking innovations entering the market might be rejected because they do not fit with existing industry structures and decision-making processes that have emerged in co-evolution with the dominant design.
- Dominant *technologies and infrastructures* form a (material) selection environment, for example, through articulated technical standards and infrastructural arrangements, which are imposed on new innovations. Path-breaking innovations may require different standards and infrastructures in order to perform (technically and economically) optimally, and as a result are perceived as not feasible.
- Guiding principles and socio-cognitive processes in the established *knowledge base* are paradigmatic and geared towards incremental knowledge development rather than paradigmatic shifts. Path-breaking innovations are rejected because insufficient resources are attributed to new knowledge development, R&D and so on, and academic and private research institutes perceive disincentives because of a lack of dedicated journals, conferences and research groups.
- *Markets and dominant user practices* form a selection environment through stabilized market institutions, supply and demand, price mechanisms, user preferences and routines. Path-breaking innovations have a hard time entering the market, for example, because external environmental costs are not represented in end-user prices, or because they require inconvenient user practices compared with accustomed habits.
- *Public policies and political power* form a selection environment through, for example, prevailing regulations, policy networks and relations with incumbent industries. Political power is exercised to maintain the status quo, in terms of jobs, tax base, and votes, which is a disadvantage for path-breaking innovations, because they require different policies and regulations, and even new political economies.
- The *cultural significance attached to a specific regime* forms a selection environment through, for example, its widespread symbolic representation and appreciation. Path-breaking innovations are put at a disadvantage, because they represent different cultural values and lack widespread stabilized representations.

Known mechanisms that shield promising niches from such outside pressures include geographical or institutional isolation. Think of wind turbine experiments on small Danish or Greek islands isolated from large power grids, the use of PV in space applications, and the institutionally isolated space of the environmentalist milieu whose members are willing to trade off higher cost and lower technical performance for environmental and social justice benefits. Other shielding mechanisms seek to actively change the odds of the new technology. Active shielding mechanisms include measures for counteracting cost differentials or performance characteristics such as tariffs, taxes, and direct and indirect subsidies. In the world of solar power, Germany's PV feed-in tariff system is a well-known example of an active niche space strategy. Other measures aiming to alter preferences include quotas, public purchasing, information campaigns, and the like. Also private actors can engage in active shielding strategies. Firms may place radical innovations in incubator programs to temporarily shield them from short-term decision-making processes within mother-firms. Civil society communities may set up renewable energy cooperatives. Either way it is important to note that the success of shielding strategies is not entirely in the hands of niche actors. From previous niche case studies we know that niche technology ups and downs also follow processes and events outside the spheres of influence of its advocates. In good times, niche actors can find it easy and attractive to argue for active shielding measures to successfully propel niche development forward. In adverse times, the same actors might find it difficult to convince others to support shielding measures. In such cases, they may resort to the more passive spaces mentioned above to continue niche development "under the radar".

When developments in the shielded niche take off due to internal niche processes (which we discuss below), niche protection should be followed-up by niche empowerment. Niche empowerment is about changing mainstream selection environments in ways favorable to the path-breaking innovations; to further scale up the niche, we need to change the selection criteria of the dominant energy regime. SNM distinguishes two types of empowerment:

- *Removing* the protective shields in relation to performance improvements of the path-breaking innovation, so that the niche learns to compete under conventional regime selection pressures. In this case energy innovation can happen without radically changing the dominant energy regime. Rather it evolves to fit and conform to prevailing regimes. Radical innovation, while changing regime properties, is then domesticated into a conservative, regime-preserving innovation. Whether or not this is a problem depends on your point of view. For instance, the transformation of the back-garden wind turbine trajectory with its environmentalist and anti-centralist values ("power to the people") into a wind-park trajectory run by conventional power companies implied an accommodation of wind power in existing regimes. It preserved conventional power plants to balance unstable wind energy production, yet it did change the regime's carbon-emission properties [17, 18].
- *Institutionalizing* the protective shields and niche practices as part of a new, reconfigured regime largely based on new (sustainability) criteria that stem from the niche. In this case the innovation stretches and transforms the prevailing regimes, so to speak. Consider the remarkable case of phasing out incandescent lighting in Australia and the European Union, in which policy makers and industry lobbies shifted the evaluation criteria from technical and economic performance to ecological performance.

14.3.2 Niche nurturing

Three niche internal processes seem highly related to successful niche development: the dynamics of expectations, social network building, and learning processes.

Expectations are of great importance for the development of technologies as they mobilize, stimulate, steer and coordinate stakeholder commitment, investment and other actions [10, 19]. Expectations, visions and beliefs can obtain the dynamic of self-fulfilling prophecies, because they guide R&D activities that work towards realizing them. They may lead to the formation of *promise–requirement cycles*: stakeholders articulate expectations and promises, which in turn translate into common goals, specifications, requirements, and tasks divisions, and hopefully the freeing of resources to start concrete projects. The evaluation of these projects may lead to adjustment of the former promises and expectations into more realistic ones, which in turn may attract additional supporters (e.g. investors).

Most niche proponents actively work on their expectations during the implementation phase. SNM proposes the active use of experiments to uncover and test existing (sometimes implicit) expectations. Important dimensions of the dynamics of niche expectations include:

- Specificity: Expectations often start with macro-expectations that are very general. In the case of desert solar power plants, early visions hinged on a very long-term promise: fossil fuels and uranium will eventually run out, but the sun's energy in desert regions is more than enough to fill the gap. Such very general promises can be translated on increasingly concrete levels: on the prospects of the PV sector in the medium term, or the prospects of certain technological breakthroughs or pilot projects in the short term. To be convincing, expectations should experiment and find a proper mix of arguments on all these levels.
- Quality: Another criterion of expectations is the fit with the real-life situation. Expectations for a niche can be strengthened when they fit with an increasing set of developments in the outside world, but are undermined by clear frictions with empirical reality. For instance, claims on the profitability of new energy technologies may be swept away in the face of a few well-published failures. Niche experiments should be designed to test and improve the reality-fit of niche promises. Consequently, niche expectations should be constantly revised in the light of new developments and insights to remain convincing.
- Robustness: Expectations do better when they are shared between a larger group of stakeholders. This does not mean that one should impose a common vision and consensus from above, however. Rather, niche players should try to make explicit their different expectations, search for common ground, and design experiments to test differences that cannot be bridged.

Next to improving niche expectations, the creation of a social network is crucial to niche success and failure. Collaboration between stakeholders facilitates the pooling of resources (knowledge, money, personnel, facilities) and sharing of risks. At the same time social niche networks are dynamic and difficult to manage. Stakeholders enter and leave, and the roles of different actors, like researchers, governments, industry, users and representatives of civil society, will change along the way. Based on existing studies, we observe the following social network aspects of successful energy niches [20, 21]:

Diversity and openness: The niche may benefit greatly from project partners with a diverse range of societal roles (e.g. industrials, scientists, NGOs, policy makers, end-users). This diversity may avoid a lock-in on a few actors, interests, or visions; such a lock-in may be vulnerable to setbacks or resistance and lock-out attempts by regime players. In addition the presence of different perspectives and interests may greatly assist the linking of the niche to the dominant regime. For instance, the solar desert power plant niche has greatly benefited from the presence of industrial partners next to university researchers and visionaries.

- As a consequence, the openness of the niche can be an important factor: Potential collaborators should feel welcome when joining the niche. Moreover, outsiders and criticasters may fulfill important roles as well, because they articulate problems on which niche proponents can react by problem solving. On the other hand the network should not become too broad as that will result in a loss of flexibility, and criticasters can also frustrate group processes. Diversity and openness therefore are not law-like predictors, but dimensions on which niche players can experiment to find a working collaboration.
- Alignment: Social network alignment refers to the coherence in the group of niche stakeholders. For instance, their vested interests may converge or diverge, and they may bring trust and experience in collaborating from earlier projects in which they worked together.
- Understanding opposition: Niche players should analyze the kind of opposition they meet. We distinguish *conscious opposition*, i.e. the withholding of political, financial or social support because the opposing actor does not see the interest of the project. *Damaging strategy* refers to deliberate actions that directly delay or damage niche experiments by draining project resources. This strategy often stems from a deep frustration or extremely differing values. Examples include legal fights and physical damaging of technological assets. Finding another location for the experiment or excluding certain stakeholders from decision-making might be necessary for dealing with this type of opposition. *Strategic opposition* influences the niche experiment by bringing in arguments and alternatives. This strategy is focused on negotiation. Often this strategy is used for making underlying frustrations explicit. SNM aims at understanding and anticipating this sort of worry. For instance, think of opposition to VLS-PV by NGOs for protection of desert wildlife or local people that face displacement, which might be met by slight planning changes.

Finally, the quality of the learning process is key to niche dynamics. As stated above, radical innovations are in constant change. Technology, its use, its regulation and so on are not fixed and stable, but still subject to major changes. Therefore SNM proposes to design experiments for learning on a broad range of issues. Next to the technological learning that dominates most niche processes, this should involve learning about economic, regulative, and institutional issues. To develop this issue further, SNM distinguishes between two levels of learning that both should be included. Most niches focus overwhelmingly on "first order learning", i.e. the accumulation of facts and data pertaining to the expected function, uses, and selection criteria for novel technology, such as technical performance and costs. A good learning process, however, also queries its own assumptions about the future functions and uses of the technology in question. This is called "second order learning". For instance, in the case of desert solar power plants one may learn that success or failure do not depend so much on the cost price of a kilowatt-hour, but on prospects for local economic development or sea water desalination. One may even speculate on a hypothetical scenario, in which desert solar power plants turn out to be drinking-water technologies rather than energy technologies. The point here is that at present we do not know how our novel technology will be used in the future; we can only design experiments so as to learn about the ongoing dynamics of potential uses and functions. Cars and bicycles, it should be remembered, were designed first as racing machines before they obtained a transport function, let alone the commuting function they have today. Major SNM lessons about learning processes from previous research include [21]:

- Include parties that are strongly concerned about the societal challenge (e.g. NGOs, government agencies, citizens) in the learning process;
- Couple learning experiences to the right persons who are able to anchor them into the regime (e.g. formulation of new laws and regulations);
- Create and stimulate diversity between experiments so lessons can be learned on various designs in various use environments;
- Aim to learn at many different dimensions including technology, infrastructure, regulations, user preferences, and cultural and societal acceptance. In addition, aim to create alignment between these dimensions to create a better fit between technology and societal context;
- Apply a reflexive attitude (i.e. questioning underlying assumptions such as social values and world views) when experiments are compared and lessons are transformed into new designs. In concrete projects a reflexive attitude is less necessary;
- Create possibilities to share lessons between experiments, e.g. by platforms, regular meetings, symposia or a newsletter;
- Create possibilities to share lessons between different types of actors;
- Systematically document learning experiences;
- Use lessons to adjust and reformulate your vision and expectations.

14.4 AN SNM PRACTITIONER TOOL: ESTEEM

SNM is a strategic management framework that operates at the level of the global niche, which may host many local experiments and projects. What SNM has to say about such individual projects follows from its overall niche concerns. In order to address more prominently the issues that entrepreneurs face in individual projects, we now turn to two project-level approaches. The first, ESTEEM (Engage STakeholdErs through a systEmatic toolbox to Manage new energy projects) [6, 7], is about taking seriously and managing social resistance to local projects. The second is about how to design and experiment with viable business models for radical innovations.

Local resistance to novel energy technologies can take various forms, ranging from a simple lack of interest and denial of cooperation to emotional stress and actual sabotage. ESTEEM is a practitioner toolkit to deal with societal acceptance issues in the implementation stage of local projects, when adaptation of the project design is still possible. ESTEEM is typically facilitated by consultants and consists of six steps (Figure 14.3 and Box 14.1).

14.5 NEW STUDIES OF BUSINESS MODELS

A second approach that deals with the project level of radical innovation stems from the business models literature. Business models are about describing "the rationale of how an organization creates, delivers, and captures value" [22] and are increasingly seen as a source of competitive advantage for a company. It is also said that "a company has at least as much value to gain from developing an innovative new business model as from developing an innovative new technology" [23]. More important to us, business models are key instruments for bringing new technologies such as VLS-PV to the market [24]. Several scholars are currently working on combining SNM with business model development into one framework, but this work is still in its initial phases. Here we limit ourselves to outlining selected business models insights and thereby mobilize these for our thinking about the implementation of desert solar power plants.

Like SNM and ESTEEM, recent work on business models generation acknowledges the importance of rapidly changing business conditions in unstable contexts, and bases itself on learning processes. It could be said that the business models approach addresses the project level, but creates a sort of global niche infrastructure to facilitate a joint exchange and learning process [22]. Either way, in terms of content it is used for practitioner experiences to distill nine building blocks that a good business model should address. Table 14.1 illustrates that such a business model not only deals with finances, it also defines for instance collaboration with partners and customer involvement. This applies to commercial companies as well as non-commercial organizations such as NGOs or government institutes that aim at creating environmental or social value (so-called social business models).

Like SNM, recent work has been done on business models to include the context in its design strategy. Then, four key business model contexts are defined: key trends, market forces, macro-economic forces, and industry forces (Table 14.2). In the model, entrepreneurs and project managers should consider these contexys as factors when constructing the nine building blocks of their business plan.

Critics find this reference model too static. For instance, it is argued that projecting currently known business

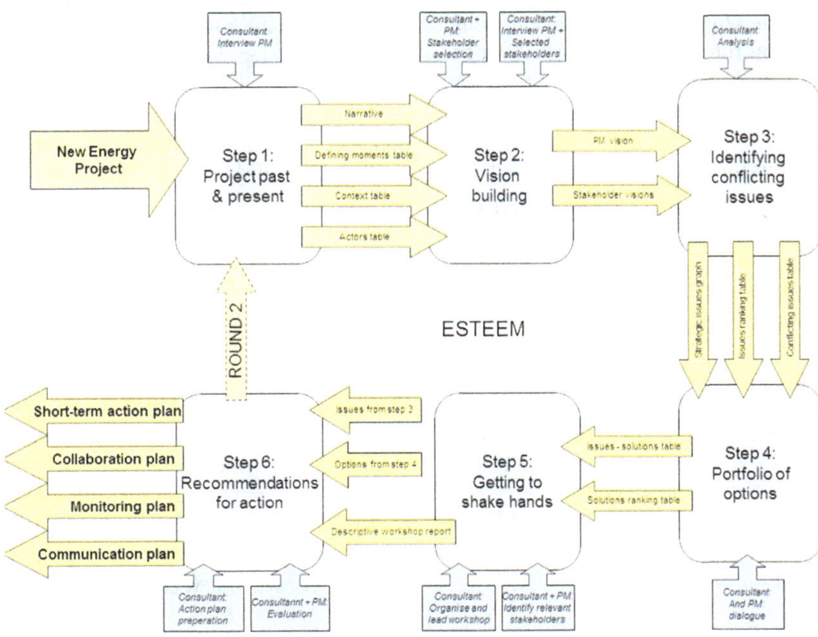

Figure 14.3 ESTEEM and the six steps to develop action plans for societal acceptance. PM stands for "project manager" [6]

Box 14.1 ESTEEM (according to Raven *et al.* 2009).

STEP 1: PROJECT PAST AND PRESENT

The first step is to document and collect information about the project for later analysis. Because the method is based on interaction between a consultant and a project manager, the consultant clearly needs to get a thorough understanding of the project. For the project manager it is a good opportunity to systemize and take a step back from the project. Step 1 is performed through a series of meetings between the consultant and project manager. Together they a) write a project narrative about the past of the project; b) derive moments that were crucial in project development ("defining moments"); c) make a context table that shows current barriers and opportunities as perceived by the project manager; and d) make an actors' table that shows which actors have been, are currently, or might be relevant in the future of the project.

STEP 2: VISION BUILDING

How to anticipate potential societal acceptance problems before they even occur? ESTEEM follows the logic of articulating future expectations and vision building to identify potential conflicts. In step 2 the consultant, project manager, and stakeholders draft several social network maps and narratives, but they are not yet confronted. First, in a semi-structured interview the consultant and project manager draft a social network map of the present, showing how the project manager sees and interprets the project and the current world surrounding it. In another semi-structured interview they also draft a future network which shows how the project manager thinks (or would like) the world (and the project) will look like in about 10 years' time. These maps are accompanied by one or more fictional newspaper articles to be published in 10 years' time in a local newspaper, in which the project manager reflects on project developments. A title summarizes the main theme of the project manager's vision. The vision (consisting of the maps, narrative, and title) are then discussed in a series of semi-structured interviews with selected stakeholders. Prior to the interview the consultant sends the vision of the project manager to the interviewee. In the interview the stakeholder is asked to react on the project manager's vision and to articulate on what issues they agree or disagree with and how they envision a different future for the project. After the interview the consultant draws a new narrative, network map, and vision title for each stakeholder. The consultant sends them for review back to the stakeholder. Step 2 thus produces a number of different visions on the future of the project.

STEP 3: IDENTIFYING CONFLICTING ISSUES

The next step in the process confronts the collected expectations and visions and analyzes points of convergence and divergence. What are the major points of agreement/disagreement that could potentially lead to further support or opposition to the project in the future? In step 3 the consultant addresses these questions by listing and describing the issues at stake in a "conflicting issues table". The table is discussed with the project manager in a face-to-face meeting. The project manager is then asked to rank the different issues in an "issues ranking table" and an "issues ranking graph". The table and graph are representations of the project manager's view on the importance and urgency of the different issues.

STEP 4: PORTFOLIO OF OPTIONS

Step 4 is concerned with devising how to improve the project acceptance by reviewing a variety of solutions for the issues identified in step 3. Which pieces of the project or project context can be modified to raise the acceptance level of the project? In this step the consultant and project manager engage in a dialogue in one or two meetings to identify potential solutions with two instruments. The first one is the "issues-solutions table", which provides a semi-structured way to think about different types of adaptations to the project or context for each issue identified in step 3. The second instrument is a "solutions ranking table", in which the project manager is elicited to rank the different solutions.

STEP 5: GETTING TO SHAKE HANDS

The aim of step 5 is to open up the process to a larger number of stakeholders and discuss in a wider setting the different issues and solutions identified. This task is performed in an interactive workshop, in which a stakeholder group of about 20–30 people can react to and vote on the different issues and solutions, but also bring new issues and solutions to the discussion. The latter is extremely important, because many stakeholders in the workshop are able to participate in the project for the first time. They can have different expectations about the project, which have not yet been included. The workshop is therefore designed to be as open as possible to allow for any discussions or feedback to emerge. Different types of workshops are distinguished within ESTEEM, depending on the requirements and characteristics of the new energy project. After the workshop the consultant produces a descriptive workshop report and sends it to all participants.

STEP 6: RECOMMENDATIONS FOR ACTION

The final step of ESTEEM is oriented towards acting and planning. In the first part of step 6 the consultant compares the outcome of step 5 (solutions identified by the stakeholders) with the outcome of step 4 (solutions identified by the project manager). As a general rule all solutions mentioned by both the project manager and the stakeholders as well as solutions mentioned by the stakeholders alone are taken into account for further processing. In the second part of step 6 several action plans are derived in cooperation between the consultant and the project manager including a short-term action plan, a collaboration plan, a monitoring plan, and a communication plan.

The ESTEEM process ends with an evaluation of the process, in which the project manager is interviewed to articulate what he or she has learned from the process. The process of six steps is straightforward and linear. While this was initially thought of for analytical reasons and for structuring the process, our actual experience is that the steps are indeed performed in a linear fashion. There is no major "going-back-and-forward" between the steps. There is however no inherent obstruction in ESTEEM to take a more dynamic approach when the project demands it. Also, when a project manager and consultant have gone through the first round of ESTEEM they might also decide to do another full round or redo part of the process. This can be useful after parts of an action plan have been implemented to verify if the project has gained societal acceptance or when new controversies have emerged.

Table 14.1 Building blocks of a business model [22]

Building block	Definition
Customer segment	The different groups of people or organizations an enterprise aims to reach and serve
Value propositions	The bundle of products and services that create value for a specific customer segment
Channels	How an organization communicates with and reaches its customer segments to deliver a value proposition
Customer relationships	The types of relationships an organization establishes with specific customers segments
Key resources	The most important assets required to make a business model work
Key activities	The most important things a company must do to make its business model work
Key partnerships	The network of suppliers and partners that make the business model work
Cost structure	All the costs incurred to operate a business model
Revenue streams	The cash a company generates from each customer segment

Table 14.2 Business model contexts [22]

Industry forces	Suppliers and other value chain actors Stakeholders Competitors (incumbents) New entrants (insurgents) Substitute products and services
Key trends	Technology trends Regulatory trends Societal and cultural trends Socioeconomic trends
Macro economic forces	Global market conditions Capital markets Economic infrastructure Commodities and other resources
Market forces	Market segments Market issues Needs and demands Switching costs Revenue attractiveness

model dimensions and their contexts into the future is not enough. Instead existing models should be supplemented by active searching and learning: "it is more sensible to engage in experimentation and discovery than to try to assume the relevant information is all known" [25]. However, this is not easy. For example, managers may feel that business model experiments are a threat to the ongoing business; cognitive barriers may prevent them from even considering the possibility [23]. Most often managers and entrepreneurs use information from established business models to make decisions about new ones. For this reason project entrepreneurs and managers need to build a learning organization, and develop mechanisms through which lessons are not lost but feed back into the project business model. Of course, a broad literature on learning organizations exists in the management sciences. We will not go into that here, but merely observe that business model experimenting should be part of project-level activities of sustainable energy niches.

14.6 CONCLUSION

Solar power plants in desert regions are on the brink of finding their way into the world's energy systems. This could mean that preparatory work, done by the International Energy Agency VLS-PV group and others, is currently being "overtaken by market developments" and can be concluded. From the innovation science perspectives described in this chapter, this would be a false conclusion. As policy makers and entrepreneurs experiment with projects and implementation models, now is the time for systematic learning about VLS-PV on a global niche level. For only the confrontation with real-life conditions can reveal the technological, business, socio-economic, and geopolitical fits and misfits of solar power plants in desert regions with the demands of the modern energy world. It is now that the main (potential) stakeholders need to learn about expected and unexpected bottlenecks that emerge in individual local projects, accumulate such lessons on an overall niche level, experiment with ways to navigate these real-world opportunities and barriers, and invent viable innovation pathways. Importantly, socio-economic or geopolitical bottlenecks might very well feed back into the technology design process: the redesign of solar power concepts for desert regions, originally conceived to serve long-distance international energy flows, to accommodate local economic development and sea water desalination illustrate technology adaptations in the (early) market implementation phase. In sum, the coordination between individual local projects and the overall desert solar power niche community is more important now than ever.

The VLS-PV niche requires continued reflection on strategic management on a niche level, and a continued search for social acceptance and business models on a project scale. To facilitate this process we have tapped into innovation sciences work on sustainability transitions. This chapter outlined the approach of SNM and briefly introduced ESTEEM and business model experimenting. This initial mapping of the territory, however, has merely opened up the topic. A next step would be a more VLS-PV specific and fine-tuned inquiry. A follow-up study should systematically examine existing or ongoing projects for current niche dynamics (in terms of articulating expectations, building social networks, organizing learning processes, shielding and empowering the niche, and emerging business models) and initial encounters with real-life contexts (in terms of technological, business, socio-economic, and geopolitical opportunities and barriers, and expected/unexpected bottlenecks that pop up). With the practical tools outlined in this chapter, we may translate these lessons into a learning-based implementation strategy for policy makers and entrepreneurs.

Note: We thank the invited writers of this chapter, Boukje Huijben, Erik van der Vleuten, Rob Raven and Geert Verbong of the System Innovations and Sustainability Transitions group, School of Innovation Sciences, Eindhoven University of Technology, for their valuable contribution to our work.

REFERENCES

1 Kurokawa, K. (2003) *Energy from the Desert: Feasibility of Very Large Scale Photovoltaic Power Generation (VLS-PV) Systems*, James & James, London.
2 Kurokawa, K., Komoto, K., Van Der Vleuten, P. and Faiman, D. (2007) *Energy from the Desert: Practical Proposals for Very Large Scale Photovoltaic Systems*, Earthscan, London.
3 Komoto, K, Ito, M., van der Vleuten, P., Faiman, D. and Kurokawa, K. (2009) *Energy from the Desert: Very Large Scale Photovoltaic Systems: Socio-economic, Financial, Technical and Environmental Aspects*, Earthscan, London.
4 Grin, J., Rotmans, J. and Schot, J. (eds) (2010) *Transitions to Sustainable Development. New Directions in the Study of Long Term Transformative Change*, Routledge, London.
5 Verbong, G., and Loorbach, D. (eds) (2012) *Governing the Energy Transition. Reality, Illusion, or Necessity?* Routledge, London, 2012.
6 Raven, R., Jolivet, E., Mourik, R. and Feenstra, Y., (2009) 'ESTEEM: Managing societal acceptance in new energy projects: A toolbox method for project managers', *Technological Forecasting and Societal Change* 76: 963–77.
7 E. Jolivet *et al.* (2008) *ESTEEM manual. Create Acceptance deliverable 5. Document ECN-E-08-031* Energie Centrum Nederland.
8 Raven, R. (2012) 'Analysing emerging sustainable energy niches in Europe. A strategic niche management perspective', in Verbong and Loorbach, *Governing the Energy Transition* Routledge, London.
9 Schot, J., and Geels, F. (2008) 'Strategic niche management and sustainable innovation journeys: theory, findings, research agenda and policy', *Technology Analysis and Strategic Management* 20: 537–54.
10 Geels, F. and Raven, R. (2006) 'Non-linearity and expectations in niche-development trajectories: ups and downs in Dutch biogas development (1973–2003)', *Technology Analysis and Strategic Managemen* 18 (3–4): 375–92.
11 Van der Vleuten, E., and Högselius, P. (2012) "Resisting Change? The Transnational Dynamics of European Energy regimes", in G. Verbong and D. Loorbach, *Governing the Energy Transition* Routledge, London.
12 Geels, F. (2002) 'Technological transitions as evolutionary reconfigurations processes: a multi-level perspective and a case study', *Research Policy* 31: 1257–74.
13 Van der Vleuten, E., and Raven, R. (2006) "Lock-in and change. Distributed generation in Denmark in a long term perspective", *Energy Policy* 34: 3739–48.
14 Van der Vleuten, E. and Högselius, P. (2012) *Resisting Change? The Transnational Dynamics of European Energy regimes*, Routledge London.
15 Raven, R.P.J.M., Schot, F.W., and Berkhout, F. (forthcoming) *Breaking out of the national: A transnational perspective of socio-technical transitions*.
16 Smith, A., and Raven, R. (2012) 'What is protective space? Reconsidering niches in transitions to sustainability', *Research Policy*48 (6): 1025–36.
17 Hirsh, R. and Serchuk, A. (1996) 'Momentum shifts in the American electric utility industry: catastrophic change – or no change at all', *Technology and Culture* 37: 280–311.
18 Heymann, M. (1999) 'A fight of systems? Wind power and electric power systems', *Centaurus* 41 (1–2): 112–36.
19 Bakker, S., Lente van, H., and Meeus, M. (2011) 'Arenas of expectations for hydrogen technology', *Technological Forecasting and Societal Change* 78: 152–162.
20 Raven, R., Heiskanen, E., Raimo, R., Hodson, M. and Brohmann, B. (2008) 'The contribution of local experiments and negotiation processes to field-level learning in emerging (niche) technologies. Meta-analysis of 27 new energy projects in Europe', *Bulletin of Science, Technology & Society* 28(6): 464–477.
21 Laak van der, W. , Raven, R. and Verbong, G. (2007) 'Strategic niche management for biofuels: Analysing past experiments for developing new biofuel policies', *Energy Policy* 35 (6): 3213–25.
22 Osterwalder, A., and Pigneur, Y. (2010) *Business Model Generation. A handbook for visionaries, game changers and challengers*. Hoboken, NJ: John Wiley and Sons.
23 Chesbrough, H. (2010) 'Business model innovation: opportunities and barriers', *Journal of Long Range Planning* 43: 354–363.
24 Zott, C., Amit, R. and Massa, L. (2011) 'The business model: Recent developments and future research', *Journal of Management*37 (4): 1019–1042.
25 McGrath, R. (2010) 'Business models: a discovery driven approach', *Journal of Long Range Planning* 43: 247–261.

PART VI
Conclusions

CHAPTER FIFTEEN

Conclusions and recommendations

15.1 UNDERSTANDINGS

Solar energy resources, PV technologies, and renewable energy will help to realize important economic, environmental and social objectives in the twenty-first century, and will be critical elements in achieving sustainable development.

In order to promote renewable energy, the diversity of challenges and resource options, as well as the financing and market conditions among and within regions and countries, implies that different approaches are required. Establishing policies for developing markets, expanding financing options, and developing the required capacity have been discussed in order to initiate and adopt policy changes and to incorporate the goals of sustainable development into their policies.

Since our previous work [1–3], solar power plants in desert regions have increasingly begun to count as a realistic energy option. Very large scale solar power plants have started to appear around the world. In order to accelerate the current transition towards renewable energy, the development of VLS-PV power generation systems that can be deployed on a massive scale is one of the ways of achieving this transformation.

Based on this background, this report has presented an overview of relevant issues and the potential of VLS-PV, guidelines for VLS-PV systems, and technical as well as strategic options for implementing VLS-PV.

Our conclusions are summarized as follows.

PROJECTION FOR THE FUTURE

Many scenarios for future global energy supply exist, but the roles of PV and other renewable energy options vary greatly between the scenarios. These are heavily dependent on underlying assumptions and projection approaches. In general, the share of PV is expected to increase drastically in most of the scenarios, but it may take a few decades for PV to be a main energy source in the global energy mix.

Our study using a simplified model for sustainability issues suggests that:

- VLS-PV can contribute greatly to global sustainability, but the main contribution will come after 2050.
- The contribution of VLS-PV is heavily dependent on the 'business as usual' scenario assumed. Energy efficiency and other renewable energy measures are also of major importance.
- Taking into account the energy depletion issues (resulting in increasing energy costs), VLS-PV could be a cost-effective alternative, although it will still need continuous improvement.
- It should be kept in mind that such innovative changes in the energy supply system require a certain lead time. Also, delay in actions would lead to more serious problems since the driving forces are gaining importance.
- Although the contribution of VLS-PV will be marginal in the next 10 years, immediate action is very important from a sustainability perspective.

PV has already obtained a higher share in the energy mix than some projections presented in the past (for example, by *IEA and others*), and should continually beat the projections in order to become a main energy source. Meeting the projection by *IEA: PV Roadmap* [4] and *Energy Technology Perspectives* [5] until 2050 should be the minimum level. Only then we can visualize the VLS-PV roadmap toward 2100 as we proposed in our previous edition [3].

SOCIO-ECONOMIC AND ENVIRONMENTAL BENEFITS

Solar electricity generation at utility scale can offer the possibility for socio-economic growth in the region.

Because of local availability of energy in desert regions, clean water can be provided, irrigation can be optimized, local food can be produced and local sources of raw materials can be exploited. The generated electricity can directly be used for lighting, communication, education, entertainment and industrial purposes. Access to electricity can be exported to neighbouring regions or beyond. By adapting a regional policy for solar electricity generation at utility scale, a sustainable local market for solar electricity modules and other related components and services will be created and this market can be served by local production. Technology transfer from industrialized countries can provide and improve education and learning, research and development, and manufacturing and sales. Local solar electricity generation also provides for security of energy supply in a sustainable way, provides fair access to energy for everybody and reduces the threat of climate change. If managed in the right way, such a closed circle between local resources and local needs offers extensive possibilities for further development of the region.

Needless to say, PV technology is environmental friendly. From a life-cycle viewpoint, it is clarified again that VLS-PV could contribute to save and mitigate fossil fuel consumption and greenhouse gas emissions. In addition, to apply a VLS-PV policy for decreasing our ecological footprint reveals that VLS-PV has significant potential to implement a sustainable ecological balance.

ENGINEERING AND FINANCING APPROACH

Assessment and evaluation of VLS-PV takes place with economic and financial considerations in mind. These are always in coherence with a basic understanding of technical issues.

The quality of execution of the planning is mainly dependent on the qualification and experience of the engineers and site managers involved. Beside electrical know-how, basic knowledge of related fields like testing procedures, mechanic and static as well as soil science and meteorology are required. VLS-PV systems contain a large number of connected conversion devices, a lot of power electronics, measuring systems etc. Depending on the region, more and more VLS-PV installations are located far away from living environments. Therefore the issues of safe, secure and trouble-free operation, maintenance and service issues should increasingly be considered in the course of plant design. From these viewpoints, the technical engineering guidelines developed in this volume facilitate the access to a basic technical understanding of VLS-PV.

The development of a VLS-PV system is a long process where almost all parameters change until the financial closing is reached. The preliminary technical design will continuously evolve during the development process, taking into account the economic, financial and legal issues that are raised all along the way. In a VLS-PV transaction, the development costs and risks are quite high and the studies are complex. In this respect, the support of the governments and development agencies should be directed toward removing the major barriers. Complete studies should point them out and suggest a pluri-disciplinary approach and management of the PV sector support to guide these players. The study should also anticipate the requirements of the credit and investment committees and the contents of technical and legal due diligences. A high quality level, pluri-disciplinary feasibility study is therefore necessary to secure the financing. It should address, together with the technical aspects, all the key financial and legal issues.

TECHNICAL POTENTIAL

Solar energy resources are available in all populated regions in the world. Looking at the technical variation of PV systems, fixed optimally tilted PV systems are the standard PV power plant system type today. However, depending on local conditions, adapted tracking systems will increase the annual yield per installed unit of capacity and lower the power generation costs.

Considering the technical potential of hybridization between VLS-PV and other renewable energy technology, the weaknesses of such technologies might be solved by engineering ingenuity. For example, due to the complementarity of solar PV and wind power, market expectations for PV may not need to be lowered. Moreover, PV together with wind power might become the backbone of the global energy supply in the coming decades. Hybrid PV-Wind systems are likely to start to strongly influence the global power system in the next decade by sophisticated hybrid concepts, like the renewable power methane technology.

In addition, a worldwide grid of ultrahigh voltage DC power lines (or perhaps ambient temperature superconducting cables) fed by the various VLS technologies may be envisaged. Not only conventional PV but also CPV, CSP and wind, each located in climatic regions appropriate to their most favourable economic performance, may be included in the future.

On the other hand, thinking about simple utilization of VLS-PV into the grid, predicting intermittence and energy storage will be important issues. It could be concluded that VLS-PV systems power output intermittence and prediction can be performed at various time scales, with the use of different instruments, important databases, and mathematical and statistical tools to predict the power output variations. There are some technical options for energy storage. Supporting integration through flexibility enhancement with 'storage on grid' can be done with batteries. 'Onsite storage' should be conjointly done with batteries for short-term uses and large-scale storages, such as pumped-hydro or compressed-air, for power shifting purposes. These may be the best candidates at this moment, but further discussion should be done.

INTERNATIONAL COLLABORATION

Despite the continuously declining cost of PV, a VLS-PV system of several hundred MW or GW scale capacity will require an increasing amount of investment. In order to promote VLS-PV systems worldwide, especially in developing countries located in deserts, innovative and favourable financing approaches and active international collaboration are required. Major international collaboration is needed to expand R&D in all areas of the VLS-PV concept, especially in policy, regulation, standards and business model development. International cooperation in these areas stimulates the strengthening of international collaboration of existing institutions and activities, as well as the creation of new joint initiatives at regional and global level.

In the near future, VLS-PV systems should become an economically viable option to meet the electricity needs of communities in remote desert regions around the world where conventional power plants cannot be built. The rate of deployment of VLS-PV is greatly influenced by the perception of general public and utilities, local, national and international policies, as well as the availability of suitable standards and codes to govern it.

In the long term, with a solid strategy for implementing of VLS-PV systems, further expansion of large scale PV generation can not only provide for increasing energy demand in sustainable way but also will stimulate local PV industry and create enough need to justify local manufacturing of PV system components. In addition, diffusion of the various kinds of high-tech knowledge, experience and know-how will be transferred to local renewable energy institutes, utilities and energy companies in order to keep track of worldwide technology developments, technology exchange between universities and scientific institutes.

The future expansion of VLS-PV depends on successful international cooperation to advance knowledge on development of the VLS-PV concept. In addition, the clear political commitment of governments and a stable regulatory framework towards creating a long-term policy is essential for further deployment of the VLS-PV concept worldwide.

STRATEGIC NICHE MANAGEMENT, SOCIAL ACCEPTANCE, AND BUSINESS MODELS

Solar power plants in desert regions are on the brink of finding their way into the world's energy systems. This might mean that our preparatory work [1–3], is currently being "overtaken by market developments". However, from the innovation science perspective, this would be a false conclusion, and now is the time for systematic learning about VLS-PV on a global niche level. Only the confrontation with real-life conditions can reveal the technological, business, socio-economic, and geopolitical fits and misfits of solar power plants in desert regions with the demands of the modern energy world. It is now that the main stakeholders need to learn about expected and unexpected bottlenecks that emerge in individual local projects, accumulate such lessons on an overall niche level, experiment with ways to navigate these real-world opportunities and barriers, and invent viable innovation pathways. Importantly, socio-economic or geopolitical bottlenecks might very well feed back into the technology design process. The redesign of solar power concepts for desert regions, originally conceived to serve long-distance international energy flows, to accommodate local economic development and sea water desalination, illustrates technology adaptations in the market implementation phase. In sum, the coordination between individual local projects and the overall desert solar power niche community is more important now than ever.

The VLS-PV niche requires continued reflection on strategic management on a niche level, and a continued search for social acceptance and business models on a project scale. To facilitate this process we have tapped into innovation sciences work on sustainability transitions, outlined the approach of strategic niche management, and briefly introduced ESTEEM and business model experimenting. This initial mapping of the territory, however, has merely opened up the topic. A next step would be a more VLS-PV specific and fine-tuned inquiry.

15.2 NEXT STEP AND STRATEGY OF TASK8 ACTIVITY

Since it was established in 1999, IEA PVPS Task8 has been examining and evaluating the potential of VLS-PV systems from various points of view and with the consideration of different aspects concerning feasibility and practicability. With the present and the past publications [1–3], a comprehensive summary of these analyses has been made available. The work of Task8 has shown that VLS-PV can be a viable and competitive building block in the world's future energy supply.

Task8 has observed and analysed PV, CPV and CSP with the same agnostic interest. The technological development of the three technologies, along with the cost reduction of the energy production especially for PV, leads to the conclusion that PV is or may become the predominant technology to harvest electricity from the power of the sun.

Throughout the last two decades we have experienced a slight but continuous paradigm shift in economies, which gains more and more momentum as the topic gains more social perception. A more sustainability-oriented energy mix with an important role of renewable energy sources is socially and politically favoured in most societies around the globe. With this, as well as with the accompanying maturing of the PV industry, the ideas and proposals presented in

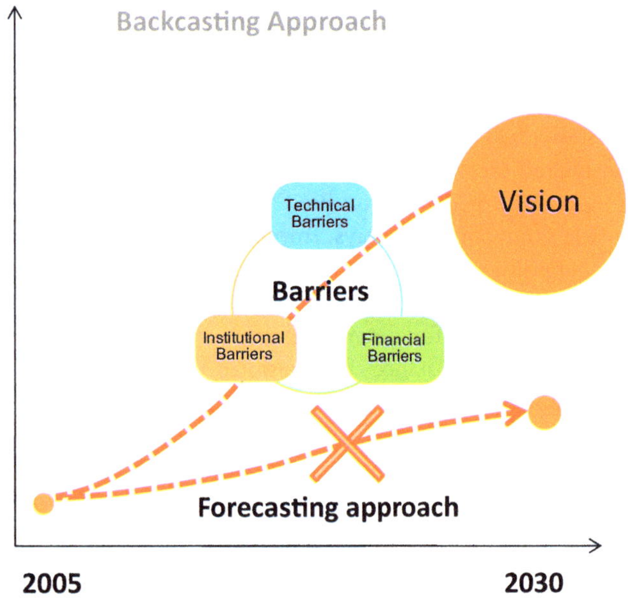

Figure 15.1 Back-casting approach

the Task8 publications have made their way into the common knowledge of all stakeholders – politicians as well as industry players, institutional investors, transnational and non-governmental organizations. The technological aspects in terms of feasibility as well as potential applications worldwide have been and are discussed in depth.

With this in mind, we will conclude, that the focus of our work should be refined to add further value to the discussion about VLS-PV. Therefore we decided to widen the scope of our evaluation and to focus on one question: What has to be done besides aspects of feasibility and practicability, to ensure that solar power plants will be a viable and sustainable option for the future electrical power supply by introducing a back-casting approach (see Figure 15.1)?

It became clear that besides the public and the governmental/administrative sides, the way of thinking and the approaches of the national and transnational electricity utilities have to be understood.

Why are the utility companies so important?

- From a global point of view, utilities currently play at best a marginal role and at worst a retarding role in the expansion rate of PV;
- Currently the utility scale PV market is dominated by investors and other non-utility players;
- As PV markets mature, utilities will remain the strong players with large-scale and long-term interests;
- The tipping point will be when PV LCOE (levelized cost of energy) reaches generation parity (peak load).

In any case, the utilities have the responsibility to ensure the electricity supply for whole countries or regions.

Therefore, they have certain interests and requirements to fulfil this mandate in the best possible way. The following list gives a first idea of what is of interest for them:

1 General
- What is the specific role of PV in the energy production portfolio of the utility company?
- What do the utilities see as the main challenges of PV integration in the energy mix?
- What share of PV integration in their energy mix do they see as realistic/feasible long term?
- What role does PV play in their regional footprint considerations?
- What is the optimal size in the eyes of a utility? What is the minimum size?
- Are there any preferences regarding distributed or central projects?

2 Financial/contracts
- What financial framework is required? What is the percentage IRR target for PV invests?
- Does it have to be same, higher or lower than corresponding fossil investment?
- According to what business models do utilities prefer purchasing traditional power plants? What can the PV industry learn from this?
- What are the preferred business models / contract types?
- Do utilities prefer long-term framework agreements or a single-project basis?
- What are the preferred supplier relationship models?
- What do the utilities actually want to purchase? Electricity? Power plants?
- What time to market requirements to exist?

3 Technological aspects
- What do utilities seek in term of maturity and professionalism of industry/technology?
- What role does storage technology play?
- What technological features do they require, e.g. standardized PV building blocks a x MWp?
- Do they need controllability of plant output? Autonomous or manual?
- Are they considering virtual power plants, by combining wind, gas, PV?

4 Service packages
- Is a yield guarantee required?
- What commercial compensation schemes are required in case of deviation from contract?

We intend to continue our work to gain further understanding of the requirements for VLS-PV from the utility point of view. Then, we will make a detailed and worldwide analysis of the requirements and expectations of major utilities. The outcome will be a competitive list of suggestions and recommendations for governments as well as for PV manufacturers to fulfil the needs of the utilities as important stakeholders for the future of VLS-PV.

Apart from utilities, we see governments and international organizations as important stakeholders and will make further efforts to understand their requirements and concerns. Eventually, conclusions, suggestions and recommendations of how to overcome hurdles/barriers from technical and non-technical viewpoints, will be proposed for potential VLS-PV owners/developers.

Also, we have started to learn the approach of strategic niche management. As our next step, a follow-up study focusing on VLS-PV specific and fine-tuned inquiry will systematically examine existing or ongoing projects for current niche dynamics and initial encounters with real-life contexts. With this practical approach, we may translate these lessons into a learning-based implementation strategy for policy makers and entrepreneurs.

As announced in January 2012, the International Energy Agency (IEA) and the International Renewable Energy Agency (IRENA) have signed a partnership agreement which will strengthen cooperation between the two organizations. There will also be increased collaboration between the two agencies at the Secretariat level, and in energy technology networks, including the IEA PVPS. Our activity, Task8, would be able to contribute to further advance international collaboration to stimulate the global deployment of large scale PV power generation systems.

REFERENCES

1 Kurokawa, K. (2003) *Energy from the Desert: Feasibility of Very Large Scale Photovoltaic Power Generation (VLS-PV) Systems*, James & James.

2 Kurokawa, K., Komoto, K., Van Der Vleuten, P., and Faiman, D. (2007) *Energy from the Desert: Practical Proposals for Very Large Scale Photovoltaic Systems,* Earthscan.

3 Komoto, K., Ito, M., Van der Vleuten, P., Faiman, D. and Kurokawa, K. (2009) *Energy from the Desert: Very Large Scale Photovoltaic Systems: Socio-economic, Financial, Technical and Environmental Aspects*, Earthscan.

4 International Energy Agency (2010) *Technology Roadmap Solar Photovoltaic Energy*.

5 International Energy Agency (2010) *Energy Technology Perspectives 2010 Scenario & Strategies to 2050*.

9780415639828